ENCYCLOPAEDIA OF
MOLECULAR BIOLOGY

ENCYCLOPAEDIA OF MOLECULAR BIOLOGY

(Vol. 5)

A. Tyagi

ANMOL PUBLICATIONS PVT. LTD.

NEW DELHI - 110 002 (INDIA)

ANMOL PUBLICATIONS PVT. LTD.
H.O.: 4374/4B, Ansari Road, Darya Ganj,
New Delhi-110 002 (India)
Ph.: 23261597, 23278000
B.O.: No. 1015, Ist Main Road, BSK IIIrd Stage
IIIrd Phase, IIIrd Block,
Bangalore - 560 085 (India)
Visit us at: www.anmolpublications.com

Encyclopaedia of Molecular Biology

First Published, 2006

ISBN 81-261-2677-9 (Set)

PRINTED IN INDIA

Published by J.L. Kumar for Anmol Publications Pvt. Ltd., New Delhi - 110 002 and Printed at Mehra Offset Press, Delhi.

Contents

Preface

Smallest fundamental unit of an element or a compound is called a molecule. The rapid progress and multi-faceted development of Science, in general; and of Biology, in particular, has ushered in an absolutely new era in the history of mankind, regarding the physio-chemical department of life.

There has been an urge to know more and more from the day one. It is due to this urge that man has been scaling ever new heights day-after-day, be it related to Science or Commerce or Politics and so on. Is this whole endeavour, which has proved to be a real boon for man, or in other words, the greatest gift, the twentieth century gave to humanity ? It really brought about a revolution in the whole human life. The emergence of Molecular Biology, as a branch of Science for study, could also be possible by virtue of this very present of the past.

The discipline is very vast and with the passage of time, big breakthroughs are in the offing. Feeling the pulse of the time, it deemed suitable, rather, necessary to bring out a comprehensive and compact encyclopaedia on this subject in order to cater to the maximum requirements of the lovers of Molecular Biology.

Comprising five volumes, this trivial attempt is expected to prove to be an asset for Science elite and students alike. Now, savour it molecule-by-molecule.

Editor

1

Expression of Gene

The synthesis of protein from mRNA involves translation of the language of nucleic acids (available in the form of mRNA) into language of proteins. This is much more complex than transcription and involves initiation, elongation and termination of polypeptide chains. The steps involved in translation, are diagrammatically represented in Figure for prokaryotes and in Figure for eukaryotes, and will be discussed in this chapter. However, for initiation and elongation of a polypeptide, the formation of aminoacyl transfer RNAs (aa-tRNAs) is a prerequisite, which will be described first, before describing the steps involved in translation.

Formation of Aminoacyl tRNA

Activation of Amino Acid : This reaction is brought about by the binding of an amino acid with ATP and is mediated by specific activating enzymes known as amino acyl tRNA synthetases or aaRs (amino acid is abbreviated as aa). As a result of this reaction between amino acid and adenosine triphosphate, mediated by specific enzyme, a complex (amino acyl-AMP- enzyme complex) is formed. It should be realized that amino acyl-RNA synthetases are specific with respect to amino acids. Therefore, for different amino acids, different enzymes would be required. Moreover a particular enzyme has a double specificity in so far as it can recognize its own amino acid on the one hand and its own tRNA on the other hand.

$$aa_1 + ATP \xrightarrow{(Enz1)} (aa_1\text{-}AMP)\ Enz_1 + PP$$

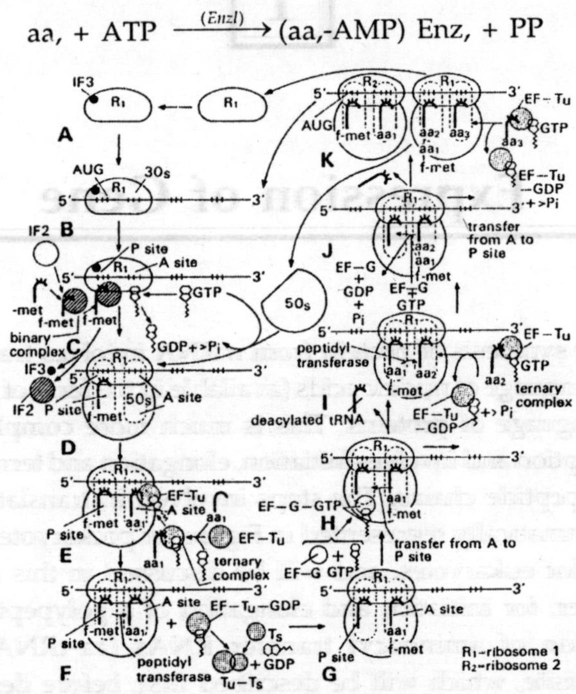

Different steps (A-K) in the translation of mRNA molecule in bacterial systems.

The Transfer of Amino Acid to tRNA : The amino acyl-AMP-enzyme complex, formed during the step outlined above, reacts with a particular tRNA and transfers the amino acid to the tRNA. This Reaction is accompanied with the liberation of AMP and the enzyme. It should be noticed in this reaction that with EnZ_1 and $tRNA_1$, 1 is used as suffix (subscript) to emphasize the specificity, which exists between them. A particular amino acid would require a particular enzyme and a particular species of tRNA. This would mean that for 20 amino acids, atleast 20 different enzymes and also atleast 20 different tRNA species would be required (while there are cases where more than one kind of tRNA may be available for the same amino acid, no cases are known, where the same tRNA can recognize two or more different amino acids).

$$(aa_1 - AMP)\ Enz_1 + tRNA_1 \rightarrow aa_1 - tRNA_1 + AMP + Enz_1$$

Although, the above reaction is generally catalysed by a protein (aaR enzyme), in recent years (1995), it was possible to use a specific selected RNA molecule, to catalyse the above reaction. It is an example, where RNA molecule can perform the same catalytic function, which is generally performed by a protein, *i.e.* aminoacyl-transfer RNA synthetases (aaRs). This gives support to the hypothesis of ancestral RNA world, where RNA performed the functions of nucleic acids (genetic information) and proteins (catalysis). For more information, on enzymatic role of RNA.

Initiation of Polypeptide

The initiation of polypeptide chain is always brought about by the amino acid methionine, which is regularly coded by the codon AUG, but rarely also by GUG as the initiating codon, very rarely initiation may also be directed by UUG (- 1%) or AUU (only one case reported). In *E. coli,* there are two tRNAs for methionine, $tRNA_m^{met}$ (non-formylatable) and $tRNA_f^{met}$ (formylatable). The methionine molecule carried by $tRNA_m^{met}$ cannot be formylated, while that, carried by $tRNA_f^{met}$ can be formylated. In prokaryotes, formylation of initiating amino acid methionine is essential requirement, so that $tRNA_f^{met}$ is meant for depositing methionine as the first amino acid, while $tRNA_m^{met}$ deposits methionine at the intercalary positions only. In eukaryotes, formylation of initiating methionine is not brought about due to the absence of $tRNA_f^{met}$ in plants and due

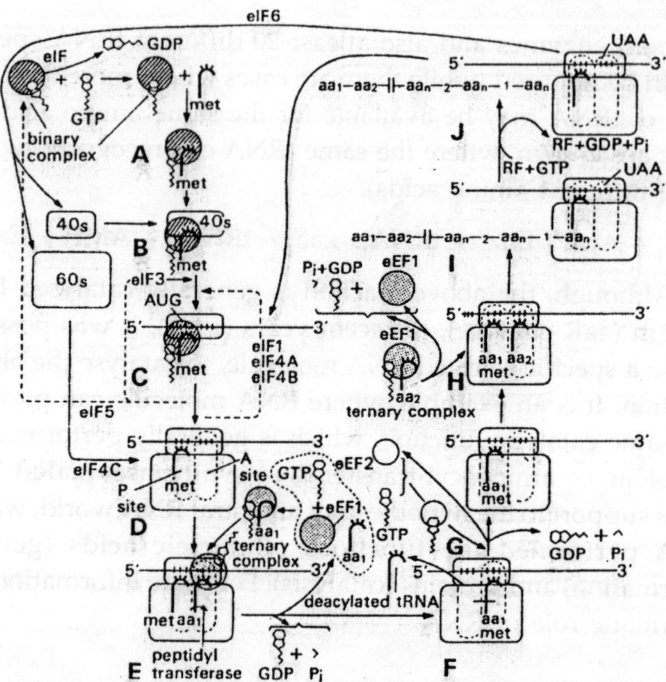

Different steps (A-J) in the translation of mRNA molecule in eukaryotes (reticulocytes).

to the absence of enzyme transformylase in animals. Initiation in higher organisms will, therefore, take place without formylation. Differences between prokaryotes and eukaryotes in the mechanisms of initiation of translation are summarized in Table.

Initiation in Prokaryotes

Ribosome-mRNA Interaction **:** The initiating codon always functions in combination with a Shine-Delgarno sequence (SD region of mRNA), which has complementarity with the 3' end of 16S rRNA (the ASD region) found in 30S subunit of ribosome. SD region is a part of ribosome binding site (RBS), which itself is a part of a longer region called translation initiation region or TIR found in mRNA (Fig.). The TIR helps in binding of mRNA with ribosome, and also accepts formylated methionyl tRNA at the initiating codon.

Differences between Prokaryotes and Eukaryotes in the Mechanism of Translation.

Prokaryotes	*Eukaryotes*
1. Initiating amino acid (inethionine) needs to be fonnylated (therefore two tRNAs for methionine *i.e.* $tRNA_f^{met}$ and $tRNA_m^{met}$ are found)	1. Initiating amino acid (methionine) is not formylated (only one tRNA for methionine *i.e.* $tRNA_f^{met}$ is found)
2. Ribosomes enter the mRNA at AUG codon or at nearby Shine-Delgarno site.	2. Ribosomes enter at the capped 5' end of mRNA and then advance to AUG codon by linear scanning.
3. No initiation factors or co-factors are required for initial contact between ribosomes and mRNA.	3. ATP and a number of protein factors are needed for ribosomes to engage the mRNA.
4. Small ribosomal sub-unit (30S) can engage mRNA before binding of initiator met-$tRNA_f^{met}$.	4. Small ribosomal subunit (40S) binds stably to mRNA only after initiator met-tRNA is bound.

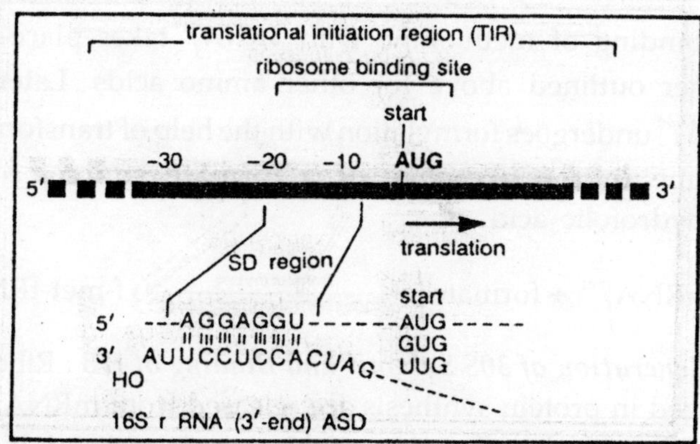

The prokaryotic translational initiation region (TIR), and ribosome binding site (RBS); the Shine-Delgarno region relative to start codon (4-12 bases away) is also shown, which has homology to 3' end of 16S rRNA (ASD region).

Ribosome-tRNA Interactions : In addition to the ribosome-mRNA interaction involving only 16S rRNA, tRNA-rRNA interactions are also known and involved in the tRNA binding to the ribosome. Bases have been identified in the 16S rRNA molecule (G926, m^2G966, G1338, G1401), which are essential for binding of the anticodon stem loop end of tRNA to the P site of the small subunit of the ribosome. Similar bases in the A site will become known in future.

The tRNA also interacts with the 23S rRNA of the larger subunit of the ribosome. For instance, the conserved CCA terminus of tRNA interacts with bases in the 23S rRNA at the P site in the larger subunit of the ribosome. The nucleotides of the 23S rRNA involved in this interaction include conserved nucleotides G2252 and G2253 (guanosines at positions 2252 and 2253) in the hairpin loop of this rRNA. It has been shown that C74 (cytosine at position 74) at the acceptor end of a tRNA interacts and forms G-C pair with G2252 in the hairpin loop of 23S rRNA at the P site of the larger subunit.

Steps Involved in Initiation : The initiation of translation involves the following steps in *E. coli*.

Formation of formyl-methionyl- $tRNA_f^{met}$ ***(f-met-*** $tRNA_f^{met}$ ***) :*** The binding of methionine with $tRNA_f^{met}$ takes place in the manner outlined above for other amino acids. Later, met-$tRNA_f^{met}$ undergoes formylation with the help of transformylase enzyme in the presence of a formate source, formyl-tetrahydrofolic acid.

$$\text{net-tRNA}_f^{met} + \text{formate} \xrightarrow[\text{(formyl-tetrahydrofolic acid)}]{\text{transformylase}} \text{f-met-tRNA}_f^{met}$$

Generation of 30S Subunit and Binding of IF3 : Ribosomes engaged in protein synthesis are released from mRNA as 70S (or 80S) particles. These particles are dissociated into two subunits (30S, 50S) with the help of the protein called initiation factor 3 (IF3). IF3 then binds to 30S subunit to stabilize it, to be ready to bind to mRNA (Figures).

$$70S \xrightarrow{IF3} 50S + 30S;\ 30S + IF3 \rightarrow IF3\text{-}30S$$

***Formation of 70S Initiation Complex*:** Several components, which participate in the formation of 70S initiation complex (70S ribosome + mRNA + f-met-$tRNA_f^{met}$) include the following:

(i) 30S and 50S ribosomal subunits,

(ii) the initiator tRNA (f-met-$tRNA_f^{met}$),

(iii) the TIR of the mRNA,

(iv) the three initiation factors (IF1, IF2, IF3) and

(v) a GTP molecule.

The probable events during initiation are depicted in Figure. The 30S subunit of ribosome is complexed with one molecule each of the three initiation factors and then binds, in random order, the mRNA and f-met-$tRNA_f^{met}$. The initiation factor IF2, in association with f-met-$tRNA_f^{met}$ forms a binary complex, which enters the 30S-mRNA complex (f-met-$tRNA_f^{met}$ does not make use of EF-Tu, see later; properties and functions of three initiation factors in *E. coli* are given in Table). This leads to the

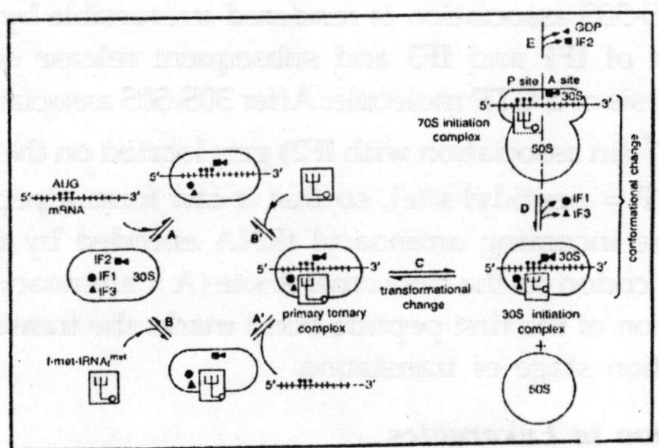

Use of initiation factors IF1, IF2 and IF3 in the formation of initiation complex with 30S subunit and the release of these factors on union with bigger 50S subunit in bacterial systems; A, B, A' and B' are in rapid equilibrium; step C is the most important rate controlling step.

Main Properties and Functions of three Initiation Factors in E. coli.

Factor	*Mass (daltons)*	*Function*
IF1	8,100	Stimulates activities of IF2 and IF3; increases the affinity of the 30S subunit for other factors.
IF2	97,300	Kinetic effector of 30S initiation complex formation; favours binding of aminoacyl-tRNAs with blocked α NH2 groups (*e.g.* f-met-$tRNA_f^{met}$); positions f-met-$tRNA_f^{met}$ in P site.
IF3	20,700	Kinetic effector of 30S initiation complex formation; shifts equilibrium 70S $\rightleftharpoons$ 30S + 50S to the right, ensuring provision of free 30S for initiation.

formation of a pre-ternary complex, which undergoes rearrangement to form the *bonafide* 30S initiation complex. This complex can either dissociate into its original components or will associate with 50S subunit to form a 70S initiation complex. The 30S-50S association is rendered irreversible by the initial ejection of IF1 and IF3 and subsequent release of IF2 after hydrolysis of a GTP molecule. After 30S-50S association, f-met-$tRNA_f^{met}$ (in association with IF2) gets located on the ribosomal P site (P = peptidyl site), so that it can form a peptide bond with the incoming aminoacyl tRNA encoded by the second mRNA codon on the ribosomal A site (A = aminoacyl site). The formation of the first peptide bond marks the transition to the elongation stage of translation.

Initiation in Eukaryotes

Initiation of polypeptide chain in eukaryotes is similar to that in prokaryotes, except the following minor differences, (i) In eukaryotes there are more initiation factors and at least ten have been identified in red blood cells. They are named by putting a prefix 'e' to signify their eukaryotic origin. These factors are eIFl, eIF2, eIF3, eIF4A, eIF4B, eIF4C, eIF4D, eIF4F, eIF5 and eIF6 (Table). The eIF2, eIF3, eIF4A and eIF4F contain multiple polypeptide chains, but others are single polypeptides; eIF2 and eIF3 are analogous to IF2 and IF3 of prokaryotes. (ii) In eukaryotes, formylation of methionine does not take place, (iii) In eukaryotes, smaller subunit (40S) associates with initiator tRNA known as $tRNA_i^{met}$ (because methionine is not formylated), without the help of mRNA, while in prokaryotes, generally the 30S-mRNA complex is first formed, which then associates with.f-met-$tRNA_f^{met}$.

Structure and Function of Nine Initiation Factors Isolated from Reticulocytes in Cukaryotes.

Factor	*Structure (mass in daltons)*	*Function*
eIF3	> 500,000 (multimer)	Maintains free 40Ssubunits; helps binding mRNA with 40S
eIF4F	200,000 (multimer)	Binding mRNA 5' end with 40S subunit; unwinding
eIFl	15,000 (monomer)	Assists mRNA binding
eIF4A	15,000 (multimer)	Assists mRNA unwinding (also bind ATP)
eIF4B	15,000 (monomer)	Assists mRNA binding & unwinding
eIF6	23,000 (monomer)	Prevents 40S-60S joining
eIF5	150,000 (monomer)	Releasing eIF2 & eIF3; eIF5A hypusine (unique aa) joining 60S subunit eIF4C 15,000 (monomer) Binding 60S subunit
eIF2	Trimer (α β γ)	Binding met-tRNA
eIF4D	Monomer	Unknown

Following steps are involved in the initiation of eukaryotic polypeptide synthesis :

(i) As shown in Figures, GTP binds to eIF2, which increases its affinity for met-$tRNA_i^{met}$; met-$tRNA_i^{met}$ associates with eIF2-GTP complex forming a ternary complex, *i.e.* met-$tRNA_i^{met}$ -eIF2 -GTP.

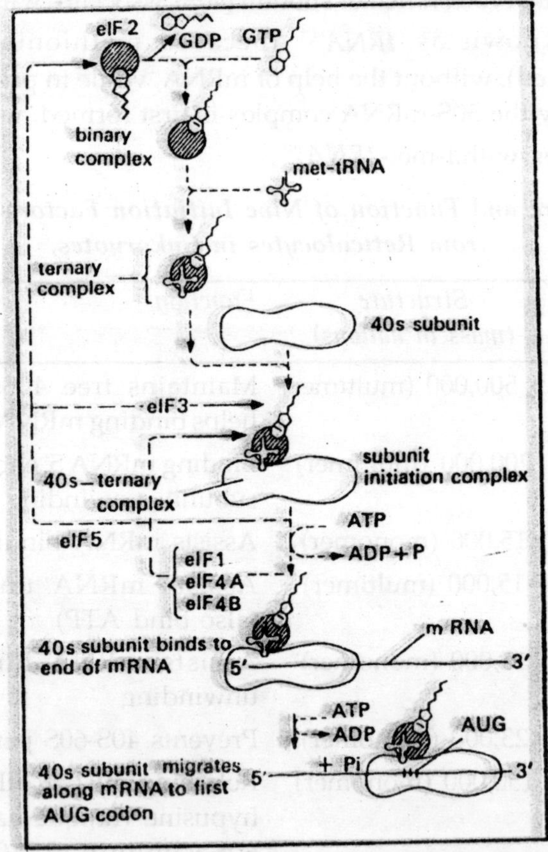

Use of a number of initiation factors in eukaryotes for binding of 40S subunit and mRNA to form a complex. 40S subunit with ternary complex then moves to AUG codon (modified from Lewin's 'Genes').

(ii) The ternary complex associates with 40S subunit, to form 43S initiation complex. (The factor eIF2 in mammals and plants like wheat has three subunits, namely α (35,000), β (38,000), γ (55,000); eIF2α binds to GTP, eIF2γ binds to met-tRNA and eIF2β may be a recycling factor (Table).

(iii) The mRNA at its 5' end binds with 43S initiation complex. This reaction depends on eIF3 and the binding of mRNA is assisted by eIF4F, eIF4A, eIF4B and a high energy bond of ATP (Table; Figures). As can be seen in Figure, the association of mRNA takes place on 5' end and not at the initiation codon AUG as in prokaryotes.

(iv) After association of the 5' end of mRNA, initiation complex moves towards 3' end of mRNA in search of initiation codon AUG, and then also associates with 60S subunit (you may recall that in prokaryotes, the initiation complex directly attaches to AUG codon and not to the 5' end of mRNA).

Association of 60S subunit with the initiation complex requires the factor eIF5, because it helps in releasing eIF2 and eIF3 (60S can not join, if these two factors are not released). eIF2 is released as a binary complex, eIF-2-GDP (Fig.). The details of the role of eIF2 and its reversible phosphorylation in the initiation of translation in eukaryotes Figure. The 40S-60S joining reaction really depends on eIF4C, and the GTP of the initiation complex is hydrolysed, when 60S subunit joins.

Structure and Function of three Subunits of Eukaryotic Initiation Factor eIF2.

Subunit	*Mass (daltons)*	*Function*
α	35,000	Binds GTP
β	38,000	May be recycling factor
γ	55,000	Binds met-$tRNA_f^{met}$

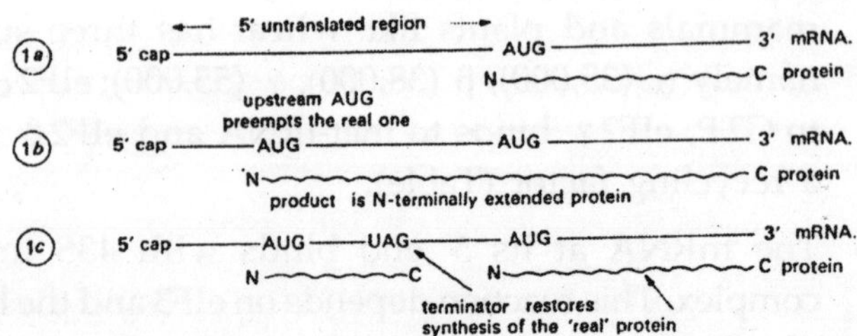

Kozak's scanning hypothesis explaining synthesis of mRNA in eukaryotes note the role of false start and stop signals preceding the Coding region).

It should be realized that the initiation process described above is characterized in prokaryotic system by *E. coli* and in eukaryotic system by red blood cells, where major work has been done. Variations may by found in other systems.

Translation in the Nucleus in Eukaryotes : In a report published in August 2001, it was shown that 10-15% of protein synthesis in a eukaryotic cell perhaps occurs within the nucleus, although earlier it was believed that translation occurs only in the cytoplasm. If translation in the nucleus is a reality, future research will confirm it and reveal details of this translation system, which does not seem to differ from that in the cytoplasm.

Kozak's Ribosome Scanning Hypothesis for Translation in Eukaryotes : In 1983, Marilyn Kozak proposed a hypothesis for initiation of translation by eukaryotic ribosome. According to this hypothesis, 40S smaller subunit of a eukaryotic ribosome with its associated met-tRNA moves down the mRNA from 5' end, until it encounters the first AUG. At this point, the 60S subunit joins and the translation begins. The 80S ribosome, after reaching termination, releases protein and dissociates in two subunits.

Kozak's hypothesis received support from cases, where one or more AUG codons were present or inserted upstream of the starting codon of a mRNA.

In these cases it was found that these extra or false AUG

codons were always read, so that if a false AUG codon was in the reading frame of the coding region, it would give an extended protein, and if followed by a termination codon, a small peptide is released (Fig.). However, during early 1990s, Kozak's hypothesis was challenged, when it was shown that in some mRNAs, the ribosome can bind to the initiation codon in the internal ribosome entry sites (IRES) without any prior scanning from the 5' end of these mRNAs. However, these internal binding sites are often the characteristic features of mRNAs that make use of initiation codons other than AUG, and amino acids other than methionine (as discussed below, sometimes codons other than AUG may also be translated into methionine).

Translation Initiation with Methionine using Codons other than AUG : Many exceptions to general outline of the mechanism of translation initiation described earlier are now known. For instance it was shown that in prokaryotes, rarely the codons GUG and UUG were translated into methionine and used for initiation with the help of mutant $tRNA_f^{met}$. Similarly, in eukaryotes, codons ACG, CUG, AUU and AGG are rarely used as initiating codons, and translated into methionine with the help of mutant $tRNA_i^{met}$.

Translation Initiation without Methionine and/or its tRNA: In the year 1990 and later, cases were also reported, where amino acids other than methionine were used for initiation. For instance in *E. coli*, UAG codon (a termination codon) was used for initiation with formylglutamine, and without the help of initiator $tRNA_f^{met}$. Later (in 1998, 2000), similar results were reported for the translation of viral RNA in mammalian and insect cells, where the initiating codons were CAA translated into glutamine and GCU/GCA translated into alanine. In most such cases with abnormal translation initiation, the messenger RNA was characterized by the absence of 5' cap and by the presence of an internal ribosome entry site

(IRES) associated with a pseudoknot, in the non-coding region, immediately upstream of the codon used for initiation. The pseudoknot may be important for IRES dependent translation, since any disruption of this pseudoknot has been shown to abolish translation.

Elongation of Polypeptide

In *E. coli,* after the formation of 70S-mRNA-f-met-$tRNA_f^{met}$ complex, elongation of polypeptide chain is brought about by the regular addition of amino acids. At 37°C about 15 amino acids per second are added to a growing polypeptide chain, so that in 20 seconds a protein with 300 amino acids can be synthesized. The following three steps are important in the elongation process.

Binding of AA-tRNA at Site 'A' of Ribosome (Classical vs Hybrid State Models for Translation) : In earlier classical model, each ribosome had two cavities, in which tRNA could be inserted. These were 'P' site (peptidyl site) and 'A' site (amino acyl site). However, later a third cavity called 'E' site (exit site) was suggested. It is not known whether f-met-$tRNA_f^{met}$ comes on P site or on A site, but ultimately it has to be present on P site, to make 'A' site available for the next amino acyl tRNA (AA-tRNA).

Another site called 'R' site located on smaller subunit of ribosome, was proposed. The presence of 'R' site was confirmed and its role was discussed. It was proposed that 'R' site plays a role in the improvement of accuracy of translation. The aminoacyl tRNA first binds to R site involving codon-anticodon pairing (Fig.). Later aminoacyl tRNA is flipped to A site using energy from GTP molecule (Fig.). During this flipping, tRNA is held only by codon-anticodon pairing. If this pairing is weak due to mismatching of wrong tRNA, aminoacyl tRNA will fly off the ribosome. R site thus provides a correcting measure.

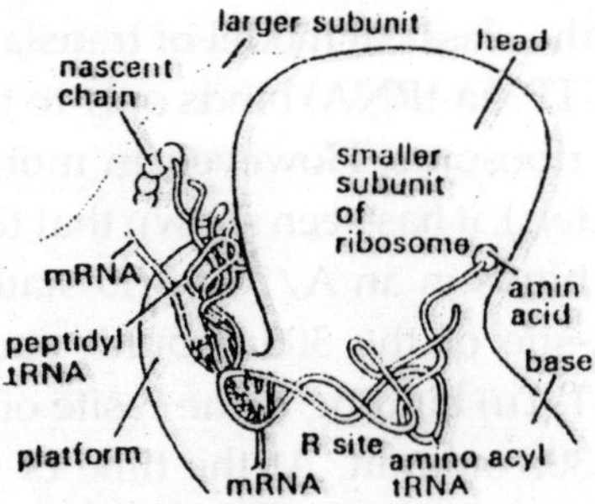

Binding of amino acyl tRNA to R-site, when its anticodon is bound to the next unread codon.

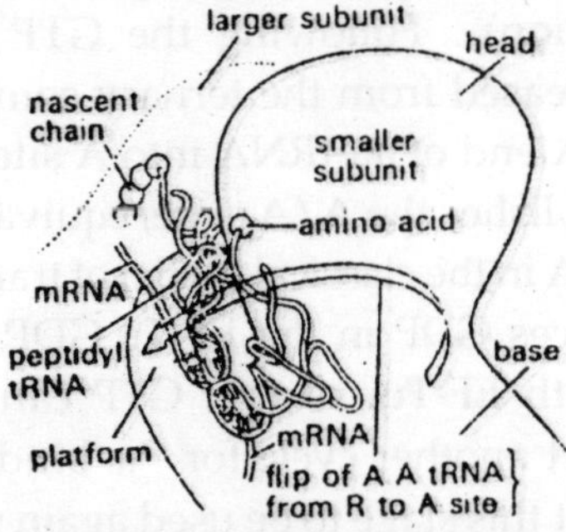

Amino acyl tRNA flips into a position to help its binding on 'A' site.

After formation of 70S initiation complex, the next amino acyl tRNA (aa-tRNA) enters 'A' site. Factors responsible for this entry include the elongation factors EF-Tu (u means unstable when heated) and EF-Ts (s means stable when heated). Although aa-tRNA can bind to ribosome even without the help of elongation factors (EF-Tu, EF-Ts), proper binding occurs only in the presence of EF-Tu, which is responsible for placing aa-tRNA on the A site. The elongation factor EF-Tu (a.G-protein) first combines with GTP and changes to an active binary complex (EF-Tu.GTP) which binds with aa-tRNA (but it can not bind to f-met-tRNA$_f^{met}$, which instead binds with IF$_2$ to form a binary complex) to form a ternary complex (Fig.).

$$\text{EF–Tu–Ts} + \text{GTP} \xrightarrow{\text{binary complex}} \text{EF–Tu.GTP} + \text{EF–Ts}$$

$$\text{EF–Tu.GTP+aa–tRNA} \rightarrow \text{EF–Tu..GTP.aa–tRNA}$$
(ternary complex)

According to the classical model of translation, the ternary complex (EF-Tu.GTP. aa-tRNA) binds only to the A site in both the subunits of a ribosome. However, in more recent models (hybrid states models), it has been shown that the above ternary complex actually binds in an A/P hybrid state, the anticodon binding to the A-site of the 30S subunit and the CCA end (sequestered by ET-Tu) binding to the P-site of the 50S subunit as well as to the 30S subunit. At the time of the entry of the ternary complex, the 'P' site is already occupied by f-met. $tRNA_f^{met}$ (at the initial stage) or by a peptidyl tRNA (at later stages of elongation). Following the GTP hydrolysis, EF-Tu.GDP+P are released from the ternary complex, permitting movement of CCA end of aa-tRNA into A site of the large 50S subunit. This establishes the A/A state, equivalent to the A site binding of aa-tRNA in the classical model of translation. Further, EF-Ts now displaces GDP in the EF-Tu.GDP binary complex and associates with EF-Tu, so that GTP can again associate with EF-Tu to start another cycle for the binding of aa-tRNA. EF-Ts is released at this stage to be used again with EF-Tu.GDP complex (Fig.). EF-Ts is believed to be a molecule which is also described as guanine exchange factor (GEF). Such a GEF is also known in the eukaryotic systems.

EF-Tu.GDP + EF-Ts $\rightarrow$ EF-Tu.EF-Ts + GDP

EF-Tu.EF-Ts + GTP $\rightarrow$ EF-Tu.GTP + EF-Ts

In eukaryotes as typified by red blood cells, there are two elongation factors eEFl and eEF2, the former being considered analogous to EF-T (both EF-Tu and EF-Ts) of *E. coli* and the latter analogous to EF-G (see later in the section). However, eEFl may have several polypeptides in addition to the two which may be analogous to EF-Tu and EF-Ts. The elongation factors in chloroplasts and mitochondria may be completely independent and have not been characterized so far.

Formation of Peptide Bond : This is a catalytic reaction during which a peptide bond is formed between the free carboxyl group (-COOH) of the peptidyl tRNA at the 'P' site and the free amino ($-NH_2$) group present with amino acyl tRNA, which is available at the A site. The 50S rRNA seems to have peptidyl transferase activity, so that the ribosome is described as a ribozyme. The peptidyl transferase reaction involves peptidyl-transferase-coupled translocation of aa-tRNA. It has been shown that the tRNA moves relative to ribosomal proteins S21 and L11. A displacement model has been proposed for the peptidyl transferase reaction, which is consistent with the hybrid states model discussed above. According to this displacement model, peptidyl chain remains in a constant position relative to ribosome, while tRNA moves during the peptidyl transferase reaction. After the formation of peptide bond, the tRNA at 'P' site is deacylated and the tRNA at 'A' site now carries the polypeptide.

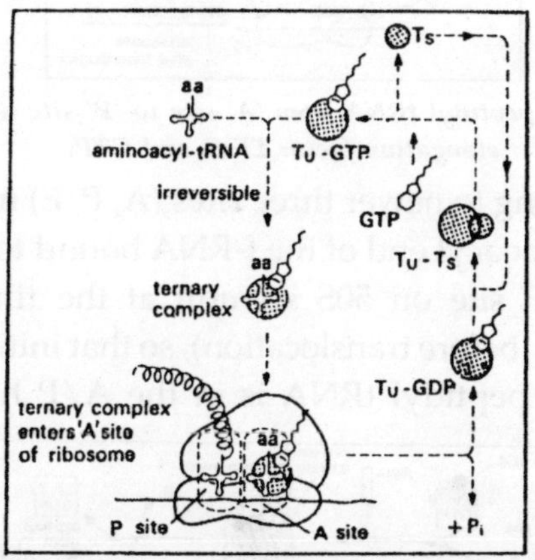

Formation of EF-Tu-GTP-tRNA ternary complex, which helps entry into 'A' site, and the release of EF-Tu-GDP, which forms Tu-Ts.

Translocation of Peptidyl tRNA from 'A' to 'P' Site : The peptidyl tRNA present at 'A' site is now translocated to 'P' site. For translocation of peptidyl tRNA from 'A' site to P site, there are two models available :

(i) According to two sites (A, P) model, deacylated tRNA is liberated from 'P' site, and with the help of one GTP molecule and an elongation factor EF-G (called EF-G, since it was isolated as GTP requiring factor; also earlier called transfer factor, TFII or translocase), the peptidyl tRNA is translocated from 'A' to 'P' site (Fig.). Thus according to this model, tRNA is either entirely in the A site or entirely in the P site,

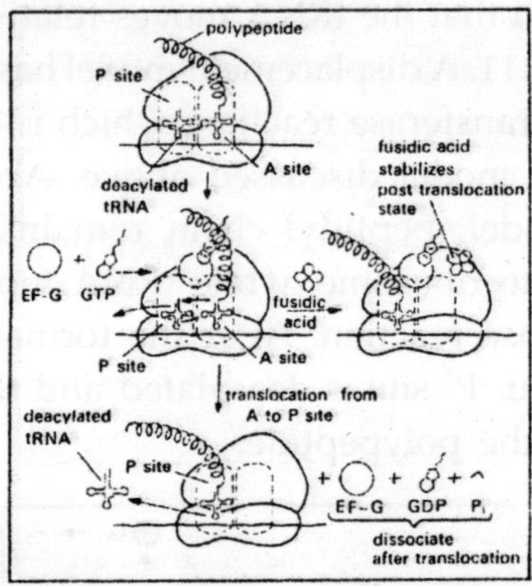

Translocation of peptidyl tRNA from 'A' site to 'P' site with the help of elongation factor EF-G and GTP.

(ii) According to newer three sites (A, P, E) model, initially the aminoacyl end of the t-RNA bound to A-site moves to the P site on 50S subunit at the time of peptide transfer (before translocation), so that initially the newly formed peptidyl tRNA is in the A/P hybrid state.

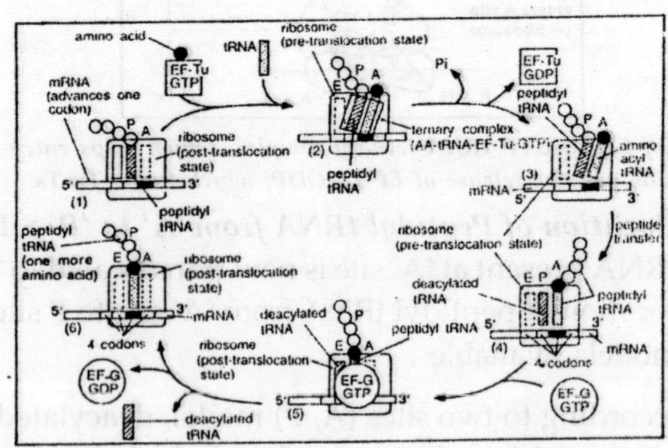

Three site (A, P, E) model for translocation of peptidyl tRNA from A to P site.

Only later during translocation, the anticodon end of this tRNA moves from A to P site on 30S subunit so that the peptidyl tRNA is now in the P/P state. Only this later step requires action of elongation factor EF-G. In this model thus, there is an intermediate state, when anticodon end of tRNA is still on A site (on 30S subunit), while the aminoacyl end occupies P site (on 50S subunit; see Fig.). A third tRNA binding site, 'E' (Exit) was also recognized, through which tRNA leaves the ribosome. E site is found mainly on 50S subunit, and deacylated tRNA interacts with it through its CCA sequence at its 3' end. The movement of deacylated tRNA from P to E site also occurs in two steps : the aminoacyl end moves to E site during peptide bond formation (deacylated tRNA is now in P/E state), and the anticodon end leaves the P site (on 30S) only during translocation (deacylated tRNA is now in E/E state). This three site model suggests that tRNAs interact mainly at their ends. It also assumes that a tRNA remains bound to ribosome at one or the other of its two ends (aminoacyl end, anticodon end), and does not leave ribosome at both its ends simultaneously (either on A or P site). The two step transfer of tRNAs from A to P site and from P to E site could result from reciprocating motions of the two subunits of a ribosome. It means that 50S and 30S subunits move alternately and not simultaneously.

The hybrid states model of translation (in comparison of the classical model) suggests the following features of the role of t-RNA in translation (described in greater detail by Green and Noller in a recent review Ann. Rev. Biochem., 1997, 66: 679-716):

(i) the peptidyl moiety remains stationary and only the tRNA moves;

(ii) movement of tRNA (translocation) takes place in two steps–in the first step, both aminoacyl tRNA and peptidyl tRNA move with respect to the larger subunit and in the second step both these tRNAs (along with mRNA) move with respect to the smaller subunit of ribosome.

These features suggest that the transfer RNA binding can have atleast five states (A/A, A/P, P/P, P/E, E/E) and perhaps a sixth state, described as A/P state.

The requirement of EF-G and GTP for translocation was revealed by the use of antibiotic *puromycin,* which closely resembles aa-tRNA in structure (Fig.) and can, therefore, occupy 'A'-site and form peptidyl puromycin. The latter can not form further peptide bond and is released prematurely causing lethality. This puromycin reaction is dependent on the presence of EF-G and GTP, suggesting their significance in teanslocation process.

The elongation factor EF-G binds to ribosome and is released on hydrolysis of GTP, which is a ribosomal function. EF-G and EF-Tu can not bind to ribosome simultaneously, so that the entry of a fresh aa-tRNA on 'A' site and the translocation of peptidyl tRNA from 'A' to 'P' site have to follow each other and can not occur simultaneously. For instance, EF-Tu.GDP must be released before EF-G can bind and similarly EF-G should be released before EF-Tu.GTP.aa-tRNA can bind to 'A' site. This was confirmed by the observation that if ribosome-

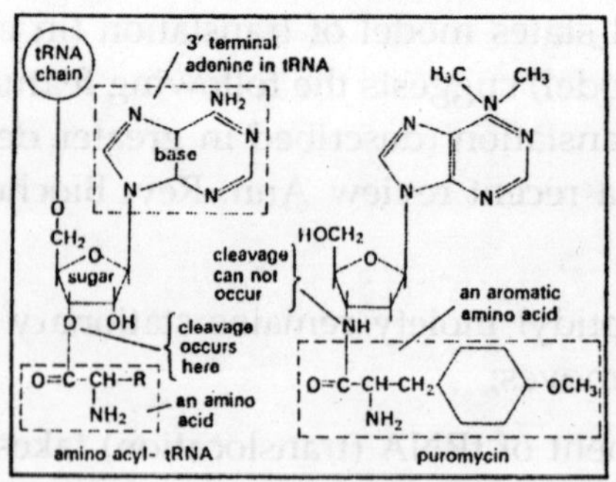

A similarity between amino acyl tRNA and puromycin, so that puromycin can occupy 'A' site, but can not be cleaved due to the presence of N in place of O in RNA.

EF-G.GTP complex is stabilized by a steroid antibiotic fusidic acid, so that EF-G is not released after translocation, further process of translation stops, due to inability of EF-Tu.GTP.aa-tRNA complex to enter the 'A' site. This exclusion of one elongation factor by the other allows orderly translation in protein synthesis. The details about EF-Tu, EF-Ts and EF-G are given in Table.

In eukaryotes, the elongation factor needed for translocation is called eEF-2. Unlike prokaryotes, a stable complex of eEF-2 with GTP can be isolated and this complex can bind to ribosome.

The above account of elongation of polypeptide suggests that for the formation of one peptide bond, one ATP molecule (for activation of amino acid) and two GTP molecules (one for transfer of aa-tRNA to 'A' site and the other for translocation of peptidyl tRNA from 'A' to 'P' site) are required. However, during 1990s evidence became available, which suggested that the number of GTP molecules for the formation of one peptide bond can be three on even more. For instance, it is suggested that for the transfer of one aa-tRNA, two molecules of the binary complex, EF-Tu.GTP, may be required, so that the total GTP molecules' for one peptide bond will be three (two GTP for transfer of aa-tRNA and one GTP for translocation of peptidyl tRNA).

Elongation Factors EF-Tu, EF-Ts and EF-G from E. coli.

Factor	*Genes coding*	*Mass (daltons)*	*Molecules per cell*	*Function*	*Inhibitor*
EF-Tu	*tufA tufB*	43,225	70,000	binds AA-tRNA & GTP	kirromycin
EF-Ts	*tsr*	74,000	10,000	binds EF-Tu displacing GDP	
EF-G	*fus*	77.444	20,000	binds ribosome and GTP	fusidic acid

Recently, the crystal structure of the ternary complex EF-Tu.GTP.aa-tRNA has been worked out. Following are some of the interesting features revealed by this study :

(i) The overall shape of the ternary complex (not the EF-Tu.GDP) resembles EF-G.GDP;

(ii) Three domains of the carboxy -terminal end of EF-G, which have no homology in EF-Tu, together resemble t-RNA and are positioned the same way as tRNA in the ternary complex.

The structural similarity between EF-G and EF-Tu also reflects functional similarity. Both bind to ribosome, when complexed with GTP and have GTPase activity on binding to ribosome. Further, after GTP hydrolysis, both adopt conformations, that have low affinity for ribosomes.

Particularly, similarity has been demonstrated between domain IV of EF-G and the anticodon domain of the ternary complex. During 1997, it was also shown, that EF-G dependent hydrolysis of GTP takes place immediately after binding of EF-G.GTP and before the translocation of tRNAs (peptidyl tRNA and deacylated tRNA). These recent studies led to the following conclusions :

(i) that GTP hydrolysis by EF-G drives translocation;

(ii) that domain IV of EF-G has an important role in coupling GTP hydrolysis to translocation and

(iii) that domain IV is essential for the release of EF-G. GDP from the ribosome after translocation.

Termination of Polypeptide

Termination in mRNA with Stop Codon : Termination of the polypeptide chain is brought about by the presence of any one of the three termination codons, namely UAA, UAG and UGA. These termination codons are recognized by one of the two release factors RF1 and RF2. Of these release factors, RF1 recognizes UAA and UAG and RF2 recognizes UAA and UGA. They help the ribosome to recognize these triplets. The release factors seem to act on 'A' site, since suppressor tRNA capable of recognizing termination codons can compete with release factors by entry at 'A' site. A third release factor RF3 seems to stimulate the action of RF1 and RF2 in a GTP-dependent and codon independent manner (one GTP molecule is hydrolysed during release of a polypeptide, when RF3 stimulates RF1 and RF2. For release reaction, the polypeptidyl tRNA must be present on 'P' site and the release factors help in splitting of the carboxyl group between the polypeptide and the last tRNA carrying this chain. Polypeptide is thus released and the ribosome dissociates into two subunits with the help of ribosome release factor or RRF (IF3). The identity of RRF in eukaryotes is unknown, where so far only one release factor is known *i.e.* eRFl. GTP seems to be necessary for binding of this factor to ribosome. GTP is cleaved after the termination step has occurred which may be a prerequisite for the release of eRF1 from the ribosome.

Termination in mRNA Lacking A Stop Codon (l0Sa RNA Works Both as tRNA and mRNA) : The above termination process can not be used in the absence of an in-frame termination codon, at the 3'end of mRNA molecule or its segment. This will be the situation in many truncated mRNAs. Under this condition, the ribosome may stall at the end of the message

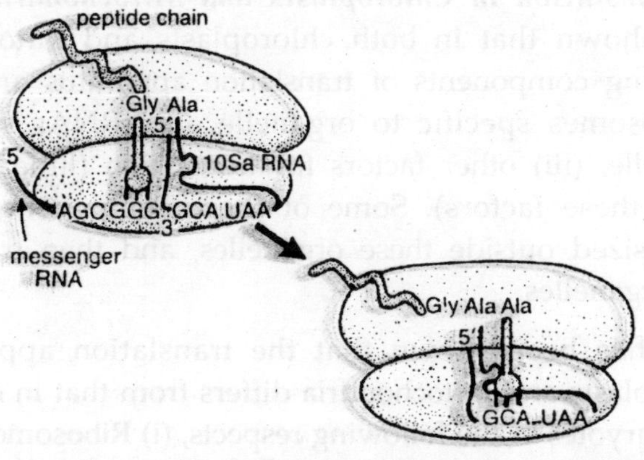

Use of 10 SaRNA for termination in mRNA lacking a stop codon.

and a newly-discovered 363 nucleotides long RNA called 10Sa RNA (encoded by the *E. coli* gene *ssrA)* is bound to this stalled ribosome, initiating a process of degradation of the incomplete polypeptide chain. This 10Sa RNA functions both as a tRNA and a mRNA. It has properties of tRNA including the G3:U70 base pair (5'-3' base pairing) that is crucial for tRNA-ala identity. Accordingly, it is charged with alanine at its 3'tRNA acceptor-like end. On binding of the charged 10SaRNA to ribosome, alanine gets incorporated to the growing polypeptide chain. The ribosome then uses 10SaRNA as mRNA and decodes a short sequence, adding 10 amino acids (ANDENYALAA) onto the C-terminus end of truncated polypeptide chain and terminates at a UAA codon available in 10Sa RNA (Fig.). The polypeptide is thus released. The C-terminal oligopeptide in this chimaeric chain is recognized by COOH-terminal specific proteases, so that the polypeptide is then degraded. This phenomenon for degradation of incomplete polypeptide remained unnoticed for a long time and was discovered only recently (1994-96). This protein degradation differs from the mechanism, where ubiquitin is added after the translation process is completed and leads to degradation in the proteasome.

Translation in Chloroplasts and Mitochondria : It has been shown that in both chloroplasts and mitochondria, following components of translation apparatus are found : (i) ribosomes specific to organelle, (ii) tRNAs specific to organelle, (iii) other factors for translation (little is known about these factors). Some of these components may be synthesized outside these organelles, and then transported into organelles.

It has been shown that the translation apparatus in chloroplasts and mitochondria differs from that in cytoplasm in eukaryotes in the following respects, (i) Ribosomes in these organelles are smaller in size (70S) than those in cytoplasm (80S), (ii) The tRNAs are specific and differ, the number of tRNAs in mitochondria being 22 as against 55 in cytoplasm, (iii) Initiation of translation takes place by formyl-methionyl tRNA both in chloroplasts and mitochondria, although no formylation takes place in cytoplasm, (iv) Translation in chloroplasts and mitochondria can be inhibited by chloramphenicol, as in bacteria since the 70S ribosomes are sensitive to chloramphenicol and not to cycloheximide; on the other hand, the translation in cytoplasm is inhibited by cycloheximide, since 80S ribosomes are sensitive to cycloheximide. There are other antibiotics like spectinomycin, lincomycin and erythromycin which also inhibit translation in bacteria and in all organelles of eukaryotes. These antibiotics help to stop the translation preferentially either (i) in cytoplasm (by cycloheximide) thus permitting protein synthesis only in chloroplasts or mitochondria or (ii) in organelles (by chloramphenicol) thus permitting protein synthesis only in the cytoplasm.

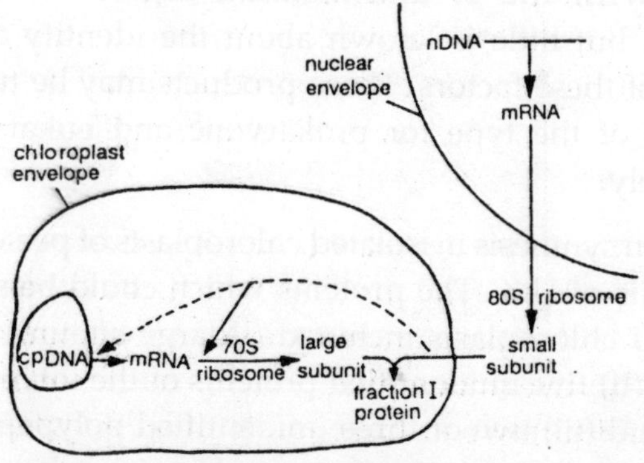

A model of cooperation between nuclear DNA and Chloroplast DNA in pea as studied by R.J. Ellis and his colleagues (redrawn from Nature Vol. 254, 1975).

Studies on , the synthesis of 'chlorophyll μ -binding proteins', which play a major role in the primary reactions of photosynthesis have also provided new information about chloroplast translation and its regulation. In many plant species, in the absence of light the mRNA of chlorophyll binding proteins, including those encoded by-chloroplast genome, remain associated with thylakoid bound polysomes, but are not translated. Transfer of seedlings to light induces synthesis of proteins, although no increase in transcription has been observed. This suggests that post-transcriptional processes play a key role in light induced chloroplast gene expression.

It has also been shown that the polycistronic mRNA encoded by the *psbB* operon in maize, need not be cleaved into tri-, di- or monocistronic forms for successful translation *(psbB* operon = *psbB* + *psbH* + *petB* + *petD).* This suggests that plastid ribosomes can bind directly to internal initiation regions for initiation of translation as in prokaryotes. However, RNA processing does take place in chloroplasts and mitochondria, and the transcripts may differ in their translatability.

There are also nuclear genes, whose products are essential for translation of specific genes in chloroplasts and mitochondria. Several of these nuclear encoded factors interact with the 5' untranslated region of chloroplast massages, but little is known about the identity and precise function of these factors. These products may be translational activators of the type for prokaryotic and eukaryotic genes respectively.

Protein synthesis in isolated chloroplasts of pea was studied by R.J. Ellis of U.K. The proteins which could be synthesized in isolated chloroplasts included (i) large subunit of Fraction I protein, (ii) five unidentified proteins of the internal lamellar system, and (iii) two or three unidentified polypeptides of the envelope. These are only few of a large number of proteins found in chloroplasts. It has also been demonstrated that while the large subunit of Fraction I protein is synthesized under the influence of chloroplast DNA, the small subunit is synthesized in the cytoplasm under the influence of nuclear DNA. Small subunit is then transported to the chloroplast where it is associated with large subunit to give rise to Fraction I protein (Fig.).

Proteins synthesized in mitochondria have also been identified by inhibiting protein synthesis in cytoplasm by cycloheximide. Although, some proteins are synthesized in mitochondria, not all proteins present in mitochondria are synthesized there. There are some mitochondrial proteins which are synthesized in cytoplasm and then transported. There are still others having dual origin, some polypeptides having cytoplasmic origin and others having mitochondrial origin. Some proteins, which are known to be synthesized in mitochondria include the following : (i) Cytochrome oxidase contains seven different kinds of polypeptides, of which three, are synthesized on mitochondrial ribosomes, the remaining four being synthesized on cytoplasmic ribosomes. (ii) Adenosine triphos- phatase (ATPase) is an important component of mitochondrial membrane, which plays an important role in coupling respiration to ATP formation. There are several polypeptides present in this complex and only two are definitely known to be synthesized on mitochondrial ribosomes. (iii) In mitochondrial ribosomes, although ribosomal proteins are known to be synthesized outside mitochondria and under the influence of nuclear DNA, but rRNA (ribosomal RNA) is transcribed from mtDNA, since both 21S rRNA and 15S rRNA (in yeast mitochondria, 21S rRNA and 15S rRNA occur instead 23S and 16S rRNAs) hybridize with specific regions of mtDNA.

Modification and Folding of Released Polypeptide

Modification of Released Polypeptide : After translation, the released polypeptide is modified in various ways. An enzyme deformylase removes the formyl group of first amino acid methionine. (Fig.). Due to the action of certain other enzymes, exo-amino-peptidases, amino acids may be removed from either the N-terminal end or the C-terminal end or both. Due to this modification, therefore, the N-terminal amino acid may not be methionine in some cases (Fig.), although in protein synthesis, the first amino acid is always methionine. The polypeptide chain singly or in association with other chains also folds into a tertiary structure. This problem of protein folding is sometimes described as 'Second Half of the Genetic code'.

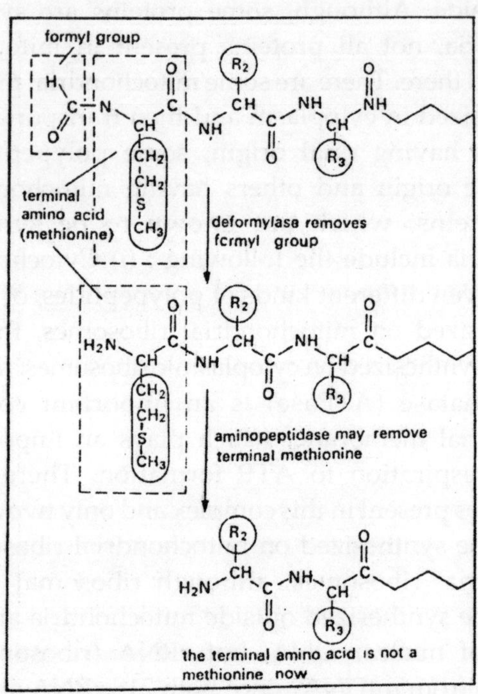

Removal of formyl group from released polypeptide and a subsequent removal of methionine, so that the second amino acid becomes terminal (modified from Lewin's 'Genes').

Inteins and their Excision from Polypeptides : We discussed split genes, which consist of intron and exon sequences. We also discussed the mechanisms used for the excision of intron sequences from pre-mRNA to give mRNA. A parallel situation sometimes exists in polypeptides that are released after translation. In other words, some polypeptides may contain sequences called inteins, that are destined to be spliced out post-translationally by a prpteolytic cleavage and ligation process. The sequences that flank inteins and are retained are called exeins. Inteins appear to autocatalyse their own excision. Some of them are site specific endonucleases, which can insert each a copy of its own DNA into its integration site in an intein-less allele. Thus, at the DNA level, like introns, intein encoding DNA sequences have also been shown to be 'mobile genetic elements'. Till about the middle of 1998, more than fifty inteins were already known from genomes of becteriophages, mycobacteria, cyanobacteria, archaebacteria, yeast and chloroplast of a red alga. Some of the proteins, known to have inteins, include DnaB, RecA, DNA polymerases. Intein encoding DNA sequences have also been shown to coexist with introns in the same genes. In an evolutionary sense, inteins are believed to be very ancient and might have been lost in other more highly evolved polypeptides. Alternatively these may represent molecular parasites, that invaded the corresponding genes in the past and merely needed to ensure the transmission of their genes. The shortest known intein is 150 amino acids long and is found in the DnaB protein of chloroplast of the red alga *Porphyra purpurea.* It lacks the endonuclease-like motifs found in all other inteins, which range in their length from 360 to 538 amino acids.

Chaperones in Folding and Assembly of Proteins : Molecular chaperones represent a diverse class of protein molecules, which play a significant role in folding of individual polypeptides and in assembly of multimeric proteins, but are not themselves components of the final functional structures. These chaperones function by inhibiting incorrect molecular interactions, which are often possible in their absence (word chaperone, rather chaperon is used in english to describe 'a person' usually a married or elderly woman who for the sake of propriety, accompanies a young unmarried lady in public, as guide and protector). Some molecular chaperones (but not all of them) are also stress proteins, so that it is possible that all stress proteins act as molecular chaperones. Another term 'chaperonin' is sometimes used interchangeably with 'molecular chaperones', but refers to only one of the different classes of molecular chaperones. (Table). Many chaperonins are ATPases,

Some Families of Molecular Chaperones.

Class	*Members*
1. Nucleoplasmins	Nucleoplasmin Protein XLNO38 Protein Ch-NO38 Nucleoplasmin S
2. Chaperonins	Chaperonin 60 (cpn60) 1. Bacterial groEL *(E. coli)* 65kDa common antigen (mycobacteria) 2. Plastid : rubisco subunit- binding protein 3. Mitochondrial : hsp 60 mitonin HuCha60 4. Eukaryotic cytosol : TCP-1 Chaperonin 10(cpnl0) 1. Bacterial groES 2. Mitochondrial
3. Heat shock proteins 70 (hsp70)	Hsp68, 72, 73; DnaK; clathrin uncoating ATPase; BiP; grp75, 78, 80; hsc 70, KAR2; SSA1-4; SSB1; SSC1; SSD1
4. Heat shock proteins 90	Hsp83, 87; HtpG
5. Signal recognition particle	SRP
6. Subtilisin prosequence	
7. *Alpha-lytic* factor prosequence	
8. Ubiquitinated ribosomal proteins	
9. Trigger factor	
10. Sec B protein	
11. Pap D protein	

whose function was demonstrated by *in vitro* folding and assembly of RUBISCO, which required groEL and groES proteins of *E. coli.* Another chaperonin encoded by a nuclear gene is hsp60 (heat shock protein) which is translated in the cytosol and imported into the mitochondrion, where it is folded and assembled into a tetradecameric structure. Some chaperonins can induce their own folding and self assembly (*e.g.* groEL protein); there are others (*e.g.* hsp60), which can fold and assemble only in the presence of other pre-existing functional chaperonins.

***Protein Translocation Across Membranes* :** The proteins synthesized in the cell have to be translocated to appropriate locations within the same cell or other cells. This process is called protein sorting or protein trafficking and involves interaction of cell membranes either with some part of a newly synthesized protein or with a protein sequence added to this new protein. These protein sequences interacting with membranes are called signal sequences or transit peptides, or leader sequences, which are recognized by receptors located within the membrane. The transfer of proteins may be coupled with translation— co-translational transfer (proteins synthesized on ribosomes attached to endoplasmic reticulum = rough ER) or may take place after the translation— post-translational transfer (proteins synthesized on free ribosomes). The signal sequences and the mechanism of intracellular protein trafficking were discovered by Gunter Blobel, who was awarded the 1999 Nobel Prize in Medicine or Physiology.

Protein translocation across membranes has been classified in the following two major systems based on whether these are transported out of the cytosol or imported into the cytosol; (i) export system and (ii) import system. Export system includes transport of proteins out of the cytosol or cytosol like compartments into extracytosolic compartments including bacterial periplasm (across bacterial plasma membrane), lumen of the endoplasmic reticulum (from the cytosol), thylakoid lumen (across the internal membrane of chloroplasts) and to the mitochondrial inner membrane (from the cytosol). The import system, on the other hand, includes transport of proteins

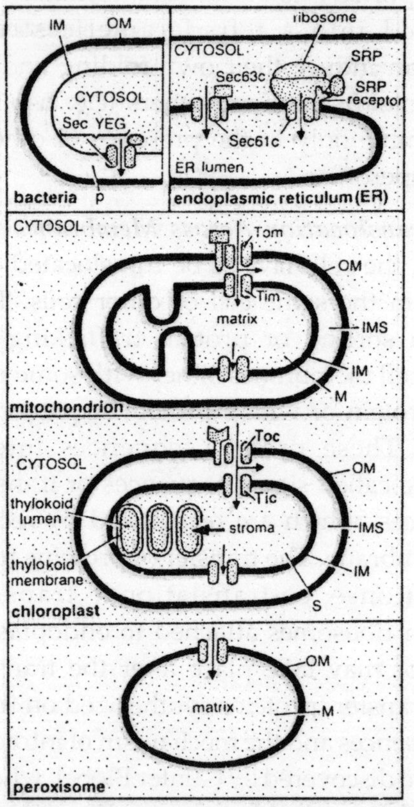

A comparison of protein trafficking among bacteria, eukaryotic cells and eukaryotic organelles.

into the cytosol or cytosol like compartments (eg. stroma of chloroplasts and matrix of mitochondria and peroxisomes). Thus the mitochondria and chloroplasts have export as well as import systems, while bacterial cells and ER have only export system, and the peroxisomes have only an import system (Fig.). This does not mean that no macromolecules leave ER, but they generally do so by vesicular traffic. Further, besides these export and import systems, there may be special systems for transport of special proteins. Similarly, nuclear pores allow nucleocytoplasmic transport. Since, these special systems are discussed elsewhere in this book, only the general export and import systems will be discussed in this section.

In these systems, following are the general features applicable to all export and import transport systems: (i) the protein to be transported carries an -NH_2' terminal signal sequence, due to which it is recognized by some cytosolic factors; (ii) cytosolic factors (*e.g.* chaperones) recognize membrane receptor and deliver the protein to the target membrane; (iii) the protein usually remains partly unfolded during transport; (iv) the signal sequence opens a hetero-oligomeric transmembrane channel that is gated both across and within the plane of the membrane (see Fig.); (v) the channel is coupled to a motor protein that is driven by the hydrolysis of ATP or GTP, and transports (by push or pull) the protein across the membrane; (vi) the protein folds on the trans-side of the membrane with the help of chaperones or folding enzymes, (trans-side means the side of the membrane away from the cis surface, where transport is initiated). These general features of export and import systems will be briefly discussed.

Targeting Signals at the NH_2 Terminus of Transported Protein : On the -NH_2 end of each protein destined to be transported, a signal sequence is always found, which is either cleaved off, once the protein reaches the trans-side of the membrane or may function as signal anchor sequence that permanently anchors the protein within the membrane generating a transmembrane protein. This signal sequence is also called leader sequence, targeting sequence, transit sequence or a presequence. The signal sequence is hydrophobic in export systems and hydrophilic in import systems. Some of the general features of these export and import signal sequences are shown in Figure.

Export Signal Sequences : In *E. coli,* the export signal sequence acts in two steps : (i) In the first step, it may retard folding of protein facilitating its binding to cytosolic chaperone, SecB, which then interacts with SecA subunit of transport machinery (in the plasma membrane). The SecB and SecA thus deliver the protein to SecYEG complex in the membrane (SecYEG is a complex of SecY. SecE and SecG). (ii) In the second step, the signal sequence does not act as a targeting signal, but only as a folding inhibitor. It interacts with and activates the membrane-associated export machinery and facilitates the recognition of protein by the transport machinery.

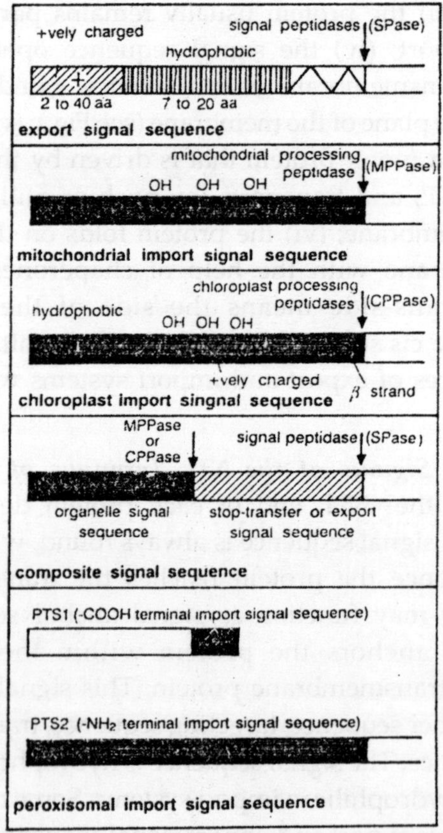

Different signal sequences known for transport of proteins across membranes.

The transported protein may either cross the membrane to reach either the trans side of a lumen, or it may become a transmembrane protein either (i) due to a 'signal anchor sequence' as mentioned above or, (ii) due to sequential function of a 'signal sequence' and a 'stop transfer sequence'. In the second mechanism, the signal sequence initiates translocation and the stop transfer sequence then triggers a lateral opening of the translocation channel, thus arresting the translocation. The stop transfer sequence then becomes the transmembrane anchor of an integral membrane protein.

Import Signal Sequences : Import signal sequences (20-35 residues long) help in transport of the proteins into the mitochondrial matrix or the chloroplast stroma. They are more hydrophilic than export signal sequences and are also rich in hydroxylated amino acids. Composite signal sequences are also known, where an import signal sequence and an export signal sequence (or a stop transfer sequence) are fused and function in a sequential manner one after the other. The import of proteins into thylakoids or to the mitochondrial intermembrane space is often affected by such a composite signal sequence. For instance, in chloroplast, the import signal sequence directs the protein to the stroma, where it is cleaved off and the export signal sequence then transports the protein to the thylakoids. Similarly, in mitochondria, the import signal sequence directs the protein towards mitochondrial matrix and a stop transfer sequence then arrests translocation either in the outer or in the inner membrane, so that the protein then becomes a transmembrane protein. The proteolytic cleavage of a protein stopped in the inner membrane may generate a soluble protein of intermembrane space.

Cytosolic Chaperones and GTPases : For the purpose of transport, a protein should remain atleast partly unfolded. Some small proteins (*e.g.* coat protein of phage Ml3) remain unfolded and traverse from ribosome to the target membrane without any help from cytosolic factors. However, large proteins with signal sequences require for their transport the assistance of cytosolic chaperones or other similar factors. Later, interaction with membrane receptor, leads to release of cytosolic factors (assisted by ATP or GTP) and directs the protein's signal sequence into the translocation channel (also called translocon). Only, rarely a protein remains folded in cytosol and gets unfolded by the membrane-linked protein transport system. Chaperones also differ in their specificity. For instance, Hsp70s interact with a wide spectrum of non-native proteins, while SecB and MSF (mitochondrial import-stimulating factor) only interact with proteins destined for transport into or across a membrane. Other factors are also needed. Release from SecB needs SecA and release from Hsp70 and Dnak requires interaction with co-chaperones DnaJ and GrpE accompanied with ATP hydrolysis. A higher level of complexity and regulation is involved in case Of cytosolic RNPs termed signal recognition particle (SRP) (see later in this Chapter).

In *E. coli*, proteins can be transported to transport machinery in plasma membrane by any of the two pathways (Fig.), one mediated by SecB and SecA and the other mediated by SRP and FtsY (both are GTPases; FtsY is known as SRα in eukaryotes). In the first pathway cytosolic SecB (a chaperone) binds the protein and delivers it to membrane bound SecA. In the second pathway, SRP (4.5S RNA + Ffh protein; Ffh a GTPase) binds to the nascent secretory protein and the protein-SRP-ribosome complex then binds to FtsY (GTP dependent reaction). SRP and FtsY increase the efficiency of transporting the protein.

SRP dependent and SRP independent translocations of proteins, destined to endoplasmic reticulum (ER), occur in eukaryotes also, where SRP dependent translocation is coupled with translation. Proteins can also be translocated to ER in a SRP independent manner as shown in yeast. It makes use of

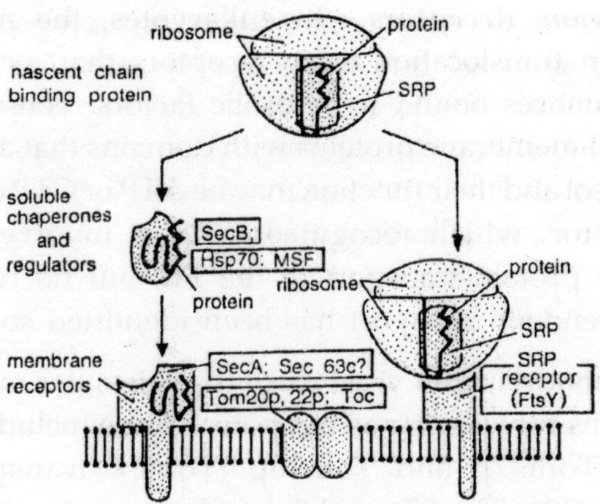

Transport of proteins to the transport machinery in plasma membrane by two alternative pathways.

Hsp70 and DnaJ homologue Ydjlp, which together help in transporting proteins to mitochondria and other organelles. The synthesis of most of the mammalian membrane proteins and secretory proteins begins on cytosolic free ribosomes. These proteins are co-translationally targeted to the membrane of the ER. They are recognized and selected with the help of three GTPases (SRP54 belonging to SRP; SRα and SRβ belonging to SRP receptor). These GTPases work as adapters between the translation machinery in the cytosol and the translocon (translocation apparatus) in the ER membrane. The translocon is mainly composed of the heterotrimeric Sec61bp. The GTPases bind to GTP and the GTP-binding activity of atleast one of them (SRP54=54kDa) is modulated by the ribosome itself. SRP54 (equivalent of Ffh of *E. coli)* is a subunit of signal-recognition particle (SRP), which is a cytosolic ribonucleoprotein (RNP), and identifies the ribosomes engaged in the synthesis of nascent chains bearing ER-directed signal sequences. SRP consists of RNA and six polypeptides, SRP54 being the most important. The amino-terminal domain of SRP54 contains GTPase activity and the carboxy - terminal domain binds to signal sequence and SRP - RNA.

Membrane Receptors : In eukaryotes, the membranes involved in translocation carry receptors that recognize the signal sequences bound to cytosolic factors. These receptors are integral membrane proteins with domains that are exposed to the cytosol and their function may be ATP or GTP dependent. SRP receptor, which recognizes SRP is involved in SRP-dependent protein transport to the ER, but no receptor for SRP-independent transport has been identified so far.

Mitochondria from yeast carry in their outer membrane 3 or 4 proteins functioning as receptors. These include Tom20p, Tom22p, Tom37p and Tom70p (Tom = transport outer membrane; 20p, 22p, 37p and 70p indicate molecular masses). The proteins Tim23 and Tim 17 found in the mitochondrial inner membrane are not receptor proteins, but are constituents of protein import channel (see Fig.). Similarly, the chloroplasts have protein import receptors called Toc 34 and Toc 86 (also called Iap34, Iap86), which have GTP binding motifs, so that the binding to chloroplast surface is controlled by GTP. Toc75 forms the channel, forming a barrel; Tic complex consists of Tic11, Tic53, Tic22, and Tic 20.

Transmembrane Protein Channels and Motors : The protein, destined to be transported through a membrane, actually moves through a hydrophilic, hetero-oligomeric transmembrane channel composed of integral membrane proteins. The movement of proteins across the membrane and within the plane of the membrane is facilitated by the following structures:

(i) the signal sequence helps in gating (opening) the channel across the membrane,

(ii) the stop-transfer sequence or signal-anchor sequence helps in gating the channel within the plane of the membrane;

(iii) translocation of the protein is powered by a peripherally attached protein translocation motor that hydrolyses

ATP or GTP and then pushes or pulls the protein across the membrane.

In bacteria and mitochondria, these motors are aided by a transmembrane electro-chemical gradient. However protein transport into thylakoids in chloroplasts can be driven by a pH gradient also.

Several transmembrane protein channels have been characterized. Some of them include the following :

(i) In *E. coli* plasma membrane, the transmembrane channel is a heterotrimer composed of the integral proteins SecY, SecE and SecG, hence the name SecYEG; this acts in concert with SecA and the membrane proteins SecD and SecF.

(ii) The channel across the ER of mammals and yeast is a Sec61 complex having three subunits similar in sequence to the three subunits of SecYEG complex found in bacteria,

(iii) In the inner membrane of mitochondria, the import channel consists of integral proteins Tim23 and Tim 17 and perhaps two additional subunits.

(iv) In chloroplasts, the transport channel across the membrane consists of two integral proteins of the inner envelope, namely lap 100 and lap 36.

Protein transport channels are not active transporters, but function like conduits (tubes) that resemble F_0 portion of F_1 F_0-ATPase. In this F_1F_0-ATPase, active proton transport is affected by coupling the gated proton channel F_o to the peripheral proton pump F_1 which is powered by the hydrolysis of ATP. Similarly, as indicated above, a peripheral attached motor protein is powered by the hydrolysis of GTP or ATP thus pushing or pulling the protein through a protein transport channel across the membrane. Following are some examples of these motors :

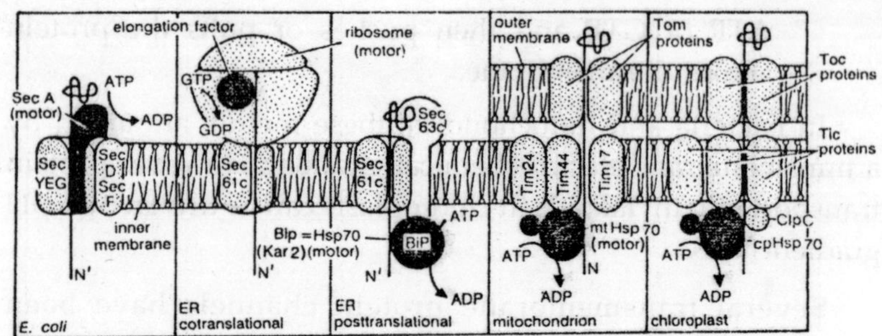

Transmembrane protein channels and motor proteins in* E. coli, *and endoplasmic reticulum (ER), mitochondria and Chloroplast of higher eukaryotes.

(i) Hsp70 (also termed Kar2 or BiP) is found on trans (inner) side of the membrane and pulls the protein into ER as a part of post-translational import;

(ii) SecA is bound on the cis (outer) side of the membrane and pushes the protein across the bacterial plasma membrane or Chloroplast thylakoid membrane;

(iii) ribosome bound to cis side of membrane may also work as a motor and may provide GTP-powered push (Fig.)

In the above discussion, it may be noted that in bacteria, the protein is pushed through the channel, while in ER and mitochondria it is pulled. It may be due to lack of ATP on the trans side of the bacterial plasma membrane, so that one may assume that the bacteria have put the motor where the fuel (ATP) is available.

The ribosome is also believed to work as a motor, so that if the ribosome is fixed to the membrane independent of the translocating polypeptide chain, then the push can be exerted by the GTP-powered polypeptide elongation machinery. Thus the growing polypeptide is extruded from the ribosome tunnel and traverses through the transmembrane protein channel. It is known that ribosome does remain tightly bound to ER with the help of Sec61 complex and some additional ribosome receptors. The ribosomal push may also need the help of Hsp70-mediated pull, or alter-natively Hsp70 helps in the removal of chain at the exit site, because Hsp70 (BiP) has been shown to be essential even for co-translational transport of freshly synthesized proteins.

The motors for transport of proteins through transmembrane channels are mainly found in (i) the inner membranes of chloroplasts and mitochondria, (ii) ER, (iii) thylakoid and (iv) bacterial plasma membrane. They seem to be absent on the outer membranes of mitochondria and chloroplasts.

Despite this, proteins are transported through these outer membranes to reach internal compartments. It is suggested that a protein with signal sequences is driven across the membrane by association with one or more binding sites on the inner side of the outer membrane. These binding sites may be provided by Tom22p in mitochondria and by Hsp70 in the chloroplasts. Thus the protein is translocated across the outer membrane of chloroplasts and mitochondria by a relay of binding sites on both faces of the membrane. (A similar, cascade of binding sites is involved in the movement of proteins through the nuclear pore complex).

Once the protein reaches the inter-membrane space after crossing the outer membrane, it may be captured by the transport system of the inner membrane, where an ATP-linked motor helps further in the translocation process. In mitochondria, this job is done by the electric potential across the membrane. This potential pulls the positively charged signal sequence across the inner membrane.

The components of the machines involved in the export and import systems of translocation differ very much and are listed in Table. The only exception is Hsp70, which is not only involved in both these systems, but also in a variety of other aspects of protein folding.

Components of Export and Import Systems of Translocation in E. coli, ER, Chloroplasts, Mitochondria and Peroxisomes.

Components of transport system	Type of system					
	Export systems			Import systems		
	E. coli	ER.	Chloroplast (thylakoid)	Mitochondrion	Chloroplast	Peroxisome
Cytosolic factors	Trigger factor; Sec B	NAC	—	—	—	Pas10p (Pas 8p); pas 7P
(chaperones or	Dnak	Hsp70	Hsp70	Hsp70	Hsp70	—
regulators)	SRP-GTP, FtsY-GTP	SRP-GTP	SRP-GTP	MSF-ATP	—	—
Membrane	Sec A	—	cpSec A	Tom 20p-Tom 22p	Toc86, Toc34	Pexl3p, pexl4p
receptors	Fts Y	SRα, SRβ	—	Tom 37p-Tom 70p		
Channels	Sec Y	Sec 61 p (Sec 61α)	Sec Y	Tom7, Tom6, Tom5	Toc75	
	Sec G	Sec lp (Sec 61β)	—	Tim 17p, Tim23p		
	Sec E	Sec lp (Sec 61γ)	—	—	—	—
		Sec 63p, 62p, 71p, 72p	—	—	—	—
	Sec D, Sec F			Tim 44p	?	—
Translocation	Sec A	—	cpSec A	—	—	?
motors	—	Ribosome	—	—	—	—
	—	Kar2p (BiP)	—	mtl Isp70	cpl Isp 70	—
Outer membrane channels of organelle				Tom 40p, 6p, 7p, 8p	Toc75; Tic110, 55, 22, 20	

Protein Folding at the End of Transport (Chaperone and Folding Catalysts) : Once the protein comes out of the transport channels, it is folded on the trans side of the membrane. This folding is mediated by a battery of chaperones and other folding helpers, whose function may be specific for a protein or may be general in nature. In the stroma of chloroplasts and matrix of mitochondria, homologues of bacterial GroEL and GroES (termed Hsp60 and Hsp10, or Cpn60 and Cpnl0) are found, which help in the folding of many but not all of the imported proteins. They also help in the folding of proteins synthesized within these organelles. Proteins exported to periplasm in bacteria are refolded with the help of several chaperones and by atleast one proline rotamase. ER lumen also contains several chaperones (eg., BiP, Hsp90, calnexin), which facilitate folding of all imported proteins.

Import of Proteins in Peroxisomes : Most proteins imported to peroxisomes carry a COOH-terminal Ser-Lys-Leu tripeptide motif collectively called PTS1 (peroxisome-targeting signal). A few other imported proteins carry a transient NH_2-terminal signal (termed PTS2). PTS1 signal is recognized by a protein Pas8p or its homologue. Pas10p (depending on which yeast species is involved). Similarly PTS2 is recognized by Pas7p. Pas8p (or Pas10p) is recognized by peroxisomal integral protein Pas20p. The system thus is comparable to other systems using cytosolic factors, coupled to membrane receptor, but no channels or motors are known yet.

2

RNA Processing

Protein coding structural genes in higher eukaryotes are transcribed in the nucleus, but the primary RNA transcripts in the nucleus differ from mRNAs used in the cytoplasm for translation. The RNA transcripts in the nucleus, due to their broad size distribution (500-20,000 bp; mean length 2000 bp), are collectively described as heterogeneous nuclear RNA (hnRNA). Primary transcripts derived from individual genes, should be better described as pre-mRNA molecules, each of which is generally much larger than its corresponding mRNA. An individual hnRNA can have one of the following fates :

(i) RNA transcripts of some genes do not seem to give rise to any cytoplasmic mRNA, but get degraded within the cytoplasm,

(ii) For each gene, only a small proportion (25%) of RNA transcripts takes part in RNA processing leading to formation of mRNA, the remaining 75% undergoing degradation in the nucleus.

Due to these two events only 5% of hnRNA (by mass) enters the cytoplasm. This also suggests that regulation of gene expression is also exercised at the level of hnRNA. The hnRNA molecules, which are destined to produce mRNA, undergo RNA processing which includes the following events:

(i) Modification of 5' end by capping (addition of 7-methyl guanine) and modification of 3' end by a tail (polyadenylation) after enzymatic cleavage;

(ii) Splicing out of intron sequences from RNA transcripts of interrupted genes. It has been shown that cleavage and polyadenylation usually precede RNA splicing, although there is no causal relationship between the two.

Besides the protein coding genes, there are other genes like those for ribosomal RNA (rRNA) and transfer RNA (tRNA), which are involved in RNA synthesis, giving rise to more than 90% of nuclear RNA. These RNA transcripts also undergo RNA processing giving rise to rRNA and tRNA. In this chapter, we briefly describe processing of hnRNA, but the discussion on RNA splicing, involved in processing of hnRNA will also include processing of precursor rRNA and precursor tRNA.

Polyadenylation of mRNA in Prokaryotes

At the 3' end of mRNA of both prokaryotes and eukaryotes, a run of adenylate (polyA) is added . This is described as polyadenylation. For sometime in the past, it was believed that polyadenylation is unique to eukaryotes, but now it is well established that polyadenylation also takes place in prokaryotes, although only in 2-60% molecules of a given mRNA species in prokaryotes as against 100% of mRNA molecules in eukaryotes (except histone mRNAs). Further the polyA tails in prokaryotes are 14-60 nucleotides long and are thus much shorter than the 80-200 adenylates long tails found in eukaryotic mRNA molecules. The sites of polyadenylation in bacterial mRNA can be diverse and can include the following:

(i) 3' end of primary transcript;

(ii) the sites of endonucleolytic processing in the 3' untranslated and intercistronic regions and

(iii) the 3' sites of degradation products within the coding regions of mRNA .

This diversity in the polyadenylation sites suggests that mRNA polyadenylation in prokaryotes is relatively indiscriminate process that can occur at all mRNA 3' ends, and does not require specific consensus sequences as in eukaryotes. The polyadenylation requires a poly(A) polymerase (PAP) and two such polymerases have been isolated in *E. coli.* Polyadenylation may play one of the following roles in prokaryotes:

(i) it may promote degradation of regulatory RNA that inhibits replication of bacterial plasmids;

(ii) it may help in mRNA stabilisation and

(iii) it may play some role in mRNA translation.

Addition of Cap (m^7G) and Tail (Poly A) for mRNA in Eukaryoes

Addition of Methylated Cap at the 5′ End : The initial RNA transcript (hnRNA), derived from genes coding for proteins, gets modified so that its 5' end gains a methylated guanine (cap) and its 3' end is polyadenylated (tail). Capping at 5' end occurs rapidly after the start of transcription and much before completion of transcription. Transcription starts with a nucleoside triphosphate (A or G), and a 5' triphosphate group is retained at this first position. The initial sequence at 5' end of hnRNA is therefore 5′pppA(or G)pNpNp...3'. To the 5' end is added a terminal G with the help of an enzyme guanyl transferase as follows :

5′Gppp + 5′pppA (orG)pNpNp → 5′Gppp5′ApNpNp + pp + p

The new G residue is in the reverse orientation with respect to all other nucleotides and undergoes methylation (using a methyl transferase) at its 7th position. The cap with a single methyl group at this terminal guanine residue is found in unicellular eukaryotes and described as cap0, but in most eukaryotes, methyl group may also be present on the penultimate base at 2' position of sugar moiety, so that with methyl groups at two terminal nucleotides, it is now described as cap1. In still other cases (10-15% of all caps), methyl group may be present at the third base also at the 2' position of sugar and this cap is now described as cap2 (Fig.). It is estimated that, at a particular moment, only - 20% of hnRNA is capped following transcription.

Capping of 5′ end of mRNA by cap 0, cap 1 and cap 2 due to methylation at several positions.

It could be shown that removal of cap leads to loss of translation activity due to loss of the formation of mRNA-ribosome complex. It suggests that the 'cap' helps in recognition of ribosome. Only in some eukaryotic mRNAs (as those for histone proteins), caps may be absent and may not be required for translation.

Polyadenylation and the Generation of 3′ End in Eukaryotes: The 3' end of mRNA is generated in two steps (Fig.) (i) Nuclease activity cuts the transcript at an appropriate location, (ii) Poly(A) is added to the newly generated end by an enzyme, poly (A) polymerase (PAP), utilizing ATP as a substrate. Estimations have shown that ordinarily 30% of hnRNA and 70% of mRNA are polyadenylated. In a region 11-30 nucleotides upstream of the site of poly(A) addition, there is a sequence AAUAAA (in all higher eukaryotes except yeast), which perhaps gives a signal for cleavage [step (i) above].

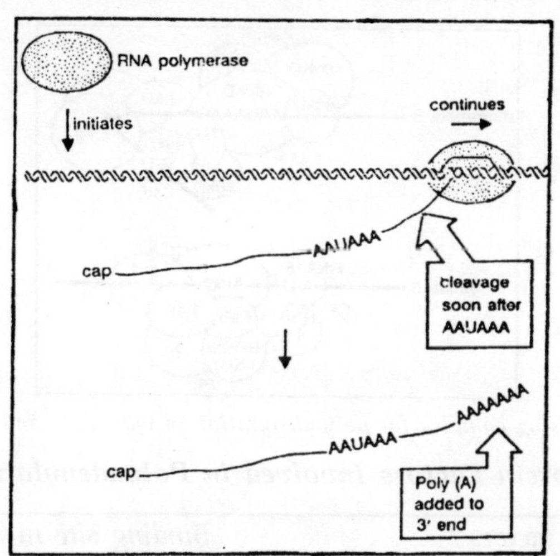

Cleavage at AAUAAA and polyadenylation at the cleaved 3′ end.

In addition to AAUAAA, there are following consensus sequences (Fig.), that are involved in polyadenylation:

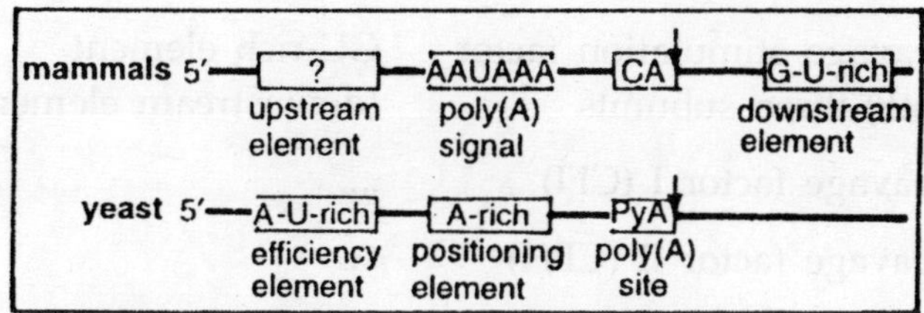

Consensus sequences involved in polyadenylation in mammals and yeast.

(i) a G-U rich element is present downstream to the site of cleavage, and is important for efficient processing for polyadenylation (this sequence is absent in yeast);

(ii) a Consensus sequence immediately 5' of the cleavage site (although this is not essential);

(iii) consensus upstream element situated 5' of a poly A signal or AAUAAA (this is found in many, but not all pre-mRNAs).

There are also protein factors, which interact with these signals. These protein factors are listed in Table, and form a processing complex for polyadenylation (Fig.).

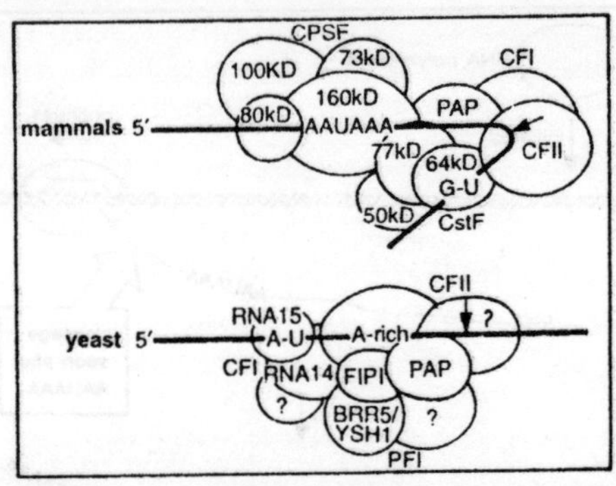

Processing complex for polyadenylation in mammals and yeast.

Protein Factors Involved in Polyadenylation

Protein factors	Binding site in mRNA
1. Cleavage and polyadenylation specificity factor (CPSF)-four subunits	AAUAAA (ooly A signal)
2. Cleavage stimulation factor (CstF)-three subunits	GU-rich element (downstream element)
3. Cleavage factor I (CFI)	....
4. Cleavage factor II (CFH)	....
5. Poly(A) polymerase (PAP)	poly (A) site (-CA)

In most mRNAs for histone proteins, no polyadenylation at 3' end is found, so that 3' ends are processed without addition of poly (A). The U7 snRNA (56 bases long and not related with other snRNAs involved in spliceosome; is involved in this processing, through extensive complementary base pairing with histone mRNAs (or hnRNAs). Other proteins may also be involved in processing of 3' end of hnRNA, whether or not it is polyadenylated.

Decapping and Deadenylation vs Instability of mRNA : Each mRNA has an intrinsic stability expressed as half life of this mRNA. A typical mRNA of *E. coli* has a half life of about two minutes; in yeast it is 10-20 minutes, and in mammals it is several hours. Even within an organism, different mRNAs may differ in stability. The half life of mRNA depends on (i) some mRNA-specific sequences and (ii) a number of mRNA binding proteins. The m^7G and poly(A) tail work as barriers to this mRNA degradation, since none of the two types of exonucleases (5' $\rightarrow$ 3' and 3' $\rightarrow$ 5') will be able to function in presence of cap and tail. Therefore, deadenylation and decapping are necessary for degradation. There are two pathways for mRNA degradation in budding yeast, one is deadenylation dependent and the other is deadenylation independent. In deadenylation dependent pathway, the 5'm^7G cap is cleaved by a decapping enzyme, Dcpl (encoded by *Dcpl* gene). Decapping is caused only after the degradation of 3'poly (A) tail and loss of a protein PABP (poly A-binding protein), that binds to poly A chain, causing removal of 10-15 adenine residues at the poly(A) tail. This is followed by rapid degradation of mRNA by the 5'-3' exonuclease encoded by *XRNl* gene. A deadenylation-independent pathway, on the other hand, is stimulated by the presence of a premature - termination codon (Fig.).

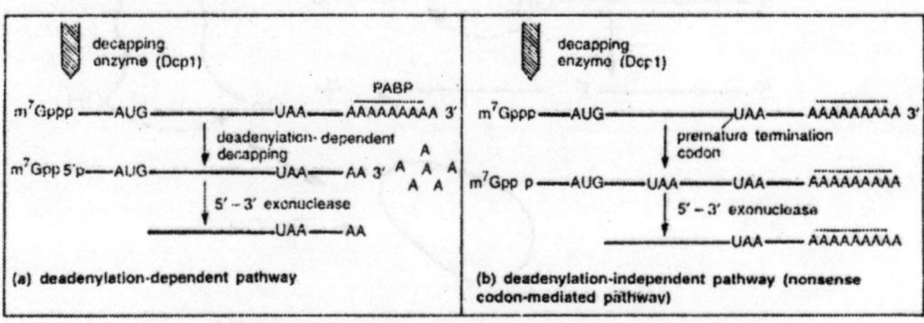

Two pathways of mRNA decapping in eukaryotes.

RNA Splicing

In eukaryotes, the RNA transcribed from DNA almost invariably undergoes RNA splicing to yield mature RNA sequences .(rRNA, mRNA, tRNA). It involves removal of sequences mainly corresponding to introns in split genes, although, other examples of RNA splicing (*e.g.* alternative splicing;) may also be available. The mechanisms available for this purpose include the following :

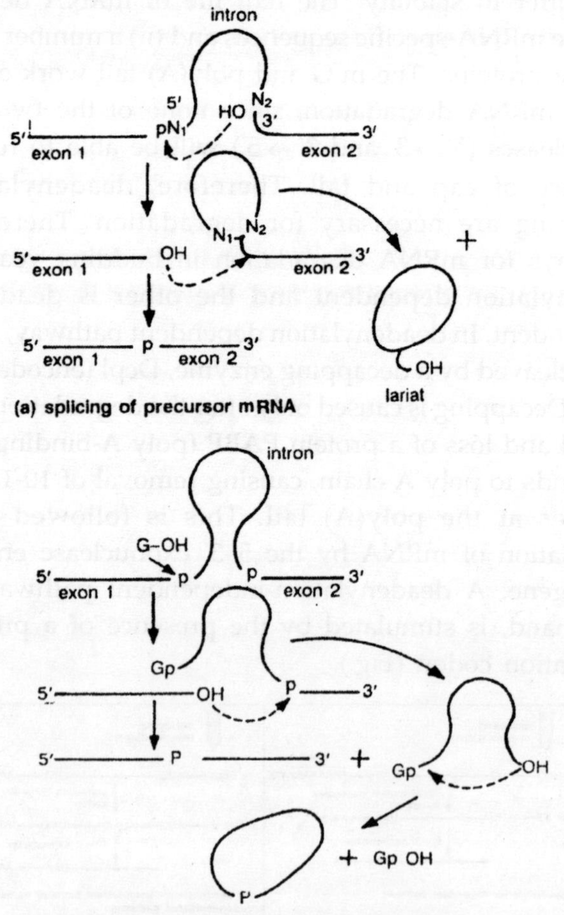

Self splicing of group I intons, by two transesterification reactions.

(i) self splicing of fungal mitochondrial and other group I introns;

(ii) splicing of higher eukaryotic nuclear introns, through the formation of spliceosomes at intron-exon junctions;

(iii) self splicing of mitochondrial group II introns through lariat formation without assistance from any proteins or spliceosomes;

(iv) splicing of yeast tRNA precursor molecules by cleavage due to endonuclease followed by fusion due to ligase.

Self-splicing of Group I Introns : Self-splicing of RNA molecules involving group I introns, found in rRNA genes of *Tetrahymena* and *Physarum* nuclei, in fungal mitochondria and in phage T4, takes place through two transesterification reactions (exchange of phosphate esters, which leave the total number of phosphodiester bonds unchanged). Group I introns are characterized by (i) the absence of conserved sequences at splicing junction, and by (ii) the presence of short conserved consensus sequences internally. In the first transesterification, the 5' splice site is cleaved, as guanosine is added to the 5' end

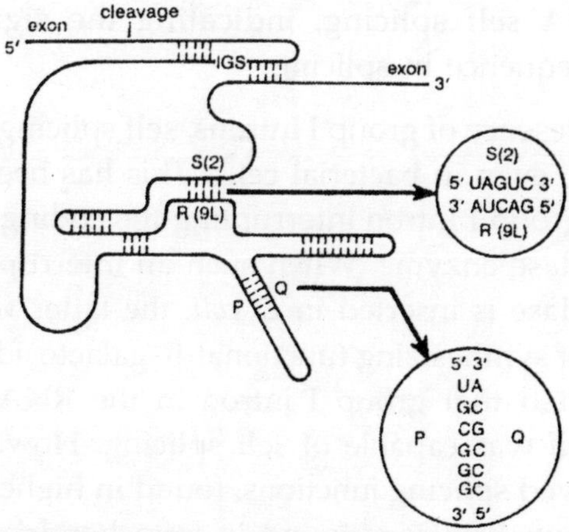

Common secondary structure of a group I intron in **Tetrahymena**, *facilitated by internal pairing due to consensus sequences (P, Q and R, S) and the pairing at the ends due to internal guide sequence (IGS).*

of intervening sequence (IVS) to be cleaved. In the second transesterification step, the 3' splice site is cleaved as the exons are joined (Fig. a). The excised IVS or intron can form a circle by cyclisation reaction and these circles can again regenerate linear molecules due to autocatalysis. It means that excised IVS RNA and truncated versions of the IVS RNA also act as ribozymes to cleave, join or dephosphorylate RNA substrates.

As indicated earlier, all group I introns have some short consensus sequences, which allow intramolecular base pairing to generate a characteristic secondary structure within an intron. A characteristic feature of this secondary structure, is the pairing between two ends of an intron, with the help of an internal guide sequence (IGS), as shown in Figure. Internal base pairing is facilitated by the presence of two pairs of elements :

(i) *P* and *Q* each 10 bases long, of which 6-7 bases are involved in pairing;

(ii) *R* and *S* elements, each 12 bases long, of which only 5 are involved in pairing (Fig.).

Any mutations in this consensus sequence have been shown to stop RNA self splicing, indicating the significance of consensus sequence in splicing.

In the presence of group I introns, self splicing of RNA can be achieved, even in bacterial cells. This has been shown by inserting a group I intron interrupting the coding sequence of β-galactosidase enzyme. When such an interrupted gene for β-galactosidase is inserted in *E. coli,* the latter was found to be capable of synthesizing functional β-galactosidase enzyme. This suggested that group I intron in the RNA transcribed within *E. coli* was capable of self splicing. However, introns with conserved splicing junctions, found in higher eukaryotes (not the group II introns found in mitochondrial genome of lower eukaryotes, plants and bacteria, need spliceosomes formed by snRNA molecules and therefore can not be spliced out in bacterial cells.

Self Splicing through Formation of Hammer-head : In case of viroids (naked infectious RNA molecules capable of replicating) and virusoids = satellite RNA (encapsidated with plant viruses, and incapable of replicating independently), a consensus sequence forms a 'hammerhead' secondary structure, which helps in self cleavage (Fig.).

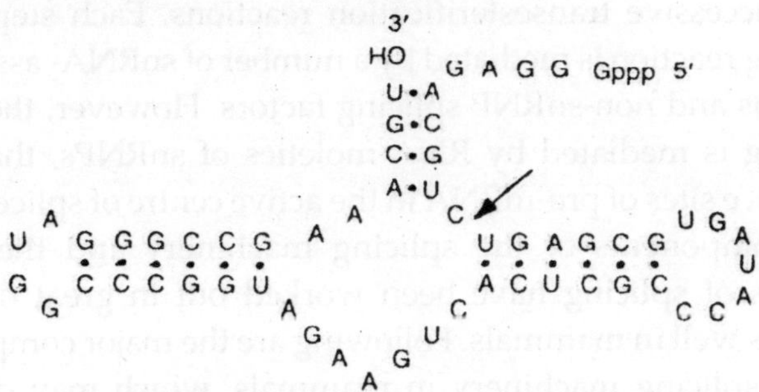

A hammerhead secondary structure in viroids and virusoids showing site for self cleavage by arrow.

Splicing of hnRNA of higher Eukaryotes through Spliceosomes : The following two features are significant in the introns of eukaryotic nuclear genes (to some extent, mitochondrial group II introns resemble these nuclear introns, although not entirely) :

(i) the two ends of an intron do not have complementary base sequences to allow formation of secondary structures as has been shown for group I introns;

(ii) the exon-intron junctions always have the same sequences, so that in more than 99% cases, an intron in RNA transcript starts with GU (5' end or left border) and ends in AG (3' end or right border); this is sometimes described as GU-AG rule (a minority of introns, are characterized by AT-AC exon-intron junctions.)

On either sides of intron-exon junctions also, there are common sequences described as consensus sequences, although these may not be found in all cases.

Splicing of Major Class of GU-AG Introns : Splicing of intron sequences of eukaryotic hnRNA (also called pre-mRNA) involves a well defined multi-step pathway. Small nuclear ribonucleoprotein particles (snRNPs = SnRNA + Sm proteins) and about 50 protein factors are essential for the formation of an active spliceosome, in which intron excision proceeds in two successive transesterification reactions. Each step of the splicing reaction is mediated by a number of snRNA- associated proteins and non-snRNP splicing factors. However, the actual splicing is mediated by RNA moieties of snRNPs, that align the splice sites of pre-mRNA in the active centre of spliceosome. The components of the splicing machinery and the actual process of splicing have been worked out in great detail in yeast as well in mammals. Following are the major components of the splicing machinery in mammals, which may differ in details worked out in budding yeast,

(i) snRNPs (or 'snurps') % consisting of snRNAs (Ul, U2. U4, U5, U6) and common Sm proteins,

(ii) SR family of splicing proteins;

(iii) polypyrimidine tract-binding proteins;

(iv) branch-site binding proteins;

(v) hnRNP proteins;

(vi) snRNP associated non-snRNP proteins;

(vii) some other non-snRNP splicing factors.

The Spliceosome Assembly : The splicing reaction involves the formation of spliceosome. For spliceosome assembly, snRNPs (Ul, U2, U4,U5,U6) associate with pre-mRNA using large number of essential protein factors. Among these essential proteins, the most important are the following :

(i) ***SR Proteins :*** (*e.g.*, SRp20, SRp35, SRp40, SRp55, SRp75); they are called SR proteins because at their C terminal region, they contain a domain, rich in serine, abbreviated as S and arginine abbreviated as R; the N-terminal region consists of an RNA recognition motif or RRM. SR proteins are among the first components that interact with pre-mRNA, thus committing pre-mRNA to the splicing pathway. They also escort snRNPs to the assembling spliceosome.

(ii) ***Polypyrimidine Tract-binding Proteins :*** (*e.g.* U2AF, PTB and PSF); they bind a polypyrimidine tract (lying between branch site and 3' splice site) and play an important role in spliceosome assembly.

Following steps are involved in spliceosome assembly. Of these steps, only the first step is ATP independent; all other steps require ATP (Fig.). (i) The first specific stage of spliceosome assembly is the binding of Ul snRNP, which requires interaction of pre-mRNA with SR proteins and also with a polypyrimidine tract binding protein, U2AF. It has been suggested that the commitment of pre-mRNA to the splicing pathway needs an interaction with SR proteins, which also have a general role of escorting snRNPs to the assembling splicosome. For instance, these SR proteins enhance the binding of Ul snRNP to 5' splice site and recruit U2 snRNP to the branch site. The binding of Ul snRNP to the 5′ splice site and that of U2AF to the polypyrimidine tract (it lies between branch site and 3' splice site) leads to the formation of complex described as commitment complex (CC) in yeast and complex E in mammals, (ii) The formation of this commitment complex is necessary for the subsequent binding of U2 snRNP to the branch site. U2AF consists of two subunits ($U2AF^{65}$ and U2AF), of which only the larger subunit is essential for binding. The 12S U2 snRNP (having associated Sn proteins) also becomes associated with a number of proteins (mainly SF3a and SF3b), with the help of a 75kDa protein, SF1, to form a 17S U2 snRNP. Only this 17S snRNP and not the 12S snRNP is competent for binding to the branch site giving rise to pre-splicing complex A. The 12S U2 snRNP is inactive in pre-spliceosome assembly; during

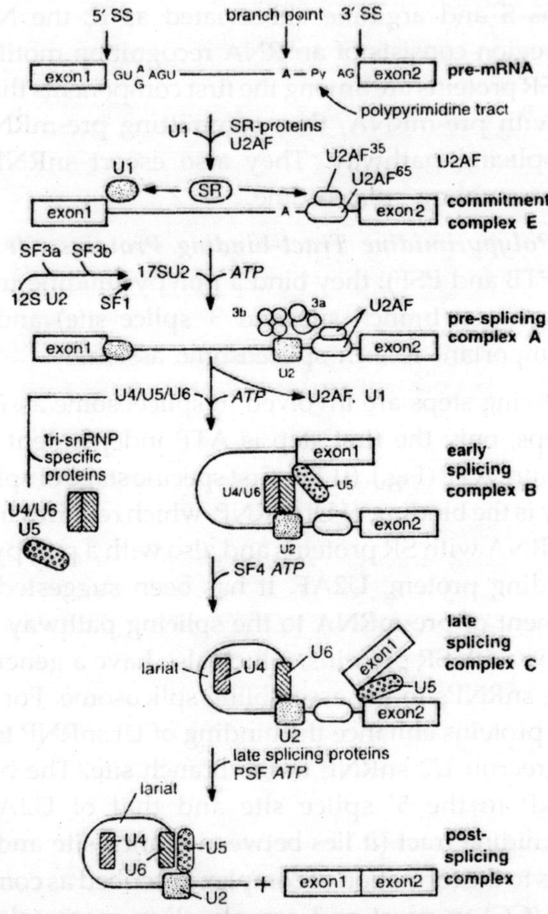

Spliceosome assembly pathway in mammals (redrawn from Schuler, 1998).

the base pairing between U2 snRNA and branch site, an adenosine at the branch point bulges out, which acts as the nucleophile for 5' splice site cleavage and lariat formation, (iii) The complex A gives rise to early splicing complex B, due to association of U4/U6. U5 tri-snRNP complex; Ul snRNP and U2AF dissociate during this step, (iv) Complex B undergoes a conformation change to form a late splicing complex C; this leads to destabilisation of base pairing between U4 and U6 snRNAs. The presence of U4 snRNA is not essential for subsequent catalytic steps.

Transesterification Reactions : The transesterification reactions are catalysed by an RNA component (U6 snRNA) of the spliceosome. In mammals, a protein called SF4 has also been described, which is not needed for the assembly of the spliceosome, but is required for the catalytic steps. The first transesterification reaction involves cleavage at the 5' splice site of an intron, which is facilitated by the nucleophilic adenosine at the branch site. For the second transesterification reaction, an intramolecular interaction between the highly conserved guanine residues at the 5' and 3' termini of an intron is essential and it also specifies the 3' cleavage site. After the two above catalytic steps, the intron is liberated, but still remains associated with U2, U5 and U6 snRNAs. The exons join together and are released from the spliceosome as a low molecular weight mRNA. Spliceosome disassembly is also an active process, which requires atleast one protein factor (PSF) and ATP.

Splicing of Minor Class of AT-AC Introns : The above mechanism of splicing through spliceosome assembly and sites have AU-AC rather than GU-AG sequences at the catalysis holds good for splicing of only those introns in pre-mRNA which have GU-AG splice sites. In metazoans, more than 99% introns conform to the GU-AG rule. However, in less than 0.1% of all known pre-mRNA introns, the splice sites. These introns are popularly described as AT-AC introns in literature (in DNA, it will be TA-TG and AT-AC on the two strands and in pre-mRNA, it will be AU-AC) and are present only rarely in metazoan genes. One such gene having AT-AC introns is P120 coding for proliferation associated nucleolar protein in several mammals. Splicing of these AT-AC or AU-AC introns involves a distinct and rare class of snRNAs , Ul 1 and U12. Three other snRNAs taking part in the spliceosome assembly in these AU-AC introns include U5, U4atac and U6atac. The sequence of the branch site also differs in the AU-AC introns and is UCCAAC. Ull interacts with the 5' splice site, while U12 interacts with the branch site, so that Ull and U12 perform the same function in AU-AC introns, which Ul and U2 perform in GU-AG introns (Fig.).

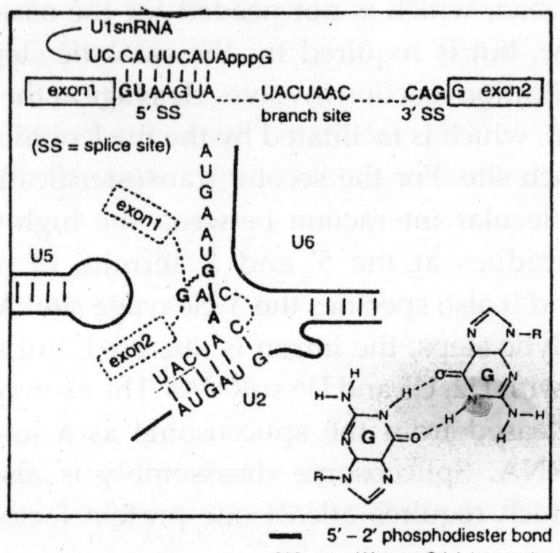

(a) major class (GU-AG introns) pre-mRNA splicing

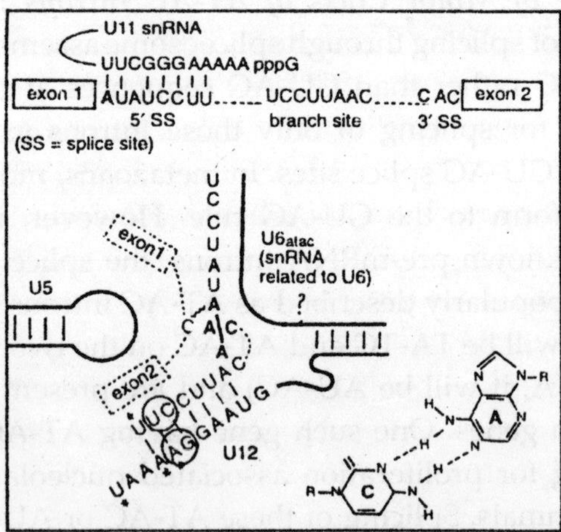

(b) minor class (AU-AC introns) pre-mRNA splicing

A comparison of pre-mRNA splicing at GU-AG and AU-AC intron junctions (redrawn from Mount, 1996).

Splicing of UA Rich Introns in Plants : The characteristic features of introns in plants and emphasized that plant introns have UA-rich elements spread throughout their length. Thus UA-rich elements help in recognition of 5' and 3' splice sites. Potential 5' splice sites are also differentiated by the degree of complementarity to Ul snRNA and by their location between AG-rich exonic elements and UA-rich intronic elements. Similarly potential 3' splice sites are differentiated by their location at the UA transition point and by a 3' splice site consensus sequence (U-rich concentration followed by GC AG). Following interactions are involved in plant intron recognition (Fig.):

(i) Ul snRNP binds to 5' splice site,

(ii) U-tract binding factors bind the U-rich sequence preceding the 3' splice site,

(iii) UA-island binding proteins bind to UA-rich elements of intron and

(iv) exon sequence element (ESE) binding proteins associate with AG elements in the adjacent exon.

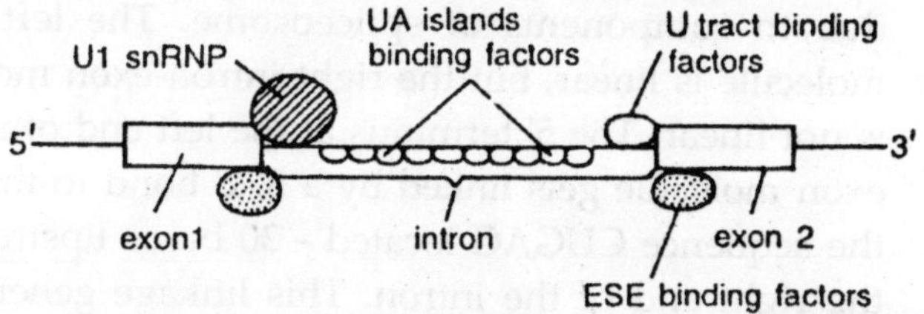

Interactions involved in recognition of plant introns.

A major difference between vertebrate and plant intron recognition systems is that while in vertebrates, the primary unit recognized for spliceosome assembly is the exon but in plants, splice site selection is primarily defined by UA-rich sequences within the intron. Consequently UA richness in plant introns is essential for efficient splicing and for 5' and 3' splice site recognition. However, the monocot splicing machinery is more permissive (less dependent on composition) than the dicot recognition machinery.

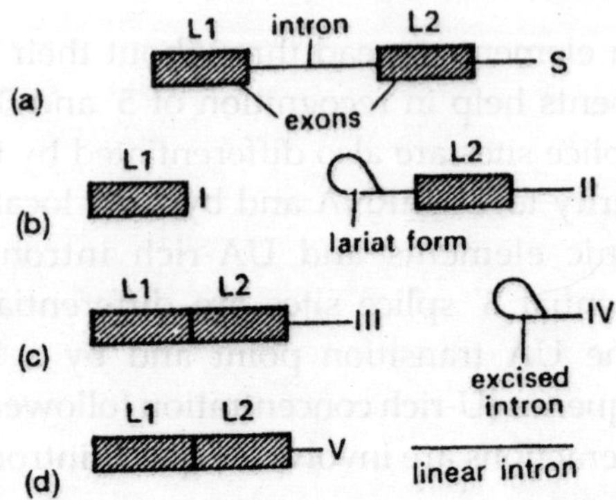

Formation of lariat type of loop configuration in excised introns in a split gene. L1 and L2 are two exons; 1-V are steps in excision.

Formation of Lariat Daring Splicing : Nuclear splicing, where spliceosome is formed, involves formation of a lariat structure. This occurs is two stages :

(i) In the first stage, a cut is made at the left end of the intron, releasing a separate RNA molecule with left exon and a right RNA molecule with intron and the right exons. The two RNA molecules are held together due to components of spliceosome. The left RNA molecule is linear, but the right intron-exon molecule is not linear. The 5' terminus at the left end of intron-exon molecule gets linked by a 5'-2' bond to the A of the sequence CUGAC located - 30 bases upstream of the right end of the intron. This linkage generates a lariat,

(ii) In the second stage, cutting at the right splicing junction releases a free intron in lariat form, and the left exon is ligated to the right exon (Fig.). The lariat is debranched to give a linear excised intron, which is rapidly degraded.

Splicing of Group II Introns : The group II introns (present in organellar genomes of lower eukaryotes, plants and also in bacteria) resemble introns of nuclear pre-mRNA or hnRNA of higher eukaryotes and are excised as lariats like those produced

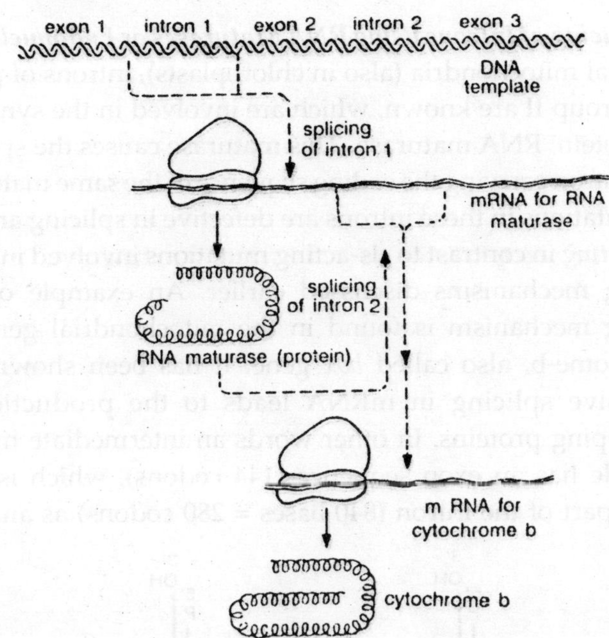

Splicing through the synthesis of a RNA maturase, using a negative feedback loop.

in nuclear pre-mRNA introns of higher eukaryotes. They have consensus sequences at the splicing junctions, GT and AP_y and a branch sequence resembling TACTAAC box. The splicing reaction is autonomous unlike that in the introns of pre-mRNA of higher eukaryotes, where trans-acting snRNPs take part in the formation of spliceosomes. Splicing in these group II introns gives rise to lariats directly, which then give rise to linear forms having no activity (Fig.). Thus, whereas in group II introns, highly conserved sequence elements juxtapose splice sites and branch sites by intramolecular base pairing interactions, in the nuclear pre-mRNA introns, the same functions are compensated by intermolecular interactions between the RNA substrate and the RNA moieties of snRNPs. This self splicing reaction of group II introns (without the help of snRNPs) may be regarded as an intermediate step between RNA mediated self splicing in group I introns and protein dependent RNA splicing of the introns of nuclear pre-mRNA of higher eukaryotes.

Splicing of Introns using RNA Maturases or Endonucleases: In fungal mitochondria (also in chloroplasts), introns of group I and group II are known, which are involved in the synthesis of a protein, RNA maturase. This maturase causes the splicing of the intron carrying the coding sequence of the same maturase. The mutations in these introns are defective in splicing and are *trans-acting* in contrast to cis-acting mutations involved in other splicing mechanisms discussed earlier. An example of this splicing mechanism is found in the mitochondrial gene for cytochrome-b, also called *box* gene. It has been shown that successive splicing in mRNA leads to the production of overlapping proteins. In other words an intermediate mRNA molecule has an exon sequence (144 codons), which is read with a part of the intron (840 bases = 280 codons) as an open

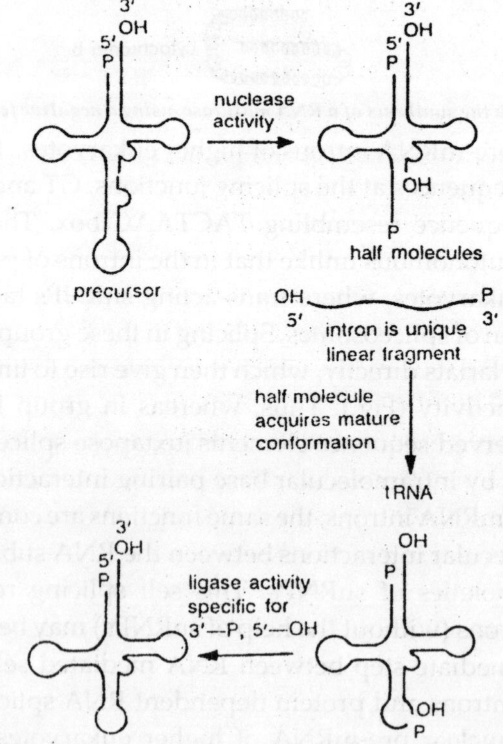

Splicing of yeast tRNA, due to nuclease and ligase activities.

reading frame, to give rise to the protein RNA maturase (the remainder of intron is not translated). This open reading frame is actually generated by splicing of the first intron. This maturase is believed to be responsible for splicing out the second intron, thus producing the mRNA, which will then disrupt the reading frame for maturase and will be translated into cytochrome b. This mechanism is illustrated in Figure. It is not necessary that the maturase has a catalytic activity, but may only recognize the splicing site, so that the splicing is achieved by a complex consisting of maturase, and other proteins and/or RNA molecules.

In fungal mitochondria, examples are also known, where an intron may code for an endonuclease, which causes a double stranded cleavage at a target site. At this cleavage site, a copy of the intron coding for the endonuclease is inserted, thus causing polymorphism for the presence or absence of introns. This is an example of DNA recombination and there may be proteins coded by introns which may have both maturase activity for RNA splicing and endonuclease activity for DNA recombination.

Yeast tRNA Splicing by Cutting and Rejoining **:** About 40 genes of approximately 400 genes for yeast nuclear tRNAs are interrupted, each with a single intron (14-46 bp), located one nucleotide away from 3' end of anticodon. This intron which may differ in different tRNA genes, has no consensus sequence, but has a sequence complementary to the anticodon. This leads to pairing between anticodon and intron giving a secondary structure, which is recognized by the splicing machinery. The splicing requires following two steps:

(i) phosphodiester bond cleavage by an endonuclease; this does not require ATP; and

(ii) ligation reaction, which requires ATP and involves bond formation with the help of RNA ligase; the splicing mechanism is illustrated in Figure.

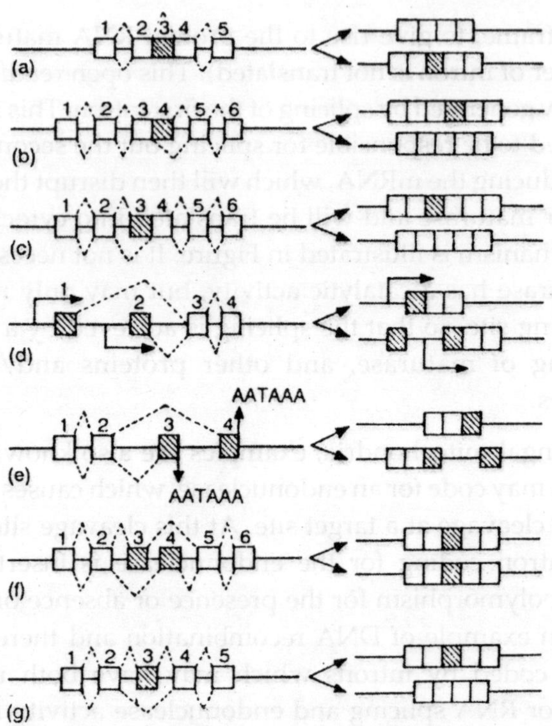

Models of alternative splicing; constitutively spliced exons are shown as open white boxes and alternatively spliced exons as hatched boxes; splicing pathways are shown by the diagonal lines; promoters and poly (A) sites are denoted as TATA and AATAAA respectively.

Trans-splicing of Transcripts in Chloroplasts and Mitochondria : Trans-splicing means that the gene is initially transcribed in pieces and after excision of introns, these pieces join to give rise to mature mRNA. Such trans-splicing reactions involving introns of Chloroplast fragmented genes *(rpsl2* in land plants and *psaA* in *Chlamydomonas reinhardtii)* have been reported.

A similar trans-splicing pathway has been reported to generate the mature mRNA for the mitochondrial gene for NAD1 (or NAD5) protein in several plant species. In such cases, coding region of the gene is found scattered among other genes and each coding segment is flanked by intron-like sequences. In these trans-splicing reactions, factors (protein as well as RNA) coded by nuclear as well as organellar genes take part.

Constitutive vs Alternative Splicing **:** In the earlier sections of this chapter, we have discussed the various mechanisms involved in RNA splicing. Generally, the pre-mRNAs undergo processing, so that the non-coding intervening sequences, *i.e.* introns, are excised, and exons correctly ligated. When this splicing takes place in all cells, without any variation, it is described as 'constitutive splicing' in contrast to 'alternative splicing', which differs in time and space. It has been shown that expression of a large number of genes is regulated by alternative splicing, which gives rise to protein 'isoforms', sharing extensive regions of identity and varying only in specific domains. Often these protein isoforms are themselves important regulatory molecules, such as transcription factors, hormone receptors and ion channels. Thus 'alternative splicing' is an adjunct to the regulation by promoter activity, gene rearrangements and by the occurrence of multigene families. This mechanism of regulation has been used in organisms extending from viruses to mammals and is an economical method of regulating gene expression at the post-transcriptional level.

In alternative splicing, splice sites that are selected for splicing in some circumstances are completely by-passed by the splicing machinery under other circumstances. The latter may lead to (i) insertion of a peptide segment or (ii) functionally different product or (iii) no functional product. Following are some examples of alternative splicing, which are depicted in Figure.

Retained Introns **:** In some genes an intron may be spliced out or retained. Examples include fibronectin, PDGFA chain and transposase of 'P' element in *Drosophila.*

Alternative 5' donor sites (or 3' acceptor sites). Introns of different lengths are spliced out due to the availability of two or more alternative donor or acceptor sites (donor or acceptor of OH group during transesterification) in the 5' region or 3' region of an intron. Examples include adenovirus Ela, SV40, ultrabithorax and transformer genes of *Drosophila,* etc.

Alternative Promoter or PolyA Sites : There are also genes having more than one alternative promoter sites (start points) at the 5' end (*e.g.* myosin light chain sites) or alternative poly A sites (cleavage sites) at the 3' end (*e.g.* a and P tropomyosins, immunoglobulin *[I* heavy chain, calcitonin CGRP genes).

Mutually Exclusive Internal Exons : In some genes, a pair of exons are present, which are neither spliced together, nor retained together, so that in one state the first is spliced out and in the other state, the second exon is spliced out giving two isoforms. Examples include genes for α and β tropomyosin, troponin-T, and pyruvate kinase.

Cassette Exons : Some genes have cassette exons that can be included or excluded, independent of others. If several such cassette exons are present, high degree of diversity can be generated in mRNA (*e.g.* genes for troponin-T, where upto 64 isoforms can be generated).

RNA Editing and Guide RNA

The phenomenon of RNA editing was discovered in mid 1980s, following some unexpected observations made by several groups of molecular biologists. These observations made particularly in the mitochondrial genes of *Trypanosoma* (which causes African sleeping sickness) and *Leishmania* (causing kala-azar), included the following :

(i) Several genes having mutations, which were supposed to render these genes inactive (due to origin of premature stop signals, or due to loss of start signals), were still active,

(ii) The sequences of mRNA molecules in several cases were such which could not have been derived from the DNA sequence of the corresponding genes.

These DNA sequences have sometimes been described as nonsensical 'cryptogenes' (cryptogenes differ from cryptic genes). These observations were later explained by RNA editing, which sometimes included addition of upto 550 and deletion of upto 41 uridine residues. In *T. brucei,* RNA editing actually inserts a total of 3030 U's into hundreds of editing sites in 12 different pre-mRNAs. It was shown that RNA editing mainly involved addition and deletion of uridines or insertion of cytidines, and that these changes took place at very specific sites and not 'randomly. Sometimes, RNA editing also involves conversion of adenine into guanine or inosine at specific sites.

RNA Editing through Guide RNA (gRNA) : Both in *Trypanosoma* and *Leishmania,* mt-DNA sequences were found (mainly in minicircle mtDNA, but rarely also in maxicircle mt-DNA), that encode small RNA molecules, less than 40 nucleotides in length. This RNA apparently carried information for uridine insertions and deletions, and was therefore called guide RNA or gRNA. In 1994, a model for RNA editing was presented, which suggested involvement of the following two transesterification reactions; (i) the first reaction joins the oligo-U3' tail of the gRNA to the downstream half of the pre-mRNA at the targeted editing site, generating a gRNA-mRNA chimeric intermediate; the step was believed to be catalysed by mitochondrial uridyl transferase activity; (ii) the second transesterfication reaction at the adjoining site, reforms the mRNA with U residues appropriately added or deleted. This model received wide support at that time. However, in 1996, *in vitro* experiments suggested involvement of endonuclease and ligase rather than the two transesterification reactions in U addition and deletion. Following enzymatic steps (rather than the two transesterification reactions) are now believed to be involved in addition and deletion of U's:

Two different models for the mechanism of RNA editing.

(i) gRNA directs endonuclease cleavage of the pre-mRNA at the targeted editing site;

(ii) addition of U residues from free UTPs (not from gRNA) at the 3' end of the upstream (5') cleavage product (catalysed by terminal-U-transferase or TUTase); deletion of U involves a 3' U-specific exonuclease (not a TUTase);

(iii) religation of mRNA with the help of ligase enzyme, where gRNA specifies edited RNA sequence by base pairing-mediated juxtaposition of the unprocessed 3' cleavage product and the processed (with added or deleted U) 5' cleavage product.

This latter model involving enzymatic reactions has been preferred over the earlier model involving two transesterification reactions, although gRNA-mRNA chimeras, earlier observed, are produced in both the models (Fig.).

RNA Editing through dsRAD : The enzyme double-stranded RNA adenosine deaminase (dsRAD) is ubiquitous in multicellular organisms and is implicated in RNA editing. It converts adenosine to inosine (a rare base found in tRNA molecules) in dsRNAs, creating unstable IU base pairs, resulting in unwinding of long dsRNAs. There are atleast two examples of this RNA editing:

(i) The mRNAs for 5 of the 16 or more subunits of glutamate receptor (GluR) channels found in the mammalian brain, are edited at several sites involving conversion of adenosine (A) to inosine (I) due to deamination Inosines are read as guanosines by translation machinery, thus causing changes in amino acid residues in the GluR subunits. For instance CAG (glutamine) is changed to CIG (arginine). Such changes cause changes in the Ca^{++} permeability of the GluR receptors,

(ii) In hepatitis delta virus (HDV) the editing in the antigenomic strand leads to conversion of A to G or I, thus changing the stop codon (UAG) to tryptophan codon (UIG or UGG). The RNA editing results in the loss of stop codon, leading to an extension of 19-amino acids at the carboxy-terminus of the protein called hepatitis delta antigen (HDAg). This longer protein inhibits replication of HDV genome and is required for packaging.

Ribozymes (RNA Splicing, DNA Cleavage and RNA Amplification) : Enzymatic catalysis was earlier thought to be the exclusive domain of proteins. As shown above in this chapter, only during 1980s and 1990s, it was shown that RNA molecules can cut, splice and assemble themselves without any outside help, thus extending the range of chemistry of enzymes. These RNA molecules working as enzymes, were called ribozymes. It has also been shown that enzymes can also be synthesized chemically, and are then described as co-enzymes. For the work on RNA molecules working as ribozymes, two scientists (Thomas R. Cech of University of Colorado, Boulder and Sidney Altman of Yale University) were awarded Nobel Prize in 1989. This discovery of RNA molecules working as enzymes has also changed our thinking about origin of life. It is now believed that four billion years ago, Earth was an 'RNA world' in which RNA moleculess carried out all the processes of life without the help of either protein or DNA.

The discovery of ribozymes can be traced back to 1982, when Thomas Cech, was trying to understand how 'sequences' (including sequences derived from introns) are spliced out of RNA molecules during processing of hnRNA into mRNA, or precursor rRNA into rRNA in *Tetrahymena.* The discovery was made when it was found that removal of proteinaceous enzymes one by one from a cell free system did not alter the splicing ability of pre-ribosomal RNA. However, these RNA molecules acted only on themselves, not on other molecules. Later in 1983 and 1984, Sidney Altman, using ribonuclease P1 (it consists of RNA + protein and is *E. coli* RNA processing endonuclease for splicing of tRNA), demonstrated that catalytic activity of ribonuclease P_1 resided only in RNA subunit, and protein only helped RNA to perform the enzymatic function. Self cleavage of viroid RNAs or virusoids (satellite RNAs) and catalytic activity of RNAase are other examples of ribozymes. Subsequently it was shown that ribozymes may also bring about (i) cleavage of DNA molecules, and (ii) selective amplification of RNA molecules.

3

Protein Synthesis

Biosynthesis of protein is under direct control of DNA in most cases or else under the control of genetic RNA where DNA is absent. Information for structure of a polypeptide is stored in a polynucleotide chain. Sequence of bases in a particular segment of a polynucleotide chain will determine the sequence of amino acids in a particular polypeptide. The relationship popularly known as central dogma (Fig.) explains how protein synthesis is controlled by nucleic acids. DNA, while undergoing self replication, also controls synthesis of specific proteins. Flow of information is one way *i.e.* from DNA, information is transferred to RNA (mRNA) and from RNA to protein.

Evidence is also available where RNA can control synthesis of DNA. Such RNA was though initially found to be genetic RNA but non-genetic RNA synthesized under control of DNA may also give rise to DNA by reverse process (Fig.). This phenomenon is utilized for the origin of pseudogenes in mammals. There is, however, no evidence whatsoever for the synthesis of RNA from protein. If this is proved, one would have to believe in Lamarckism because, if information from proteins can be transferred to RNA and then eventually to DNA, it would mean that information about any acquired character would be possibly transmitted to DNA, the genetic material. This would result in inheritance of acquired characters.

DNA —transcription→ RNA —translation→ Proteins

(a) one way flow of information (central dogma)

DNA ⇄ transcription RNA —translation→ Proteins

(b) reverse flow of transcriptional information (as discovered recently)

Two versions of the flow of genetic information.

The synthesis of DNA on RNA template in tumour viruses was demonstrated in 1970, simultaneously by two American scientists H. Temin and D. Baltimore. This discovery was considered significant for the understanding of cancer. Therefore, these two American scientists, along with R. Dulbecco, another American scientist, who worked in London, shared the Nobel Prize in medicine/physiology for 1975.

Beside other types, following three kinds of non-genetic RNAs were listed, which take an active part in protein synthesis; (i) mRNA (messenger RNA), (ii) tRNA (transfer RNA), and (iii) rRNA (ribosomal RNA). While mRNA and tRNA directly take part in protein synthesis, rRNA does so by making part of structures called ribosomes. The manners in which mRNA and tRNA take part in protein synthesis are now well defined, but the role of rRNA is only partly understood. However, it is definitely known that, while mRNA and tRNA have some kind of specificity, rRNA has a general function to perform. Besides other roles, perhaps it helps in mRNA and tRNA binding by complementary base pairing.

Two major steps are involved in protein synthesis: (i) transcription, involving transfer of genetic information from

DNA to mRNA, and (ii) translation, involving translation of the language of nucleic acids into that of proteins. Transcription will be discussed in some detail in this chapter both in prokaryotes and eukaryotes.

Transcription involves synthesis of mRNA on DNA template in such a way that the mRNA resembles one of the two DNA strands. This DNA strand which is identical to mRNA (except that uracil is present in place of thymine) is called the coding strand and the other complementary DNA strand is called the template strand. The transcription is carried out by an enzyme called RNA polymerase and a transcription unit may include more than one gene. The mechanism of transcription and the RNA polymerase enzyme used for transcription differ in prokaryotes and eukaryotes and, therefore, will be discussed separately.

Transcription in Prokaryotes

Single RNA Polymerase in E. coli : In bacterial systems like *E. coli,* a single RNA polymerase (RNAP) species is responsible for synthesis of all kinds of RNAs (mRNA, tRNAs and rRNA). The bacterial RNAP consists of a core enzyme ($\beta'\beta\alpha'\alpha''\omega$) and a sigma (σ) factor. The two largest subunits ($\beta'\beta$) of the core enzyme are involved in catalysis, and the remaining three subunits ($\alpha^{I},\alpha^{II},\omega$) take part in RNAP assembly and transcriptional regulation. During RNAP assembly, $\beta\alpha'\alpha''$ is an assembly intermediate, so that omega (ω) subunit facilitates

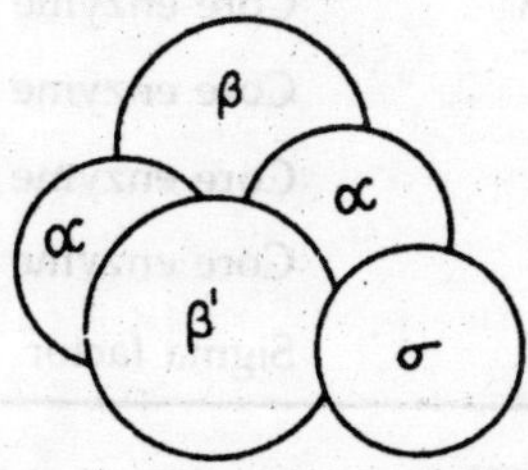

A model for the structure fo prokaryotic RNA Polymerase showing association of five polypeptides ($\alpha_2\beta\beta'\sigma$).

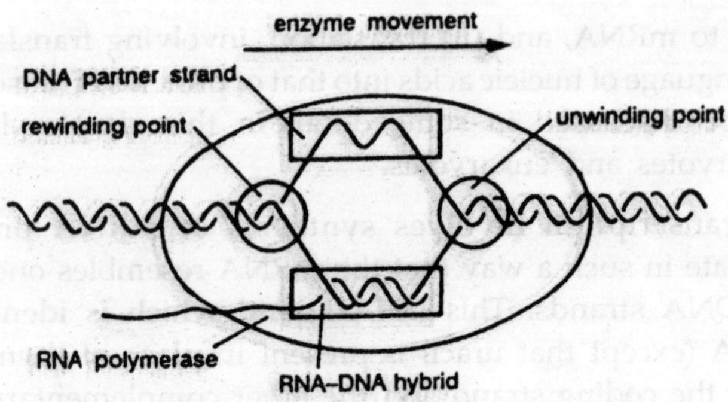

Active centres in bacterial RNA polymerase enzyme.

association of β′ with βα′α″ to form the core enzyme (β′βα′α″ω). The α′ and α″ subunits are identical in sequence, but differ in location within RNAP (α′ interacting with β, and α″ interacting with β′). The active sites of core enzyme are shown in Figure.

The sigma (σ) factor helps in recognition of start signals on DNA molecule and directs RNA polymerase in selecting the initiation sites. In the absence of sigma (σ), core enzyme initiates RNA synthesis in a random manner, suggesting the role of σ in recognition of initiation sites. Once RNA synthesis

Different Components of One Molecule of E. coli RNA Polymerase (RNA) Enzyme.

Subunit and gene	*Mass (kilo daltons)*	*Location*	*Possible function*
α_2 *(rpo A)*	40 each	Core enzyme	RNAP assembly
β *(rpo B)*	155	Core enzyme }	Catalysis
β′ *(rpo C)*	160	Core enzyme }	
ω *(rpo -)*	—	Core enzyme	RNAP assembly
σ *(rpo D)*	32-90	Sigma factor	Initiation

Note : The α and β′ have constant sizes in most bacteria; the σ varies from 32 to 92 kilodaltons.

is initiated, σ dissociates after RNA is 8-9 bases long and then the core enzyme brings about elongation of mRNA. The dissociated sigma factor may again combine with core enzyme to form RNA polymerase holoenzyme (Fig.).

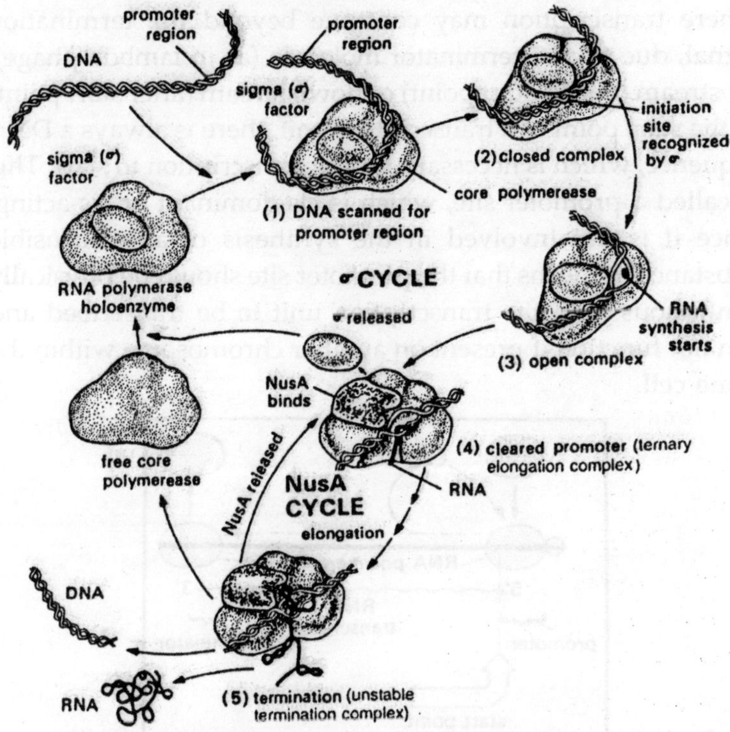

Role of sigma factor Nus A, and core enzyme of RNA polymerase during transcription.

Although, in prokaryotes like *E. coil*, all RNA synthesis is done by only one kind of RNA polymerase molecules, there may be more than one sigma (σ) factors, which associate, each with the same core enzyme at different times for expression of different genes. For example, in *E. coli*, besides σ^{70} used under normal conditions of growth, atleast four other sigma factors (σ^{32}, $\sigma^{41.5}$, σ^{54}, σ^{28}) are now known, which are used under adverse conditions like high temperature, starvation, nitrogen deficiency and for chemotaxis.

Promoter Sites for Initiation of Transcription in Prokaryotes : For synthesis of RNA on DNA template, there are transcription units, each having a starting point and a termination signal (Fig.). It is not necessary that the RNA may be released at the termination signal, since there are situations where transcription may continue beyond the termination signal, due to anti-terminator molecule (as in lambda phage). Upstream (before start point) or down stream (after start point) to the start point of a transcription unit, there is always a DNA sequence, which is necessary for the transcription to start. This is called a promoter site, which is cis-dominant or cis-acting, since it is not involved in the synthesis of any diffusible substance. It means that the promoter site should be physically continuous with the transcription unit to be transcribed and can not function if present on another chromosome within the same cell.

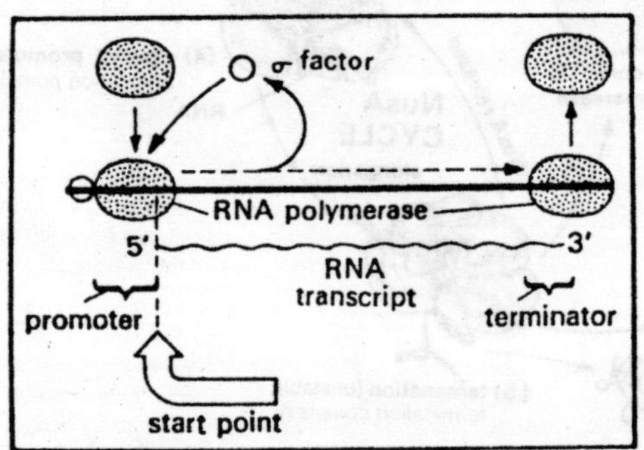

A transcription unit, showing promoter region, start point and terminator; also shown is the role of RNA polymerase with its core enzyme and sigma factor.

The binding sites for RNA polymerase enzyme lie within the promoter sequence and may be 41 to 44 base pairs in length in *E. coli,* where more than 100 promoters have already been sequenced. There is hardly any long conserved sequence over majority of the promoter regions, although short sequences within the promoters are conserved. The start point is a purine in over 90% cases and upstream from this start point is a 6bp region (TATAAT), which is called Pribnow box and is available in almost all promoters. The conservation of these six bases varies from 45% to 96% as represented by the subscripts in the following sequence (the Pribnow box for *lac* operon in *E. coli* is TATGTTG).

$$T_{80}A_{95}T_{45}A_{60}A_{50}T_{96}$$

The centre of Pribnow box lies usually 10 bp upstream and therefore it is sometimes described as –10 sequence (although the actual centre varies from –18 to –12). There is another sequence ($T_{82}T_{84}C_{78}A_{65}C_{54}A_{45}$) lying –35 bases upstream and is called recognition region. Although details of binding reaction are not known, typical promoters in *E. coli* use –35 and –10 sequences to be recognized by RNA polymerase (Fig.). It is obvious thus, that a typical promoter in bacteria has three components : (i) consensus sequence at –10bp; (ii) consensus sequence at –35bp, and (iii) start point.

Initiation and Elongation of RNA Synthesis in Prokaryotes: RNA synthesis by RNA polymerase proceeds in four steps : (i) The holoenzyme first binds at the promoter site, forming the closed promoter complex in which DNA remains double helical. (ii) The closed complex isomerises, and causes unwinding and separation of DNA strands to form open (binary) promoter complex. (iii) After unwinding, only one of the two strands is copied; this is achieved by incorporation of nucleotides, initially without movement of enzyme leading to the formation of RNA chain, upto 9 bases in length. During the incorporation of these 9 bases, at every step, there is a possibility for the release of this small RNA chain, a process described as 'abortive initiation'. A cycle of abortive initiations usually occurs generating a series of short (2-9 base) oligonucleotides, before initiation is actually successful. (iv) Once initiation succeeds, the sigma factor of RNA polymerase dissociates. (v) The dissociation of σ factor marks the entry of NusA protein, which helps elongation, and promotes pausing and termination at specific sites. Core enzyme now undergoes a major conformational rearrangement and a stable ternary elongation complex is formed. This complex moves along DNA, synthesizing RNA all along its path at a rate of about 40 bases per second at 37°C. RNA polymerases in some phages (T3, T7) are each a solitary polypeptide (< 100,000 daltons), which synthesizes RNA very rapidly (~ 200 bases/second at 37°C). Elongation of RNA transcript continues till an unstable termination complex is formed.

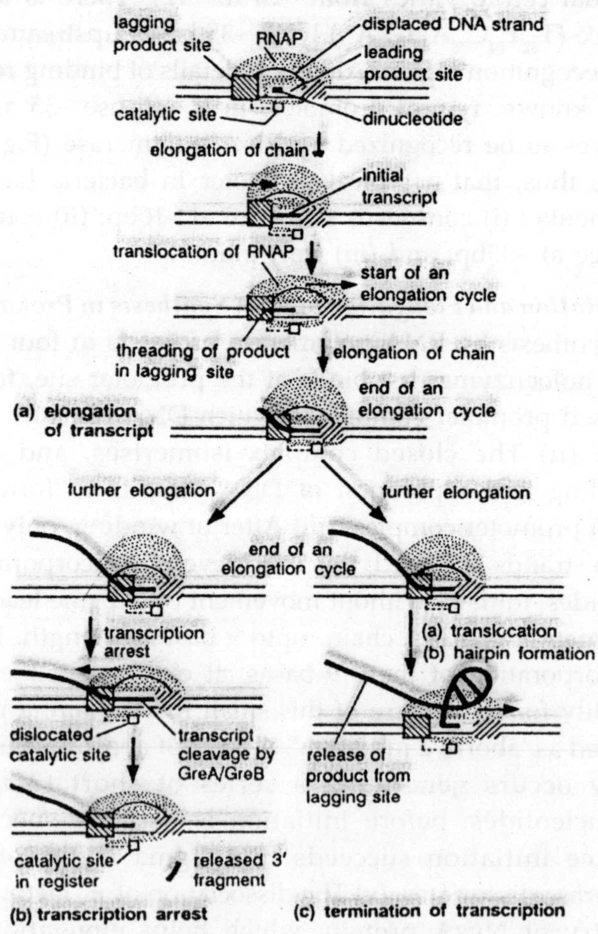

Inchworm model of elongation showing elongation cycle : (a) transcription arrest (b) and termination.

'Inchworm Model' for Elongation of Transcript : In the year 1993, M.J. Chamberlin proposed that the transcript elongation involves inchworm like translocation of RNAP (RNA polymerase) along the DNA template. RNAP is assumed to have one site for DNA binding and two sites for RNA binding (the 'leading product site' and the 'lagging product site'). While the 'leading product site' gets filled by about 10 newly added nucleotides (nt) in each successive cycle of elongation, the 'lagging product site' occupies the preceding segment of nascent RNA earlier synthesized.

RNAP translocates forward by about 10nt, when the leading site is full, so that the leading site in emptied again, thus permitting the next cycle of elongation. Stability to the elongation complex is provided by the interaction of DNA with the lagging site of RNAP. The above 'inchworm model' is based on experimental evidence and is diagrammatically represented in Figure.

Stability of Elongation Complex (RNA-DNA Hybrid vs RNAP-DNA Interactions) : The elongation complex in *E. coli* is a ternary complex containing RNA polymerase (RNAP), DNA template and nascent RNA. The classical model of transcription elongation assumes that the strength of

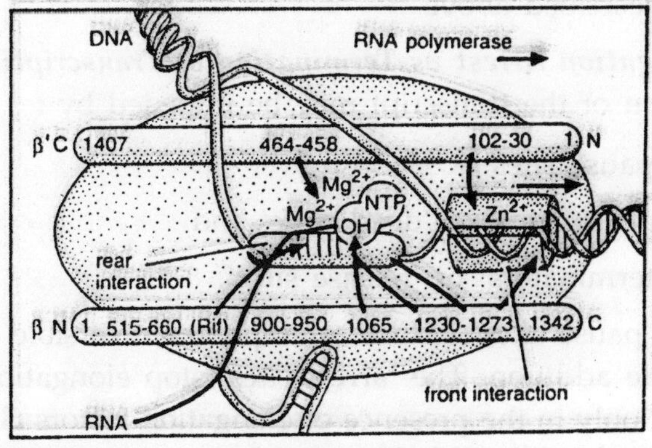

RNAP-DNA interactions for stability of elongation complex.

RNA-DNA hybrid (8-12bp) in the transcription bubble, plays a critical role in the stability of the ternary elongation complex. In a recent study, this assumption was questioned and a new model was proposed. According to this new model, stability of ternary complex depends upon the following two types of RNAP-DNA interactions (Fig.):

(i) The front (F) interaction, that occurs ahead of the growing end of RNA and is responsible for RNAP processivity. This interaction is non-ionic and requires 7-9 base pairs of intact DNA duplex, ahead of the transcription fork.

(ii) The rear (R) interaction, that is ionic and requires 5 nucleotides of template strand of DNA behind the active site and one nucleotide ahead of it; the non-template strand is not involved in this R-interaction. The F-interaction involves conserved Zn-finger motif in NH_2-terminus of β′ subunit of RNAP and R-interaction involves COOH-terminal catalytic domain of β subunit. This model allows the DNA replication complex to bypass the transcriptional elongation complex. The model is also in agreement with Chamberlain's inchworm -model for elongation, described above.

Elongation Arrest vs Termination of Transcription : The elongation of the transcript may be impeded by :

(i) 'pause sites',

(ii) 'arrest sites' or 'dead ends' and

(iii) 'terminators' or 'release sites'.

The 'pause sites' induce a temporary reversible block to nucleotide addition. The 'arrest sites' stop elongation, to be resumed only in the presence of elongation factors GreA and GreB (in eukaryotes the corresponding transcription factor is

TFIIS), which assist in a novel transcript cleavage reaction (Fig.). The 'release sites' lead to release of RNA and/or RNAP either intrinsically, or upon activation by factors like NusA, Rho, or Tau.

Although the nature of DNA sequences causing transcription arrest is not known, the mechanism of transcription arrest and the roles of GreA and GreB in restoring elongation can be explained in the framework of 'inchworm model' described above. The transcription arrest may result when the catalytic site of RNAP is misplaced from the 3' end of growing transcript. This, in turn, may result, if the 'lagging product site' slipped backward on the template, bringing the catalytic site out of register. Transcript cleavage by GreB would create a new 3' end into the right register, thus restoring elongation. Further, at some template sites, translocation may disturb the placement of catalytic site into the right register, and GreA-induced cleavage of short segments might restore the appropriate register to resume elongation. Thus an elongation barrier may be overcome by RNAP through a 'backup and restart' mechanism.

Termination and Antitermination of mRNA Synthesis in Prokaryotes : The termination of mRNA chain in prokaryotes is brought about by certain termination signals on DNA. These DNA sequences providing termination signal are called terminators. However, sometimes termination event, despite the presence of a terminator, can be prevented by some factor interacting with RNA polymerase. This phenomenon is called antitermination and the factor responsible for this phenomenon will be called antiterminator. It will be seen that the termination actually depends on the scrutiny of not only the DNA template, but also that of the product (RNA) obtained by transcription.

Rho (ρ) Dependent and Rho (ρ) Independent Termination in E. Coli : The termination signals whenever found on DNA can be of two types :

(i) those, which are recognized by a protein factor known as rho (ρ) factor, cause rho (ρ) dependent termination and

(ii) those, which do not need rho (ρ) factor for termination cause rho (ρ) independent termination. A transcription factor NusG is also needed for many rho dependent terminators inside the cells of *E. coli.* Rho (ρ) factor is a protein of molecular weight 55,000 and was isolated from *E. coli.* It causes termination of most mRNA molecules *in vitro* at relatively low ionic strength, but is not needed for termination of some of the mRNAs. However, *in vitro,* high ionic strength needs to be maintained to demonstrate *rho* independent termination. *In vivo,* distinction between *rho* dependent and *rho* independent terminators was possible with the help of mutants deficient for *rho* factor, but still showing termination at the rho independent sites. In both cases (rho dependent and rho independent terminations), near the termination sites, there is similarity in secondary structure, which is generated by palindromic sequences like the following :

5′ G G T A C C 3′
| | | | | |
3′ C C A T G G 5′

In this palindrome the sequence on both strands is the same when read in 5' to 3' direction. In this palindrome,

G G T
| | | also represents an inverse repeat giving dyod
C C A

symmetry in the sequence. This causes complementarity between bases on same strand (GGT being complementary to ACC read in opposite directions), so that the latter can base pair to form a hairpin in mRNA or cruciform in DNA.

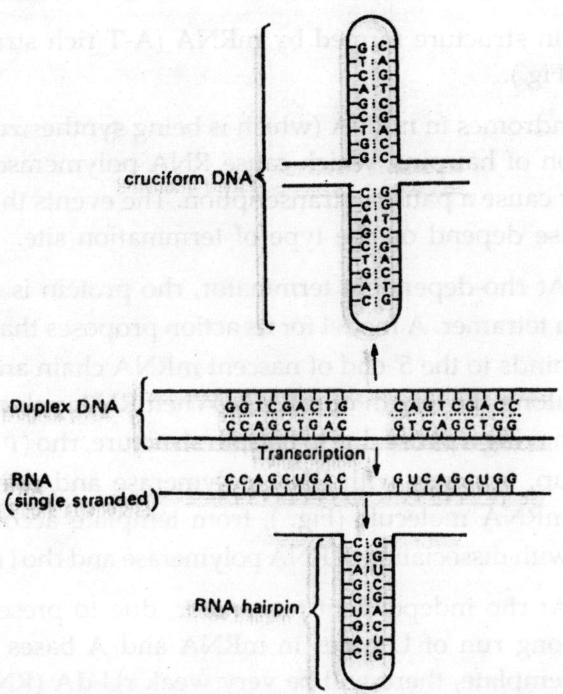

Secondary structures (cruciform DNA and RNA hairpin) formed due to inverted repeats in double stranded DNA and its RNA transcript (modified from Lewin's "Genes").

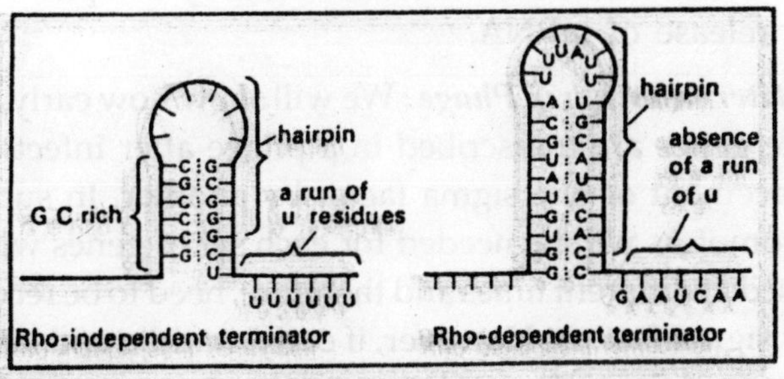

RNA hairpins followed by a run of U residues in rho independent terminator, but not in rho dependent terminator of transcription (modified from Lewin's "Genes").

Rho independent terminator differs from rho dependent terminator in the presence of a long run of U residues in mRNA (A residues in DNA) and of a G-C rich region in stem of hairpin structure formed by mRNA (A-T rich structure in DNA) (Fig.).

Palindromes in mRNA (which is being synthesized) cause formation of hairpins, which cause RNA polymerase to slow down or cause a pause in transcription. The events that follow this pause depend on the type of termination site.

(i) At rho-dependent terminator, rho protein is active as a tetramer. A model for its action proposes that rho (ρ) binds to the 5' end of nascent mRNA chain and moves along the length of mRNA. When RNA polymerase is having a pause due to hairpin structure, rho (ρ) catches up, interacts with RNA polymerase and releases the mRNA molecule (Fig.), from template accompanied with dissociation of RNA polymerase and rho (ρ) factor.

(ii) At rho independent terminator, due to presence of a long run of U bases in mRNA and A bases in DNA template, there will be very weak rU-dA (RNA-DNA hybrid) base pairs needing very little energy to break. Since RNA polymerase slows down due to hairpin structure, rU-dA bond breaks at any one point causing release of mRNA.

Antitermination in Phage : We will show how early, middle and late genes are transcribed in a phage after infection, due to replacement of one sigma factor by another. In such cases new promoters will be needed for each set of genes which are expressed at different times and therefore, need to be recognized by new sigma factors. However, if early, middle and late genes are arranged in order and termination is prevented by an antiterminator, then after early genes, the middle genes and after middle genes, the late genes will be transcribed in a sequence without the need of new sigma factors (Fig.). This is what is achieved by the mechanism of antitermination in phage lambda *(λ)*. There is an early gene *N* in phage lambda (λ)

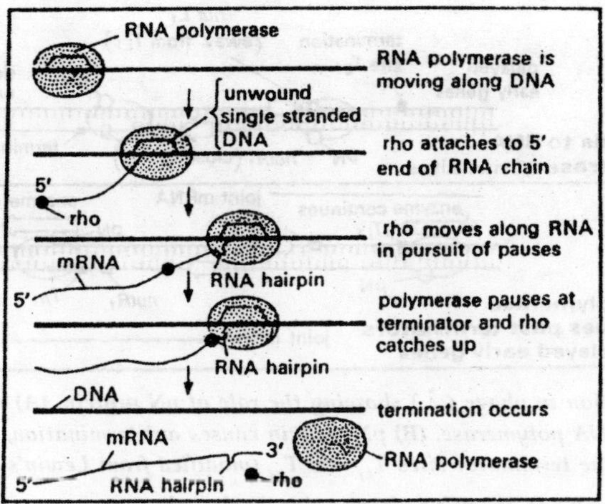

Mechanism of action of rho (ρ) factor in termination of transcription, involving its travel along RNA till it catches RNA polymerase pausing at a RNA hairpin (modified from Lewin's "Genes").

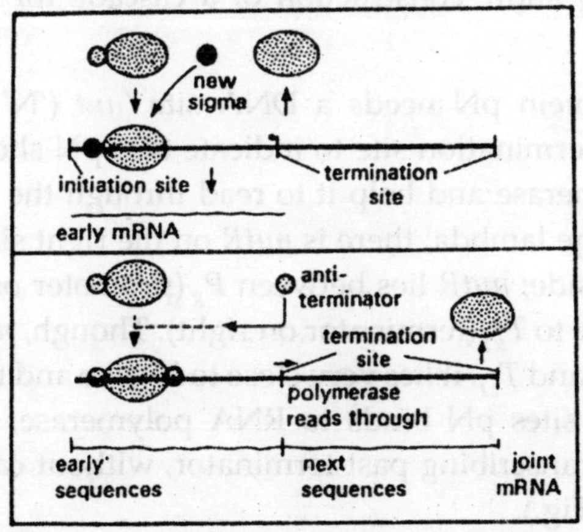

Control of termination by the absence or presence of an antiterminator molecule, which attaches to RNA polymerase enzyme (modified from Lewin's "Gene").

and its product pN is the antiterminator, which allows RNA polymerase to read past T_{L1} and T_{R1} (termination sites on the left and right of the λ DNA; the two DNA strands in λ DNA are transcribed in opposite directions; see Fig.). The product of a delayed early gene *Q* (transcribed after early genes like

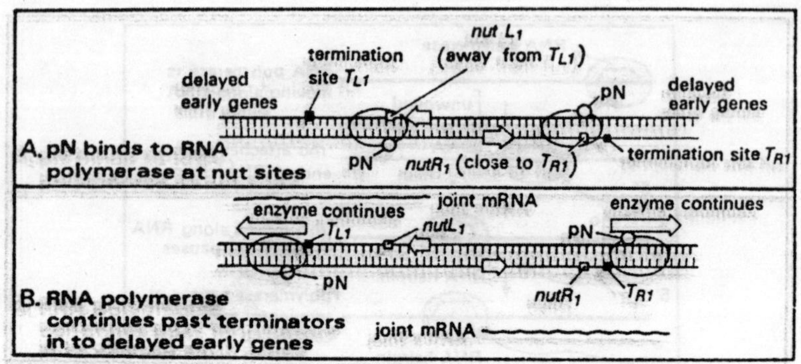

Transcription in phage (λ) showing the role of pN protein. (A) pN protein binds to RNA polymerase. (B) pN protein causes antitermination, i.e. reading through the terminator sites T_{L1} and T_{R1} (modified from Lewin's "Genes").

middle genes) is pQ which similarly allows transcription of late genes without termination of mRNA at the end of delayed early genes. In this manner, the antitermination proteins like pN and pQ allow construction of a cascade for phage gene expression.

The protein pN needs a DNA site *'nut* ('N' utilization) before the termination site to indicate that pN should bind to RNA polymerase and help it to read through the termination site. In phage lambda, there is *nutR* on the right side and *nutL* on the left side; *nutR* lies between P_R (promoter on right) and T_{R1} but close to T_{R1} (terminator on right). Though, *nutL* also lies between P_L and T_{L1}, it lies very close to *N* gene and much before T_{L1}. At *nut* sites pN binds to RNA polymerase, which then continues transcribing past terminator, without caring for the T_{R1} or T_{L1} (Fig.).

In *E. coli*, there are several *nus* loci, mutations at which will prevent antitermination protein pN from performing its function. These loci include *nusA*, *nusB*, *nusE* and *nusG* (*nus* strands for N utilization). The *nusA*, *nusB* and *nusG* are concerned solely with the termination of transcription; *nusE* codes for ribosomal protein S10. NusA is a general transcription factor that increases the efficiency of termination (Fig.), NusB and S10 form a dimer that binds RNA containing *boxA* sequence. NusG may be concerned with the general assembly of all the Nus factors into a complex with RNA polymerase. (For more details consult 'Genes VI' by Benjamin Lewin).

Transcription in Eukaryotes

Multiple RNA Polymerases in Eukaryotes : In eukaryotes there are atleast three different nuclear RNA polymerases (RNAPs), whose properties are given in Table.

Properties of three Different Eukaryotic Nuclear RNA Polymerases.

Enzyme & location	*Subunits*	*Product & abundance*	*Sensitivity to a amanitin*
RNAP I (nucleolus)	RPAs	rRNA (50-70%)	Not sensitive
RNAP II (nucleoplasm)	RPBs	hnRNA (20-40%)	Sensitive
RNAP III (nucleoplasm)	RPCs	tRNA (~10%)	Inhibited in animals at high level; not in yeast & insects

The three RNAPs can be distinguished on the basis of their sensitivity to certain inhibitors of transcription, *e.g.* α amanitin. Each of these enzymes is a large protein (~ 500,000 daltons), with two large and several (8 to 10) smaller subunits, some of them showing homology to the subunits of *E. coli* RNAP. Eukaryotic RNAPII consists of 12 subunits, RPB1-12 in a decreasing order of size. Of these subunits, RPB1, RPB2, RPB3, RPB11 and RPB6 have homology with $\beta', \beta, \alpha', \alpha''$ and ω respectively. RPB2 also has sites for ribonuclease activity, thus helping in RNA cleavage for proof reading.

Promoter, Enhancer and Silencer Sites for Initiation of Transcription in Eukaryotes : In eukaryotes there are three

RNA polymerases: RNA polymerase I or RNAPI for synthesis of pre-rRNA; RNA polymerase II or RNAPII for synthesis of pre-mRNA or hnRNA (heterogeneous nuclear RNA) and several snRNAs, and RNA polymerase III or RNAPIII for synthesis of 5S RNA, tRNAs, U6 snRNA, 7SL, EBER2 and VA (7S RNA is present in a small RNA-protein complex called SRP; EBER RNA is specified by Epstein-Barr virus; VA RNA is specified by adenovirus). Different promoter sequences have been identified for different RNA polymerases.

Promoters for RNA Polymerase I : Promoters for RNA polymerase I have at least two elements: (i) a GC-rich upstream (–180 to –107) control element (UCE) and (ii) a core region that overlaps the transcription start site (–45 to +20).

Promoters, Enhancers and Silencers for RNA Polymerase II : The promoter sequences for eukaryotic RNAPII include core promoters for basal transcription and other regulatory sequences for activator-dependent transcription. A core promoter is generally a sequence, ~40 base pairs long, that is sufficient to direct accurate transcription by RNAPII with the help of basal transcription factors, also described as general transcription factors (GTFs). There are several known core promoter motifs, which include the following : (i) *TATA* box (an A/T rich sequence) is located ~30 bp upstream of transcription start site, at which TATA binding protein (TBP) subunit of TFIID binds; (ii) *BRE* (TFIIB recognition element) is located immediately upstream of the TATA box of some TATA-containing promoters and increases the affinity of TFIIB for core promoter; (iii) *Inr* (initiator) is a conserved sequence encompassing the start site, and functions to direct accurate initiation of transcription either by itself or in conjunction with a *TATA* or DPE-motif, and (iv) *DPE* (downstream promoter element), which is a conserved sequence (as common as TATA box), located ~30 bp downstream of the start site. A typical DPE-containing promoter has *Inr* and *DPE* motifs and lacks a *TATA* box. TFIID binds to *Inr* and *DPE* motifs through its TAF

(TBP-associated factor) subunits. Since loss of TATA box can be compensated by DPE, the eukaryotic promoters can be classified into *TATA*-driven promoters and *DPE-driven* promoters. In *Drosophila, DPE*-driven promoters are used by DSTF (DPE-specific transcription factor), which is now described as dNC2 (in view of its homology to the human transcription repressor NC2) and consists of two subunits (dNC2α and dNC2β; prefix 'd' stands for *Drosophila).* dNC2 is a bifunctional transcription factor, which represses *TATA-driven* promoter and activates DPE-driven promoters (Fig.). *TATA* box is compared with *Pribnow* box of prokaryotes and further upstream of this box are present GC box (–60 or –100bP) and CAAT box (–80bP) as shown in Figure.

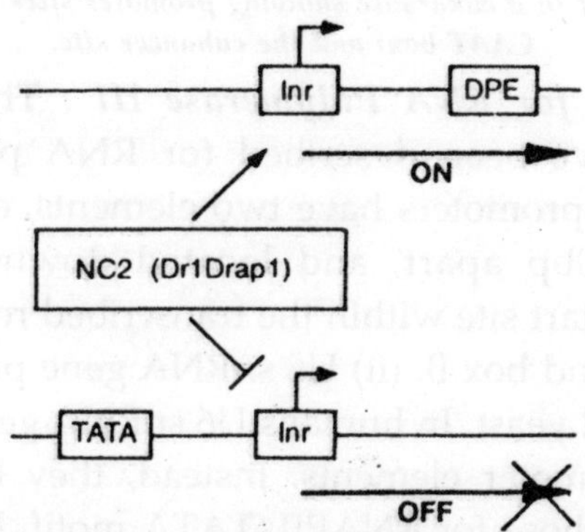

Two types of promoters (Inr-DPE and TATA-Inr) in eukaryotes and the contrasting effects of transcription factor NC2 on these two promoters.

Eukaryotic promoters often also possess elements located 100 or 200 base pairs upstream, which interact with proteins other than RNA polymerase. These elements, called enhancers or upstream activation sites (UAS) can be moved several hundred or even thousands of base pairs upstream or downstream without alteration of their activities. These enhancers may cause upto a 200 fold increase in transcription rate of an affected gene, hence the name. There are other regulatory elements known as silencers, which repress gene expression. Silencers, like enhancers, can function at great distances from genes they repress. They are sites for binding of proteins just like enhancers. The location of different promoters and an enhancer is shown in Fig. (Silencers are found in yeast and repress expression of *HML* and *HMR* loci, involved in switching of mating types).

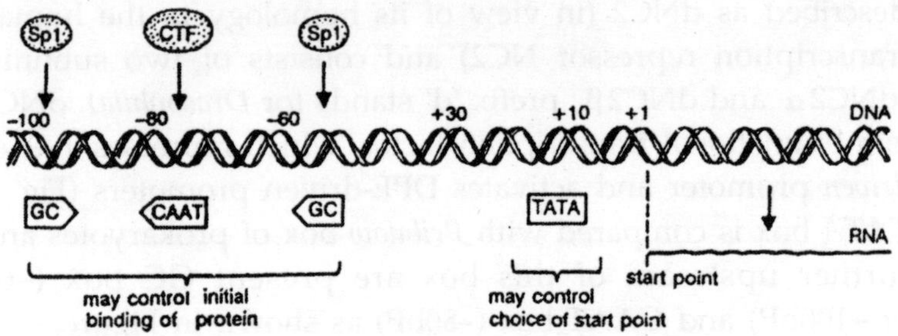

A DNA segment in a eukaryote showing promoter sites (TATA box and CAAT box) and the enhancer site.

Promoters for RNA Polymerase III : Three classes of promoters have been described for RNA polymerase III: (i) tRNA gene promoters have two elements, each l0bp long, spaced 30-120bp apart, and located downstream of the transcription start site within the transcribed region; these are called box A and box B. (ii) U6 snRNA gene promoters differ in humans and yeast. In humans U6 snRNA gene is devoid of intragenic promoter elements. Instead, they have promoter elements like those for RNAPII,(TATA motif, PSE and OCT). In yeast, on the other hand, intragenic elements, box A and box B are found along with TATA box (but no PSE). (iii) 5S RNA gene promoters have intragenic control region called box C, which lies well within the transcription unit, 50bp downstream rather than upstream from the start point (Fig.). However, the location of downstream promoters for RNA polymerase III may vary in different genes (Fig.).

One would be interested in knowing how RNA polymerase III attached to +55 to +80bp sequence can start transcription from the start point. It is believed that the enzyme is big enough to occupy start point and the promoter region simultaneously and thus may start transcribing at the start point without difficulty.

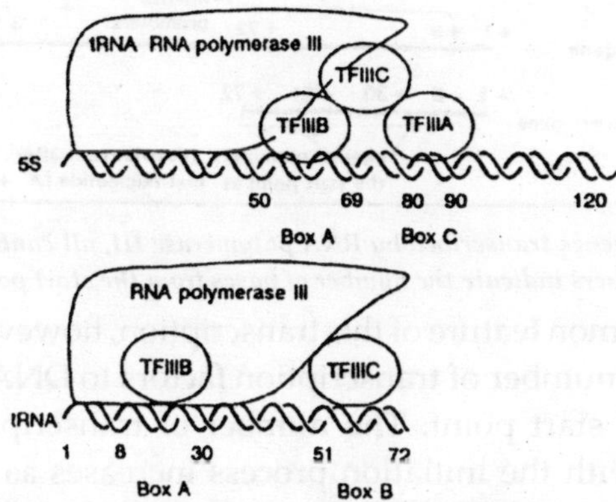

Binding of TFIIIB to downstream promoters (box A), located at different positions in 5S and tRNA genes.

Transcription Factors and Initiation of RNA Synthesis in Eukaryotes : For eukaryotic RNA synthesis, a number of transcription factors are involved in the initiation of transcription. Each RNA polymerase is believed generally to have its own set of transcription factors, although TFIID or a part of it (TBP) is required for all the three RNA polymerases. The only exception is the adeno-associated virus (AAV) P5 promoter, on which transcription can start in the absence of TBP, making use of Inr binding protein (IBP), the YY1 (Fig.). These transcription factors may be defined as 'proteins, which are needed for initiation of transcription, but are not a part of RNA polymerase'. The regulatory DNA sequences and the transcription factors used by RNA polymerase II (used for mRNA synthesis) may be either common for many genes expressed constitutively (basal transcription) or may be specific for tissue or stimulus specific expression of genes, whose expression is regulated (activator dependent transcription). It means that for transcription of different genes, no single combination of DNA sequences or factors is adequate and essential, and that the transcription may be initiated in many different ways.

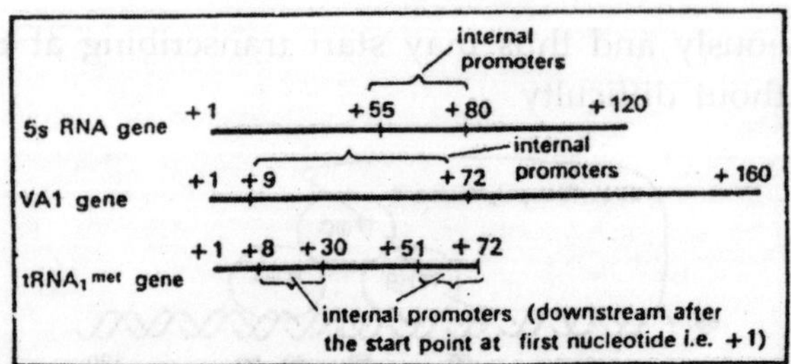

Three different genes transcribed by RNA polymerase III, all containing internal promoters; numbers indicate the number of bases from the start point of the gene.

The common feature of this transcription, however, involves binding of a number of transcription factors to DNA sequences upstream of start point. The number of transcription factors associated with the initiation process increases as the variety of genes transcribed by RNAPs increase. These transcription factors help in DNA binding of an RNA polymerase to constitute what is described as pre-initiation complex (PIC) or transcription complex. After the formation of this complex, initiation of transcription occurs. All known transcription factors may not necessarily recognize DNA sequences, and a factor may instead recognize another factor or may even recognize RNA polymerase (Fig.)

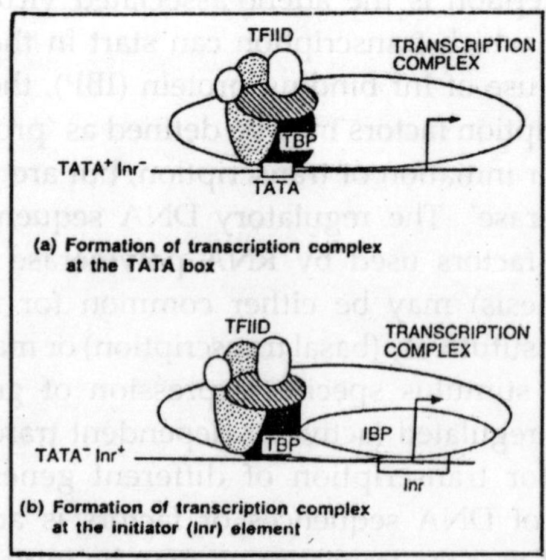

Transcription complexes of TATA box and initiator (Inr) element.

Formation of transcription complex or PIC involves the following four steps, although in case of RNAPI and RNAPIII, only the first three steps are involved (Table). (i) The first step is nucleation, which involves recognition of the promoter followed by binding of a transcription factor; (ii) the second step involves association of a factor which links RNAP with the promoter (after initial nucleation, each RNAP requires an intermediary to link the polymerase with the core); (iii) the third step involves RNAP entry and (iv) the fourth step involves complex maturation (only RNAPII requires initiation factors subsequent to the association of RNAP for complex maturation):

Formation of pre-initiation (transcription) complex or PIC with RNA polymerase II (pol II). The promoter sequence responsible for constitutive expression of common genes (also called house keeping genes) in all cells is called generic promoter, which can not bring about regulated expression (tissue or stimulus specific). Initiation of transcription on this generic promoter by RNA polymerase II is described as basal transcription and requires the action of several general transcription factors or GTFs (variously designated in humans, rat and yeast; Table) in the following defined order (Fig.):

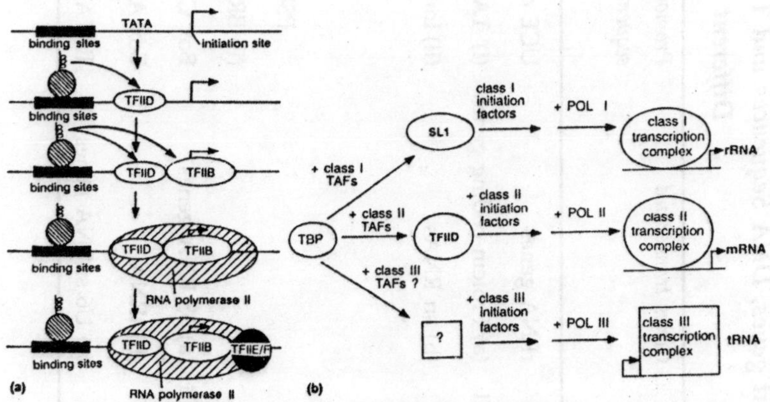

(a) Formation of transcription complex at the promoter for RNA polymerase II. (b) Association of TBP with different groups of TAFs to form transcription factors used in the formation of transcription complexes.

Different genes, DNA Sequences and Transcription Factors Involved in Basal Transcription by three Different RNA Polymerases in Eukaryotes

Enzyme	*Genes transcribed*	*Promoter sequence(s)*	*Nucleation factor (s)*	*Factor linking RNAP with promoter*	*Factor needed after RNAP binding*
1. RNAPI	rRNA genes	UCE core region	UBF + TAFs	TBP + TAFs	None
2. RNAPII	(a) Protein coding genes	(i) TATA box	TBP + TAF_{II}		
	(b) sn RNAs	(ii) Inr and/or	TATA IBP + TAF_{II}		
			IBP (TF_{II}-1 or	TFIIB, TFIIF	TFIIE, TFIIH
			YY1) + TAFs		(also ATP)
		PSE	SNAPc		
		(iii)BRE (iv) DPE			
3. RNAPIII	(a) 5S RNA genes	Box C	TFIIIA		
	(b) tRNA	Box A, Box B	TFIIIC	TFIIIB	None
	(c) U6 snRNA genes	TATA, PSE	TBP, SNAPc,		

(i) TFIID binds at the TATA box (nucleation step), which permits the association of TFIIA and TFIIB in the next step;

(ii) very early in assembly of the pre-initiation complex (PIC), TFIIA can stably associate with TFIID-DNA or TFIIB-TFIID-DNA complexes to give DA or DAB complexes, which are more stable than either the TFIID-DNA complex or the TFIIB-TFIID-DNA (DB) complex. This wards off inhibitory co-activators and enhances stimulatory effects of transcriptional activators. TFIIA has thus been defined as a co-activator involved mainly in activator dependent transcription;

(iii) only after binding of TFIIB, forming the so called DAB complex, RNA polymerase associates to promoter site; thus TFIIB provides a link between the promoter and RNAPII;

(iv) RNA polymerase II is escorted to the promoter by TFIIF to form a transcription complex: TFIIF is unique in its ability to form a very stable complex with pol II, referred to as pol/F;

(v) orderly addition of TFIIE. TFIIH and TFIIJ helps in the initiation. After pol II (RNA polymerase II) initiates transcription, it leaves the promoter, but TFIID remains bound to the core promoter supporting rapid re-initiation of transcription by recruiting all the necessary factors and pol II. The requirement of TFIIA and TFIIE has not been found universal. The function of TFIIA can be performed by TFIIG or TFII I (two factors discovered in 1990's); TFIIE, on the other hand, though needed for adenovirus (Ad) ML, H2B and H4 promoters, has not been found essential for immunoglobulin heavy chain gene (IgH) promoter. Besides the above transcription factors needed for basal transcription, hundreds of other transcription factors (transcriptional activators and co-activators) have been discovered, which are needed for tissue, developmental stage or stimulus specific activator-dependent transcription (TFIIA is, also believed to be involved only in activator dependent transcription).

Various designations (in human, rat, yeast) for transcriptions factors used as initiation factors for RNA polymerase II.

Humans	*Rat*	*Yeast*
TFIID complex	τ	d (TBP)
TFIIB	α	e
BTF2 (TFIIH)	δ	b
RAP30/74 (TFIIF, FC)	βγ	g
TFIIE	ε	a
TFIIG	—	—
TFIIJ	—	—
TFIIA	—	TFIIA

Formation of Pre-initiation Complex (PIC) with Pol I and Pol III : For the initiation of transcription of rRNA genes transcribed by RNA polymerase I (pol I), in the first step, the promoter UCE is nucleated by upstream binding factor (UBF), sometimes also described as assembly factor. The need for UBF is not obligatory, but is species dependent. In the second step, the bridging between UBF-DNA and RNAPI is brought about by SL1 (TBP+TAFs). Similarly, in case of RNA polymerase III (pol III), in the first step the nucleation of promoter is brought about due to the formation of the following complexes: BoxA, BoxB/TFIIIC complex in case of tRNA genes, PSE/SNAPc and TATA/TBP complex in case of hU6 snRNA genes and BoxC/TFIIIA complex in case of 5S RNA genes. In the second step bridging between each of these complexes and RNAPIII is facilitated by TFIIIB in case of tRNA and U6 snRNA genes, and by TFIIIC in case of 5S RNA genes.

In *Xenopus laevis,* for RNA polymerase III (pol III), two nucleation or assembly factors (TFIIIA and TFIIIC) in case of 5S RNA genes and one such factor (TFIIIC) for tRNA genes are needed to mediate the binding of transcription initiation factor, TFIIIB. These assembly factors can be removed, after binding of the initiation factor without any effect on subsequent transcription (5S RNA is an example). The different steps involved in the formation of pre-initiation complex (PIC) utilizing pol I and pol III are shown in Fig.

UBF, discussed above for the formation of PIC for RNA polymerase I, is often used as an example of architectural transcription factors, which do not have an activation domain, but function by changing DNA conformations. UBF wraps more than 180 base pairs (like histone octamer in a nucleosome) of promoter DNA into a distinct structural unit and thus brings closer two regulatory elements located far apart.

Structure and Role of TFIID and Other Transcription Factors (TBP, TAFs) : Of the above transcription factors, TFIID is the most important and is found as a multiprotein complex. A subunit of TFIID is 'TATA binding proteins (TBP)', which is actually required by all the three eukaryotic RNA polymerases (even for promoters lacking a TATA box). However, there are other 'TBP associated factors (TAFs)', which differ for different RNA polymerases (more than a dozen TAFs are known, each associated with TBP in TFIID). Therefore, it is suggested that each of the three RNA polymerases requires an essential transcription factor, which is a complex containing TBP and TAFs, the composition of TAFs differing in the three cases. The TAFs involved with RNAPII, are designated as TAF_{II}s (dTAF_{II}s in *Drosophila*, yTAF_{II}s in yeast and hTAF_{II}S in humans); the different TAF_{II}S in an organism are distinguished by their molecular weight, ranging from 250 to 15kD (*e.g.* $hTAF_{II}250$, $hTAF_{II}135$, $hTAF_{II}95$, etc.). A number of TAF_{II}s have considerable amino acid sequence similarity with histone proteins. For instance $dTAF_{II}42$ ($hTAF_{II}31$), $dTAF_{II}$ 62($hTAF_{II}80$) and $dTAF_{II}30\alpha/22$ ($hTAF_{II}20/15$) are homologues of H3, H4 and H2B respectively. No homologues of H2A are known among TAF_{II}s.

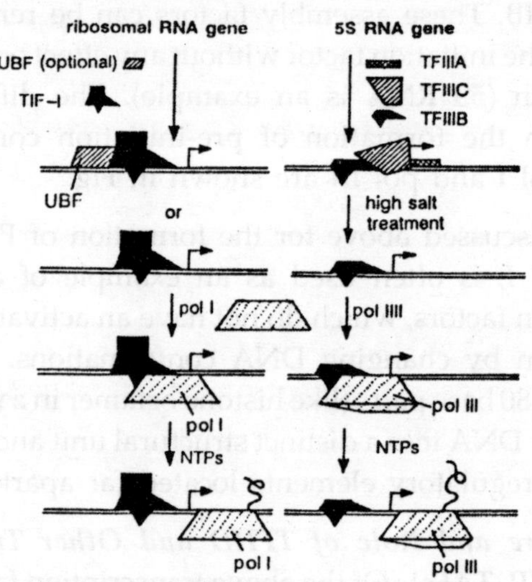

Use of transcription assembly factors (hatched) and transcription initiation factors (solid black) for binding of RNA polymerase I (pol I) and RNA polymerase III (pol III).

Another abundant complex is B-TFIID, which is involved in transcription by RNA polymerase I, RNA polymerase II or RNA polymerase III. B-TFIID contains TBP, and MotI regulates the availability of TBP. The TBP-TAFs arrangement is dynamic and TAFs can be displaced from TBP-TAFs to allow contacts by other transcription factors and activators.

TFID provides the following three functions: (i) promoter recognition through the TBP- TATA box interaction; (ii) complex assembly through interactions between TBP and general transcription factors (GTFs) and (iii) activator-directed transcription through the TAF_{II}s. However, in 1996 it was

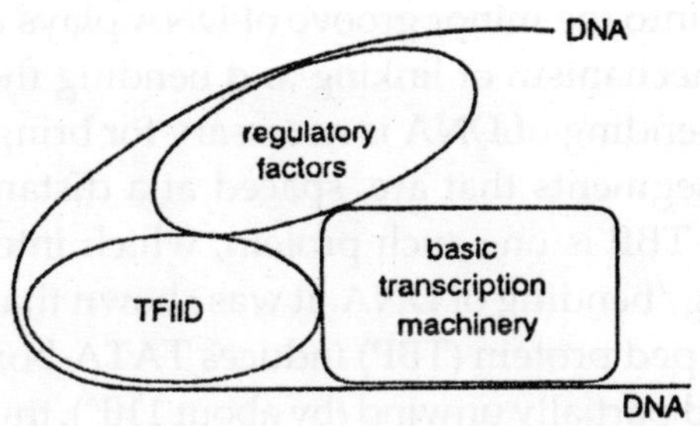

Induction of DNA bending by TFIID, to facilitate enhanced interaction between regulatory factors and the basic transcription factors.

reported that in yeast a variety of genes may be transcribed normally even in the absence of functional TAF_{II}s, suggesting that TAF_{II}s are not essential for the function of promoter specific activators. In 1998, a TBP-free TAF_{II}s-containing complex (TFTC)′, was also isolated, which supported transcription of genes containing TATA or Inr promoter sequences, and also responded to transcriptional activators. It is thus obvious that although generally TFIID (TBP+TAF_{II}s) is needed for basal as well as activator-dependent transcription, TBP and TAF_{II}S are both individually dispensable and that a variety of TFIIDs (those containing both TBP and TAF_{II}s, those lacking TAF_{II}s but containing TBP i.e. TRF-nTAF and TFTC lacking TBP) may occur to take part in eukaryotic transcription.

Binding of TFIID at the TATA box also leads to bending of DNA. This bending perhaps facilitates contact between regulatory factors bound at upstream sites and one or more components of the general transcriptional machinery (Fig.). This DNA bending due to TFIID can also be compared with CAP induced DNA bending described for prokaryotic promoters. Several other transcription factors including TFIIB are being purified and their structure being worked out.

During 1990s, three-dimensional structures of several protein-DNA complexes (including TBP-TATA-box) have been worked out. These studies, suggested that the side chain intercalation into the minor groove of DNA plays an important role in the mechanism of linking and bending the DNA. This linking and bending of DNA is necessary for bringing together small DNA segments that are, spaced at a distance from one another. The TBP is one such protein, which intercalates and causes linking/bending of DNA. It was shown that the binding of saddle-shaped protein (TBP) induces TATA-box to curve (at about 80°) and partially unwind (by about 110°), thus presenting a widened minor groove to the concave surface of TBP. The dimensions of molecular saddle of TBP are such that its concave surface could sit nicely on a DNA-double helix, at right angle to the DNA axis. The long axis of the saddle of TBP is actually parallel to DNA axis, wrapping around the DNA in the minor groove. This bending of TATA-box, brings other initiation factors together and its unwinding allows interaction of DNA with other factors leading to further unwinding towards the start site. Eventually this leads to separation of two DNA strands (at the start site) required for initiation of transcription. It is believed that, in due course of time, the three dimensional structure of the complete pre-initiation complex (appropriately designated as 'transcriptosome'), will be determined so that the mechanistic and regulatory aspects of transcription will be fully understood.

TFIIB Domains for Interaction with TFIID/TATA Complex: TFIIB has two fundamental activities: (i) formation of TFIID -TFIIB/TATA (DB-DNA) complex or TFIID-TFIIA-TFIIB/TATA (DAB-DNA) complex and (ii) basal transcription. In the year 1993, deletion mutants of TFIIB were used to identify the domains responsible for the above two activities.

Following two domains were identified for specific functional roles : (i) The C-terminal domain, containing two direct repeats and two basic residue repeats was found to be necessary and sufficient for interaction with TFIID/TATA complex (also called D-DNA complex), but not with either the

TFIID or TATA alone. In the absence of TFIID, TFIIB does not bind DNA and in the absence of TATA, it can not bind TFIID, suggesting that an interaction of TFIIB with both TATA and TFIID takes place. (ii) An extra 84-residue N-terminal region (which is not needed for the formation of D-B complex) was found to be required for basal transcriptional activity, so that in the absence of this region, despite the formation of D-B complex, transcription was greatly reduced or completely stopped. This region may, therefore, be involved in the recruitment of RNA polymerase II to the complex. It has been shown that in TFIIB and TFIID, similar structural domains are present for protein-DNA interactions (these similarities include C-terminal direct repeats, basic residue repeats, sigma homologies).

It has also been shown that an acidic activator is also needed for the basal transcription activity of TFIIB. A mutant form of TFIIB, which is unable to interact with acidic activator, could not support transcription. This suggests that TFIID and TFIIB, in the absence of an acidic activator, assemble to form only a non-productive D-B complex, but form a productive D-B complex only in the presence of this acidic activator, which transactivates the initiation complex.

The crystal structure of TFIIB-TBP-TATA ternary complex was also described in 1995. The structure revealed the following domains with defined functions : (i) core element of TFIIB (resembling cyclin A) recognizes TBP-DNA complex through protein-protein or protein-DNA interactions; the N-terminal domain of this core element of TFIIB forms the downstream surface of ternary complex permitting fixation to the transcription start site; (ii) the remaining surfaces of TBP and the TFIIB can interact with TAFs, with general initiation factors and also with transcriptional activators/co-activators.

Structure of TFIIA -TBP-DNA complex and the role of TFIIA as a co-activator in activator dependent transcription. In

1996, X-ray crystallographic studies of TFIIA-TBP-TATA (TA) complex have also been carried out. Results of X-ray crystallographic studies of TFIIB-TBP -TATA (TB) complex were published earlier in 1995, so that it was possible to combine the results of TA and TB complexes to prepare a model for a TAB (TFIIA-TBP-TATA-TFIIB) complex. TFIIB binds to the C-terminal stirrup of TBP, while TFIIA binds to the N-terminal stirrup of TBP molecular saddle. While TFIIB interacts with DNA sequences upstream as well as downstream of TATA element, TFIIA interacts with DNA backbone upstream of TATA element, but on the opposite face of the double helix (Fig.). This model of TAB complex suggests the following features of eukaryotic transcription : (i) TFIIA, TFIIB and TBP provide extensive surfaces having recognition sites for many TAFs, pol II and other general initiation factors and transcriptional activators/co-activators; (ii) TFIIB interacts with DNA downstream of TATA, stabilizes TBP-DNA complex and acts as a bridge between TBP and pol II located at the transcription start site. TFIIA similarly interacts with DNA upstream of TATA, recognizing regulatory proteins that are needed for transcription; (iii) TFIIA's impact on the activity of TFIID involves the non-conserved N-terminal portion of TBP and/or TAFs; this association of TFIIA does not lead to any conformational change in the structure of TBP-DNA complex.

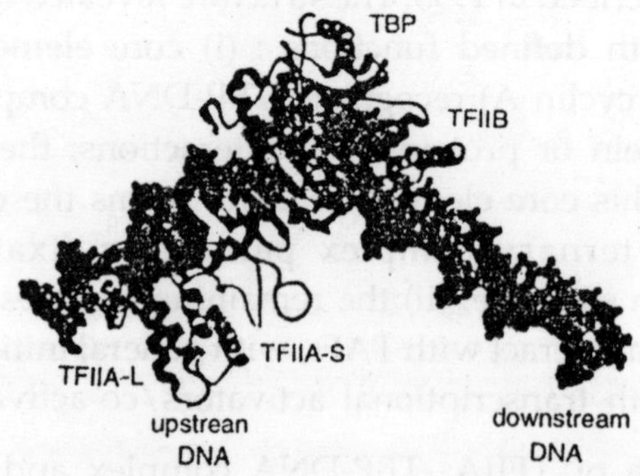

Three dimensional structure of TFIIA-TBP-DNA complex, showing interactions between three components of the complex.

Phosphorylation of the CTD of A Subunit of Pol II **:** RNA polymerase II is found in a non-phosphorylated form (called IIA) at the time of binding to the promoter. It gets altered into a phosphorylated form (called IIO), before it uncouples from pre-initiation complex to take part in the onset and elongation of transcription (Fig.). Extensive phosphorylation takes place in the carboxy-terminal domain (CTD) 'tail' of the largest subunit of RNA polymerase II. TFIIH has been shown to be the CTD kinase responsible for phosphorylation of CTD. The kinase is held in check and its activity is increased with the assembly of pre-initiation complex, and with the addition of TFIIE. TFIIH actually consists of five subunits and carries out activities other than phosphorylation of pol II. In particular, the interaction of TFIIH, TFIIE and TFIIJ may be important for transcription.

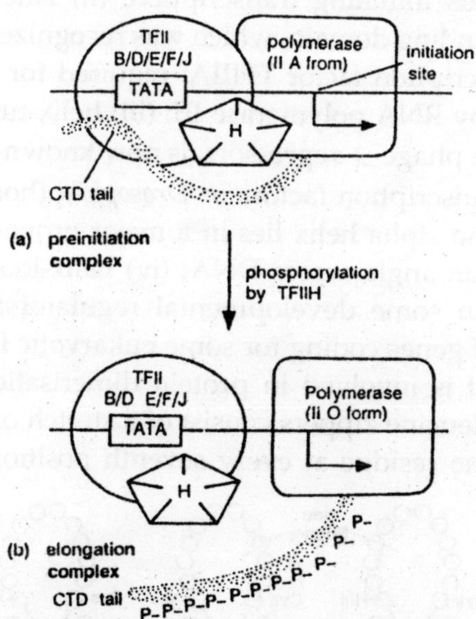

A model showing the action of transcription factor TFIIH (a kinase) on the carboxy terminal domain (CTD) tail of the largest subunits of RNA polymerase II (note that the polymerase can elongate the RNA on DNA template, without the transcription factors, once it binds to promoter and gets phosphorylated).

Separate DNA Binding and Transcription Activation Domains: Transcription factors have separate (i) DNA binding domains and (ii) transcription activation domains. Through their 'DNA binding domains, they recognize specific DNA sequences for regulating gene expression, and after binding to DNA, they may interact with either RNA polymerase or with other transcription factors. Transcription activation domains, on the other hand, function through protein-protein interaction to bring about activation of transcription.

DNA Binding Domains : Following are some of the protein motifs, which are involved in regulation of transcription through DNA binding : (i) several steroid receptors *(steroid-hormone receptor superfamily)* are activated by corresponding steroids (working as ligands), and in turn activate intracellular receptors. These intracellular receptors eventually bind specific DNA sequences thus initiating transcription; (ii) zinc linger motif has a DNA binding domain, which was recognized for the first time in transcription factor TFIIIA, required for transcription of 5S rRNA by RNA polymerase III; (iii) helix-turn-helix, first recognized in phage λ repressors, is now known to be present in several transcription factors in *Drosophila* (homeobox) and mammals; one alpha helix lies in a major grove of DNA and the other at an angle across DNA; (iv) helix-loop-helix motif was found in some developmental regulators; it regulates expression of genes coding for some eukaryotic DNA-binding proteins, and is involved in protein dimerisation and DNA binding; (v) Ieucine zippers consist of a stretch of amino acids with a Ieucine residue at every seventh position.

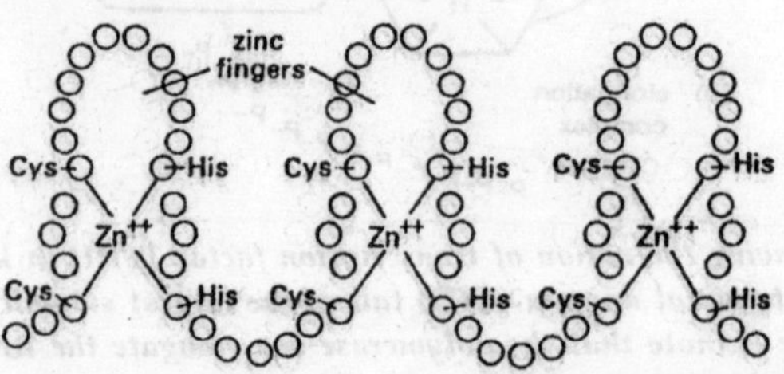

Three Cys_2/His_2 zinc fingers in the transcription factor SP1.

(i) *Steroid Receptors* : The steroid receptors include receptors for the steroid hormones, retinoids, vitamin D, thyroid hormones and a number of other compounds. These receptor proteins contain separate domains for hormone binding, DNA binding and for transcriptional activation. The DNA binding domain contains 70 amino acid residues, with eight conserved cysteine residues, forming two zinc fingers with zinc. A peptide from this domain, can fold (into two α helices) in the presence of zinc, and is utilized for recognizing the appropriate binding site in DNA.

(ii) *Zinc Proteins (with Zn fingers, Zn clusters and Zn twists):* In some proteins, there are zinc binding sites with distinct DNA binding motifs. In one such motif, a loop of amino acids protrudes out as zinc finger and form zinc binding site (Fig.). In other motifs, a zinc twist or a zinc cluster is found. More information is available about zinc fingers than for zinc twists and/or zinc clusters. The zinc fingers are described as Cys_2/His_2 and Cys_2/Cys_2 fingers.

Cys_2/His_2 fingers have the following consensus sequence: $Cys–X_{2-4}–Cys–X_3–Phe–X_5–Leu–X_2–His–X_3–His$. Each finger consists of about 23 amino acids, linked to another finger by 7-8 amino acids. Details of some transcription factors with zinc fingers are listed in Table. These zinc fingers are required for binding to DNA. On one extreme, these fingers may involve almost the entire protein as in TFIIIA (9 fingers), and on the other extreme only a small domain is involved in forming zinc fingers, as in ADR1 (2 fingers). Since zinc fingers are found in several known transcription factors, this feature also helped in recognizing proteins, that may function as transcription factors, *e.g.* TDF (testis determining factor), ZFX, ZFY (zinc finger proteins encoded by genes on human X and Y chromosomes).

Some Transcription Factors and other Proteins, which Bind to DNA by using Cys_2/His_2 Zinc Fingers.

Protein	*Sources*	*Size (daltons)*	*Number of fingers*	*Target*	*Promoter for RNAP*	*Function of protein*
TFIIIA	Mammal	37,000	9	50 bp (5S gene)	polymerase III	Factor for 5S transcription
SP1	Mammal	105,000	3	10 bp (GC box)	polymerase II	General transcription factor
ADR1	Fly	150,000	2	22 bp (*ADH2* gene)	polymerase II	Activates *ADH2* gene
TDF(?)	Mammal	?	13	?	?	Male sex determination
Kruppel	Fly	60,000	5	?	?	Embryonic segmentation
Hunchback	Fly	80,000	4+2	?	?	Embryonic segmentation

In contrast to the above Cys_2/His_2 fingers which are repetitive in nature, there are also Cys_2/Cys_2 fingers which are non-repetitive. Some examples are given in Table. As shown in the table, some of the proteins with Cys_2/Cys_2 fingers are steroid receptors, which recognize and perhaps bind to response elements (in DNA) for these receptors. Mutations altering the amino acids in finger regions have been shown to alter the DNA binding ability of these zinc finger proteins, suggesting that fingers are responsible for DNA binding. Rarely fingers

Some Regulatory Proteins, which have 1 or 2 Zinc Fingers of the Cys_2/Cys_2 Type.

Protein	*Size (daltons)*	*Number of fingers*	*Target*
Glucocorticoid receptor	94,000	2	20bp in GRE
Estrogen receptor	66,000	2	20bp in ERE
GAL4 yeast regulator	99,000	1	17bp in UAS_C
Adenovirus E1A	~30,000	1	?

may bind RNA rather than DNA or may not bind any nucleic acid at all. Translation initiation factors (*e.g.* eIF2B) have also been shown to have zinc fingers which help in recognition of initiation codons.

(iii) *Helix-turn-helix (HTH) and Homeodomain :* The HTH motif was the first DNA-recognition motif discovered. The HTH motif consists of a 20-residues segment with an α helix (1-7 residues), a turn (8-11 residues) and a second α helix (12-20 residues). It occurs in a large family of prokaryotic DNA-binding proteins including CAP, λ Cro and X Rep. The second HTH helix should lie in the major groove, and contributes to important base pair contacts for DNA binding.

Among eukaryotic regulatory proteins, the most important domain with HTH motif is the 60-residues long homeodomain motif found in the products of a number of developmental genes in *Drosophila*. The polypeptide chain representing homeodomain, folds as a 3-helix bundle, in which second and third helices form the HTH.

(iv) *Helix-loop-helix (HLH) :* The HLH proteins have some similarities with the leucine zipper family (see below). Like leucine zipper proteins, the HLH proteins have a basic region that contacts the DNA and a neighbouring region that mediates dimer formation. The dimerisation region forms an α-helix, a loop and a second α helix. The HLH proteins play an important role in differentiation and development and their activity is modulated by heterodimer formation. For instance MyoD protein forms a heterodimer with E2A protein and helps in the differentiation of muscle cells.

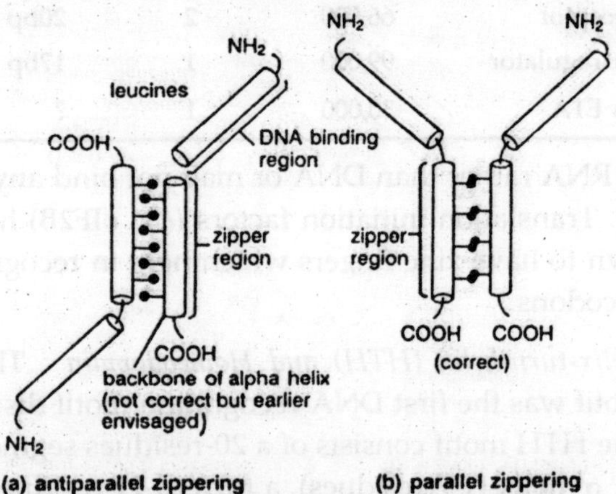

Association of leucine zippers : (a) antiparallel zippering (as earlier envisaged); (b) parallel zippering due to side-by-side overlapping (as known now to be correct).

(v) *Leucine Zippers, Dimer Formation and DNA Binding :* In 1984, Robert Tijan at the University of California (Berkeley) had identified a protein SP1, which was shown to bind at a motif repeated five times at the promoter of early genes in mammalian virus SV40.

This led to selective activation of these genes. This first report of regulation of eukaryotic genes by DNA binding proteins, led to isolation and purification of a protein C/EBP (it binds to CAAT and to SV40 core enhancer; C/EBP = CAAT/Enhancer Binding Protein) by S.L. McKnight and his co-workers at the Carnegie Institution of Washington at Baltimore (U.S.A.). The gene for C/EBP was cloned and sequenced, giving information about the sequence of 359 amino acids in this protein. A computer search revealed a 60-amino acids region in C/EBP similar to regions in two other proteins encoded in *myc* and *fos* proto-oncogenes (proto-oncogenes function normally, but become cancer causing due to mutations). Subsequently, in these and several other proteins, a leucine zipper was discovered, which is a stretch of amino acids rich in leucine, occupying every seventh position (heptad repeat) in the potential zipper. These leucine residues may form an amphipathic alpha helix (amphipathy means that in a alpha helix hydrophobic groups face on one side, and hydrophilic groups face the other side; amino acids with different properties segregate on alpha helix). The leucine zippers form dimers among similar molecules (e.g. Jun-Jun homodimer) or among dissimilar molecules (e.g. Jun-Fos heterodimer). Initially it was believed that leucines of one protein may interdigitate with the leucines of the zipper of another protein in reverse orientation. However, later it was established that zipper regions associate in parallel (leucines overlap and line up side-by-side), when they form a dimer (Fig.) in the form of a coiled coil (two right handed helices wind around each other).

Amino acid sequences were compared in atleast 11 regulatory leucine zipper proteins including four mammalian proteins (C/EBP, CREB, Jun, Fos), two yeast proteins (GCN4,

YAP1), two other fungal proteins (CYS-3, CPC1) and three plant proteins (HBP1, TGA1, OPAQUE2). This comparison revealed a consensus sequence shown in Figure. Following are the general features of these proteins :

(i) proline and glycine are absent or rarely found, thus permitting formation of alpha helix;

(ii) 2-5 heptad repeats of leucine were found, which helped in dimer formation for DNA binding;

(iii) basic amino acids, arginine and lysine (arg/lys) were found close to leucine zipper, helping in DNA binding (zippering of two molecules bring arg/lys in proper position for combining with dyod symmetric motifs in DNA);

(iv) zippering leads to formation of Y shaped dimers;

(v) an asparagine is always found at the same position and interrupts alpha helix (like proline and glycine), thus allowing bending of arms of Y shaped dimer to permit a grip on DNA motif meant for binding (Fig.).

Transcription Activation Domains : The transcription activation domains are separate from DNA binding domains and each consists of 30-100 amino acids. Different types of activation domains have been exchanged and combined with other DNA binding proteins to produce chimeric transcription factors: The transcription activation domains function through protein-protein interactions and often help in establishing contacts with components of transcription complex, which leads to activation of transcription. Following three types of activation domains, shown in Figure, have been identified :

(i) Acidic domains were first identified in yeast transcription factors GAL4 and GCN4. Later they were found in glucocorticoid hormone receptor and also in AP-l/Jun transcription factors. They have two features in common, *i.e.* there are regions of significant negative charge and they can form amphipathic alpha helical structures. The acidic domains help in association of TFIIB and TFIID.

arginine
consensus molecule
lysine
invariant asparagine
leucine zipper

A consensus sequence in protruding arms of leucine zippers, based on a comparison of several proteins; regions that aid in binding and touch the DNA are highlighted.

(ii) Glutamine rich domains were first identified in SP1, and later also_ in factors involved in *Drosophila's* antennapaedia and ultrabithorax, in yeast's HAP1, HAP2, and GAL11 and in several mammalian factors (OCT-1, OCT-2, JUN, AP-2, SRF).

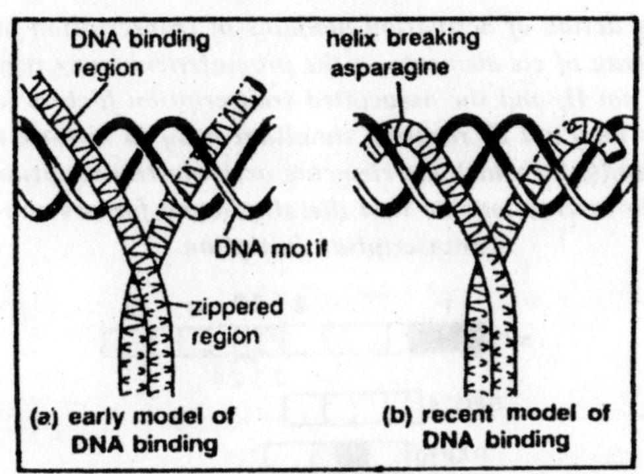

DNA binding regions of leucine zipper : (a) alpha helices protruding out from DNA (as earlier envisaged); (b) the protruding regions of alpha helices bend at asparagine residues forming a Y-shaped dimer, to establish a grip on DNA motif (as now believed).

(iii) Protein rich domains have been identified in CTF/NF-1 and in several other mammalian factors (AP-2, JUN, OCT-2, SRF). The mechanisms of action of these activation domains are shown in Figure.

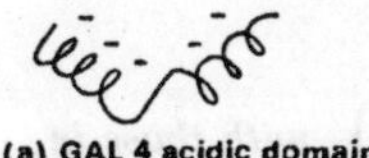

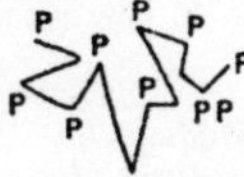

(a) GAL 4 acidic domain (b) Sp1 glutamine-rich domain B (c) CTF proline-rich domain

Three types of protein domains responsible for transcriptional activation by DNA binding factors.

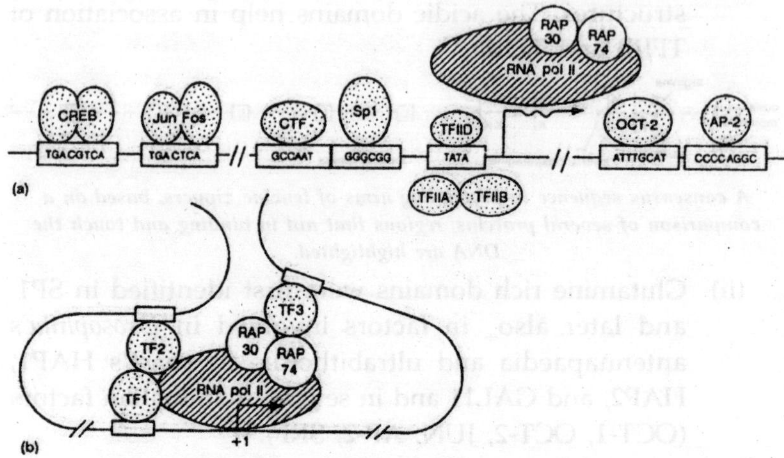

Mechanism of action of activation domains of transcription factors : (a) a hypothetical array of cis elements in the promoter/enhancer regions of a gene transcribed by pol II, and the associated transcription factors (all these DNA binding factors may not be required simultaneously to initiate transcription); (b) mechanism(s) by which cis elements activate transcription involving protein-protein interactions so that distally bound factors can take part in transcription initiation.

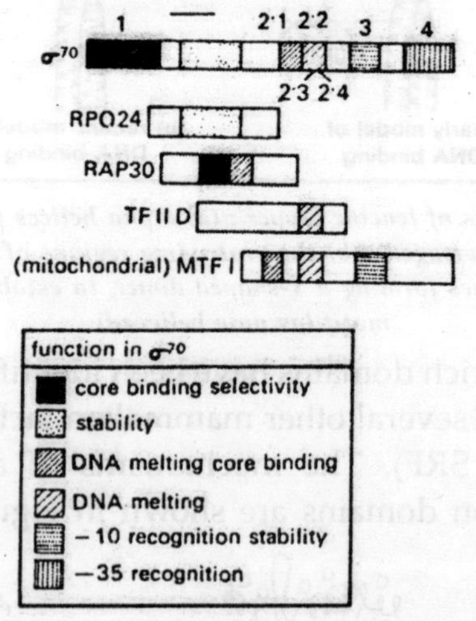

Similarity of several regions of E. coli sigma factor (σ^{70}), with those in three nuclear transcription factors, i.e. RPO24 (a subunit of yeast RNA polymerase II), RAP30 (a human transcription factor) and TFIID; and one mitochondrial transcription factor, MTF1.

Similarity between Prokaryotic Sigma Factor and Some Eukaryotic Transcription Factors : Recently similarity between sigma factor and atleast three different nuclear mRNA transcription factors has been demonstrated. These transcription factors include the following :

(i) RPO24, which is the fourth largest subunit of yeast RNA polymerase II.

(ii) RAP30, which is a part of a complex of two proteins that bind tightly to human RNA polymerase II.

(iii) TFIID, which, binds to TATA box. The similarities of certain domains of sigma with these three factors are shown in Figure.

Proteins that Influence Transcription without Binding to DNA: Several proteins, which influence transcription without binding to DNA (sometimes through recognition of another protein) include the following: (i) E1A activates Ad (adenovirus) genes, pol II and pol III genes, (ii) Tat acts at post-transcriptional level with the help of tar sequence in HIV RNA and (iii) Vmw 65 activates initial early (IE) genes in HSV (herpes simplex virus), by associating with another transcription factor.

Transcription Activators and Co-activators : In eukaryotes a network of protein-DNA and protein-protein interactions is involved in the assembly of transcription initiation complex. During this assembly, problems like the following are encountered: (i) competition among thousands of genes for a common pool of GTFs; (ii) difficulty in access of GTFs to promoter elements due to packaging of DNA into chromatin; (iii) the presence of protein factors, that inhibit formation of initiation complex; (iv) promoter elements having weak affinity for DNA-binding proteins due to deviations in consensus sequences.

In vitro, these problems are not encountered due to supply of GTFs in excess and due to the use of naked DNA. The basal transcription described earlier making use of GTFs only (using no other proteins as activators/co-activators) hardly occurs *in vivo* in eukaryotes and a given gene is silent due to (i) packaging of DNA in chromatin and (ii) presence of a variety of repressers. It has been shown that a variety of activators are actually involved in activating the transcription initiation complex. It has also been suggested that these activators do not merely activate the complex by protein-protein interactions but are actually involved in the recruitment of this complex to the promoter/ initiation site. Further, an individual activator is often incapable of performing its function unless assisted by another protein, which is therefore described as a co-activator.

Activators Targeting TBP of TFIID, TFIIB and TFIIH : Promoter proximal DNA sequences, 50-200 bases upstream of the start site, and the distal enhancer elements further upstream are the sites, where activators bind and modulate transcription. Following are the ways, in which activators/Co-activators play their role in transcription :

(i) We know that DNA, organized in the form of nucleosomes, is not accessible to GTFs and thus remains repressed due to histones. It has been shown that some activators can induce changes in the nucleosome structure, making the DNA accessible to GTFs, a phenomenon described as anti-repression. An example of these activators in yeast is provided by SWI/SNF complex, which overcomes histone mediated repression in the chromatin.

(ii) Transcriptional activators also interact with $TAF_{II}S$ (co-activators) and these activator-TAF_{II} interactions stimulate recruitment and/or stabilisation of TFIID. Two or more activators can also exert synergistic effect on transcription through interactions with multiple components of the PIC. $TAF_{II}s$ thus form a large multiprotein complex, which sits on top of TBP, thus integrating signals from many other activators and non-TAF_{II} co-activators.

(iii) Activators, other than those interacting with TAF_{II}s, are also known; they interact with other GTFs including TFIIA, TFIIB, TFIIE and TFIIH. For instance, the effect of TFIIA and TFIIB in stabilizing TFIID-promoter complex can be potentiated by interactions with various activators, leading to stimulation of transcription initiation. Activators are also known, which interact with TFIIF, TFIIH and pol II, which join PIC. Thus it seems likely that every component of PIC is the target of atleast one transcriptional activator, and mutation in these target proteins affect the function of these activators, although basal transcription is not affected by these mutations.

Co-activators **:** RNA Pol II transcription complex is composed of about 40 different subunits. As described above, it has been shown that several activators can directly interact with components of transcription complex (e.g. GTFs, TFIIA, TFIIB, TBP, TAF_{II}s). However, co-activators or co-activator complexes are sometimes used to link the upstream activators with the components of Pol II transcription complex. CBP (CREB binding protein) is the most widely characterized co-activator protein, which interacts with fZIP (basic leucine zipper) transcription factor, CREB (cyclic AMP response element binding protein). Actually CBP and a related co-activator p300 protein (CBP/p300) in *Drosophila* work as co-activators of a variety of transcriptional activators, including the following - (i) retinoic acid receptor (RAR), (ii) thyroid hormone receptor (TR) and (iii) two Rel- containing transcription factors, the p65 subunit of KB (a nuclear factor or NF) and the Dorsal in *Drosophila,* (iv) nuclear steroid hormone receptor. There are several other examples of activators, which need co-activators- to bring about activation of transcription.

CBP possesses a histone acetyltransferase activity that is believed to decondense chromatin and facilitate binding of the Pol II transcription complex to the core promoter. CBP has also been shown to be a component of large protein complexes containing additional histone acetyltrans-ferases including the following : (i) p/CAF (p300/CBP associated factor) and (ii) steroid receptor co-activators (SRCs). CRB also associates with other co-activator complexes such as ARC (activator - recruited cofactor). Since the activator/co-activator sites are nearly one kb away from start site, the role of co-activators in decondensation through histone acetylation can not be easily explained, but it seems that CBP functions in concert with other co-activators that directly interact with Pol II. It also appears that CBP has dual role in chromatin decondensation and Pol II recruitment, since it does interact with a part of Pol II, that has helicase activity. CBP has been shown to interact with Dorsal a transcription factor involved in early development of *Drosophila* embryo. Yeast SAGA co-activator complex is another example of a co-activator complex, which contains both a histone transacetylase activity (GCN5) and TAFII components of Pol II complex.

Modular/non-modular Enhancers and Enhanceosome in Activation : Enhancer sequences, briefly described earlier, have multiple binding sites for activator proteins. These enhancers can be of two types : (i) modular enhancers, which do not have special combinations of activator binding sites, so that each activator contributes to overall transcription, and (ii) nonmodular enhancers which have each a combination of sites, one site for one activator, so that several activators bind to form an enhanceosome where each activator is essential for transcription. A 65-base pair enhancer for human IFN-β gene was reported in August 2001, as a paradigm for nonmodular enhancers. It has binding sites for activators NF-kB, IRF and ATF/Jun families and for architectural protein HMGI(Y). All these proteins assemble into an enhanceosome for transcription of the *IFN*-β gene in response to viral infection, and each is essential for transcription (for details, consult Science, Vol. 293, pages 1054, 1133, 10 August 2001).

Transcriptional Repressors and Corepressors : Like activators, repressors can also have direct contact with different components of the Pol II transcription complex, particularly TBP and TFIIE. But sometimes corepressors are involved in helping the repressors in performing their function. Rpd3 (reduced phosphate dependency), which is a histone deacetylase, functions as a corepressor of several mammalian regulatory proteins including the following nuclear receptors (i) Mad/Max heterodimer (a helix—loop—helix (HLH), (ii) RAR and TR. Rpd3 seems to be a subunit of a larger complex like Sin 3 (SWI-independent) SMRT (silencing mediator for retinoid and thyroid hormone receptors) and N-CoR (nuclear receptor corepressor).

Due to its deacetylase activity. Rpd3 might mediate repression by condensing chromatin at the core promoter or at the distal enhancer, thus impeding the binding of either the components of Pol II complex or the upstream activators. In *Drosophila* also several corepressors have been identified which involve atleast 10 different sequence specific repressors; some of them are short range repressors (working over ~100 base pairs) and others long range repressors (working over one kb or more). These corepressors include Groucho and dCtBP.

Recruitment Model for Transcriptional Activation : As discussed above, tor successful transcription, a number of activators and co-activators are needed in addition to a variety of general transcription factors (GTFs) that take part in the formation of a transcriptional complex. The classical view is that the activators help transcription by protein-protein interactions between the activator and the transcriptional complex or its components already bound to DNA at the promoter site. However, more recently a recruitment model has been proposed, according to which an activator is actually involved in bringing the transcriptional machinery to DNA (rather than interacting with such a machinery already, bound to a DNA site). An activator according to this model, interacts with transcriptional machinery (not yet bound to DNA) and recruits it to DNA. In many cases, it has been shown that TAFs are required to transmit the effects of activators to the transcriptional machinery. This is why TAFs are believed to be co-activators.

The above similarities are consistent with the observation that the two largest subunits of eukaryotic nuclear RNA polymerases are homologues of the two large subunits of bacterial RNA polymerase.

Transcription Factors for Elongation of RNA Chains in Eukaryotes : Certain accessory proteins involved in transcription, and called the 'elongation factors', enhance the overall activity of RNA polymerase II, leading to increase in elongation rate. Atleast two such proteins (transcription factors) are known :

(i) The transcription factor TFIIF accelerates RNA chain growth relatively uniformly, in concert with RNA polymerase II or pol II.

(ii) The transcription factor TFIIS (also called SII) helps elongation of RNA chain, by relieving the obstructions in the path of such elongation. TFIIS is known to function by first causing hydrolytic cleavage at 3' ends of RNA chains, which are stuck and can not elongate. Thus, RNA polymerase moves backwards (hydrolytic cleavage) under the direction of TFIIS and the obstruction is removed, before it moves forward again to continue elongation (Fig.). It is possible that hydrolytic activity resides in pol II and is enhanced by TFIIS (second largest subunit of pol II has regions of homology with bacterial rihenuclease; see Fig.).

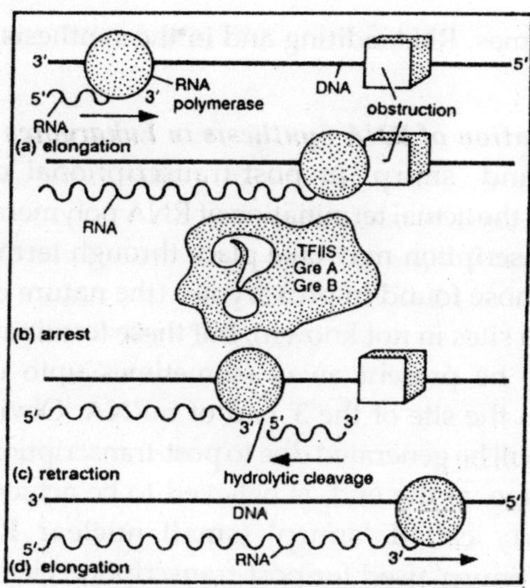

The role of TFIIS in elongation of RNA chain.

Prokaryotic elongation factors, GreA and GreB (discovered in 1992-93) also cause hydrolytic cleavage at 3' end of nascent RNA in *E. coli.*

Transcription Slippage : Usually the RNA transcripts are precisely complementary to the DNA template, but rarely RNA polymerase can undergo transcriptional slippage, so that the transcript obtained is either longer or shorter than that encoded by the DNA template. The longer RNA transcripts can result, when RNA polymerase does not move and yet incorporates additional nucleotides. Similarly, shorter products are produced due to translocation of RNA polymerase without incorporation of nucleotides. More often, transcription slippage leads to addition of extra nucleotides. This transcription slippage is more frequent, when the template consists of homopolymeric runs of a nucleotide, like $(dA)_n$ or $(dC)_n$, etc. It has also been shown that homopolymeric runs as short as two or three nucleotides can cause slippage during transcription initiation, but much longer runs are needed for slippage during elongation. Transcription slippage is a means of regulating transcriptional reading frames, RNA editing and in the synthesis of caps and tails.

Termination of RNA Synthesis in Eukaryotes : AAUAAA sequence and 'snurp' in post-transcriptional cleavage. In eukaryotes, the actual termination of RNA polymerase II activity during transcription may take place through termination sites similar to those found in prokaryotes (the nature of individual termination sites in not known), but these termination sites are believed to be present away (sometimes upto one kilobase away) from the site of the 3' end of mRNA. Obviously 3' end of mRNA will be generated due to post-transcriptional cleavage. This cleavage, at the end, is believed to be achieved by what is popularly called 'snurp' (small nuclear RNA-protein complex). 'Snurp' used for post-transcription cleavage has not been identified so far but is believed to be certainly different than the Ul *snurp*, which is believed to be involved in intron splicing in split genes. Moreover, a sequence 5'AAUAAA 3' has been found just on the 5' side of poly(A) addition site in several eukaryotic mRNAs (*e.g.* α and β globins in mouse and ovalbumin in chicken). We will notice that poly(A) tail is added to 3' end of eukaryotic mRNA after processing of precursor mRNA.

The sequence 5'AAUAAA3' in mRNA 3' end seems to be common in eukaryotic mRNA and mutations in this sequence cause elongation of mRNA. This will suggest that this sequence contains the signal for endonucleolytic post-transcriptional cleavage. This sequence, therefore, is not involved in the termination of the synthesis of mRNA, but helps in generating the 3' end later through endonuclease cleavage, in which *snurp* helps in an unknown manner.

Chromatin Decompaction for Transcription in Eukaryotes : A eukaryotic chromosome, which is barely few microns long (at metaphase), contains nuclear DNA (which can be as long as 2 m) in a very compact form. The compactness is achieved by coiling and supercoiling of nucleosome subunits to constitute what is described as chromatin. This chromatin is found in the nucleus, as nucleoproteins due to an association of DNA with proteins. During transcription, this chromatin has to undergo decompaction to allow transcription machinery's access to the DNA to be transcribed. Although some information about disruption of DNA-histone contacts at the nucleosome level is available, but very little is known about the mechanism of decompaction at higher level. At the nucleosome level, the disruption and restructuring of histone-DNA contacts in the chromatin are brought about by the combined effect of the following two mechanisms :

(i) reversible acetylation of lysine residues in the amino-terminal 'tails' of histones due to histone acetyltransferases (HATs) and histone deacetylases (HDACs); acetylation activity of TAF_{II}-250 component of TFII D also brings about enzymatic ubiquitination of H1 histone (Fig.) ;

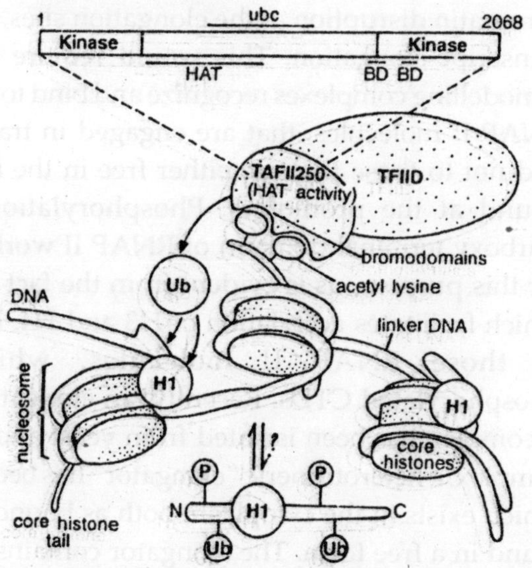

Role of TAF_{II} 250 in acetylation and ubiquitination of nucleosome histones leading to decompaction of chromatin.

(ii) chromatin remodelling due to a variety of enzyme complexes, that have been purified (e.g. SWI/SNF, HM_GIY, FACT, Spt4, Spt5, Spt6). To allow access of transcriptional machinery to the DNA recognition sites, disruption of chromatin at gene promoter sites, is brought about by a mechanism that is atleast partly known now as described above. However, little is known about this disruption during transcript elongation, which requires that the passage of RNA polymerase II (RNAP II) is coupled to a mechanism for chromatin disruption. It is believed that during transcript elongation, histone octamer does not leave DNA and the entire histone octamer is actually transferred backwards on the DNA fragment, through a transiently formed DNA loop, which allows RNAP II to travel at a rate of 25 nucleotides per second. The chromatin remodelling complexes perhaps reach the elongation sites by hitching a ride on RNAP II, as it travels on its journey through a gene. This helps in chromatin disruption at the elongation sites, leading to transcript elongation. This would require that these remodelling complexes recognize and bind to only those RNAP II molecules, that are engaged in transcription and not to those that are either free in the nucleus or bound at the promoters. Phosphorylation at CTD (carboxy terminal domain) of RNAP II works as a tag for this purpose, as is evident from the fact that HAT, which facilitates acetylation of H3 and H4, binds only to those RNAP II molecules, which have phosphorylated CTDs. Recently, an elongating RNAP II complex has been isolated from yeast and from this complex, a heterotrimeric "elongator" has been isolated, which exists in the cytoplasm both as bound to RNAP II and in a free form. The elongator contains a subunit Elp3 (60-kD) having HAT activity towards all the four histones. This subunit is believed to be recruited for acetylation during the transcript elongation process.

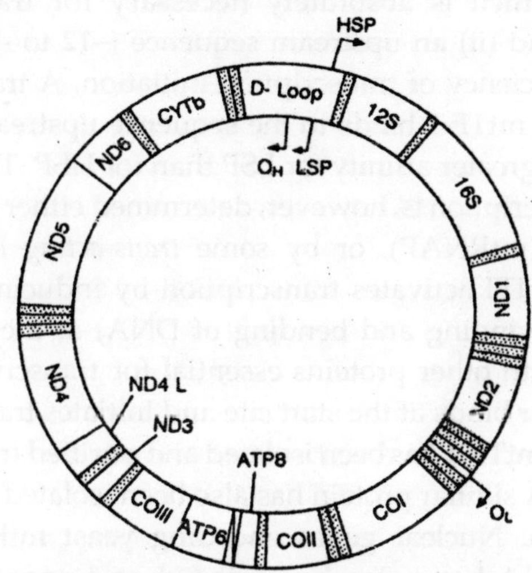

A map of vertebrate mtDNA, showing location of D loop, origins for DNA replication (O_H and O_L), promoters for transcription (HSP and LSP), genes for 22 tRNAs (shaded areas) and other protein coding genes (CO = cytochrome oxidast; ATP6, ATP8 = ATPase subunits; cyt b = cytochrome b; ND = NADH dehydrogenase); arrows indicate direction of replication or transcription.

Transcription in Mitochondria

Transcription of Vertebrate mtDNA : The transcription of each strand of vertebrate mtDNA starts at a site within the *D loop*.('displacement loop', where a short piece of nascent DNA is found at the heavy or H strand). The transcription continues uninterrupted, so that in the transcript, tRNA sequences lie among other RNA sequences. During RNA processing, tRNA species and different mature mRNAs are produced. The most abundant RNA species in mitochondria are, however, two rRNA species, which are transcribed separately and at a faster rate. In human and mouse mtDNA, the two promoters for transcription occur within the D loop, separated from one another by 150bp (Fig.). These are called H-strand promoter

(HSP) and L-strand promoter (LSP). Each promoter (50bp) has two functional domains : (i) a short sequence at the transcription start site (which is absolutely necessary for transcription initiation), and (ii) an upstream sequence (–12 to –39), which increases efficiency of transcription initiation. A transcription factor, called mtTF-1 binds to the sequence upstream to 'start site' and has greater affinity for LSP than for HSP. The species-specific transcription is, however, determined either by mtRNA polymerase (mtRNAP), or by some *trans-acting* faclor(s) or both. The mtTFl activates transcription by inducing physical changes (unwinding and bending of DNA) at the promoter, in concert with other proteins essential for transcription. The mtRNAP later binds at the start site and initiates transcription. The protein, mtTF-1 has been isolated and purified from human and mouse. A similar protein has also been isolated from yeast mitochondria. Nuclear genes encoding yeast mtRNAP and human mtTF-1 have also been cloned and sequenced. The mtRNAP is a single polypeptide, like that in the RNA polymerases of T3 and T7 phages and differs from bacterial RNA polymerase, having four polypeptides ($\alpha_2\beta\beta'$).

Although, nothing is known about the events associated with elongation of mitochondrial transcripts, but some progress in our understanding of the mode of termination of transcription has been made for human mtDNA. A 13bp sequence (5'-TGGCAGAGCCCGG-3') has been found to be adequate to direct termination. A termination protein is perhaps involved, which binds at the termination site and acts as a barrier to the transcription process.

Transcription of Yeast and Plant mtDNA : The mtDNA in yeast is 78kbp and that in higher plants (master circle) is 200-240kbp, in contrast to a small (16-17kbp) mitochondrial genome in vertebrates. In yeast, mitochondrial genes are separated by extensive AT-rich non-coding spacer sequences; a bulk of mtDNA in higher plants is also non-coding. A 5S rRNA gene is a novel gene found only in plant mtDNA. In mtDNA of both yeast and higher plants, there are multiple transcription units, each with a separate promoter (unlike vertebrate mtDNA having a single promoter on each strand (LSP and HSP).

In higher plants, only 30%-70% of mtDNA is known to be transcribed, which gives rise to 20-30 polypeptides, 3 rRNA species and 15-20 tRNAs. Currently promoters of transcription units in mtDNA of higher plants are being discovered. The RNA polymerases and the transcription factors associated with transcription of plant mtDNA, will be isolated and characterized during 1990s and in the early years of the next century.

Most genes in plant mtDNA give monocistronic transcripts, although in some cases adjacent genes do give rise to dicistronic messages. (In chloroplasts of higher plants several polycistronic transcripts are produced). The transcription initiation sites are now known for many genes in plant mtDNAs. They contain a conserved CRTA (R = A or G) motif, although the degree of conservation is low, when compared with that in yeast.

Transcription in Chloroplasts

The Chloroplast genome differs from the nuclear genome in having a much smaller information content (10^{-3} to 10^{-4} fold less), but with 100–10,000 copies per cell, making upto 15% of the cellular DNA. There are about 120 plastid genes in the chloroplast genome, about 60 coding for plastid proteins, 30 coding for a part of photosynthetic apparatus (photosystem II or PSII, Cytochrome b6/f, photosystem I or PSI, ATP synthase and RUBISCO), and the remaining genes with unknown functions. (Many subunits of photosynthetic apparatus are coded by nuclear genes). It has been shown that in *Chlamydomonas reinhardtii*, most chloroplast genes are transcribed independently of each other as monocistronic transcripts, but chloroplast genomes of maize and other higher plants contain several polycistronic transcription units, a single unit encoding subunits of different photosynthetic complexes.

Further, it has also been shown in *C. reinhardtii,* that for the expression of each of the nearly 120 chloroplast genes, factors coded by 5-70 nuclear loci are required, so that more than 1000 nuclear genes (5-10% of the nuclear genome) are involved in the expression of these genes. These nuclear encoded factors may function at the level of transcription, RNA processing, RNA maturation, translation, or during the assembly of photosynthetic complexes. While differential transcription has been documented for several chloroplast genes, regulation is manly exercised at the post transcriptional level.

The mechanism of transcription for majority of chloroplasts genes resembles that in prokaryotes in several respects including the following : (i) Promoter sequences are found at –10 and –35 positions for all kinds of genes (rRNAs, tRNAs and mRNAs), so that same RNA polymerase is used for the transcription of these three classes of genes, (ii) Termination of transcription in plastid genes is facilitated by the formation of hairpin structures (like in prokaryotes) due to the presence of short inverted repeat sequences, (iii) Chloroplast mRNAs have neither a cap at the 5' end, nor a long poly A tail at their 3' end (as in prokaryotes). However, three classes of introns are found in chloroplast genes, which need to be spliced out during RNA processing. There are also chloroplast genes (*e.g.* genes for large submit of RUBISCO and 32 kD protein of PSII) which lack introns in higher plants, but have introns in chloroplasts of plants like *Euglena.*

4

Organisation of Genetic Material

During the last few decades, the use of the techniques of molecular biology (including recombinant DNA technology) in genetic studies gave several surprises. Some of these surprises led to the discovery of the following : (i) split genes, meaning that the sequences containing actual information of the gene (exons) are interrupted by other sequences (introns) which are spliced out after transcription; (ii) overlapping genes, meaning that same DNA sequence can become part of each of two or more genes expressed at different times and in different reading frames; (iii) pseudogenes, which represent DNA sequences derived from mRNA through reverse transcription; these pseudogenes, therefore, differ from the split genes to which they belong, due to the absence of intron sequences; and (iv) cryptic genes, which are generally silent under a specific environment and may be activated (due to a variety of events) in a few individuals of a population in another environment; these genes differ from selfish DNA/selfish genes/pseudogenes and offer a selective advantage to the organism, that carries that carries them. These interesting and surprising developments in the study of cell biology will be discussed in this chapter, although for a full appreciation of this discussion.

Split Genes or Interrupted Genes

Discovery and Nature of Split Genes : After rediscovery of Mendel's laws in 1900, a gene was considered earlier to be a point on a chromosome, and later to be a continuous segment of DNA. It was also proved as late as 1964 that there was co-linearity between sequence of nucleotides on DNA and sequence of amino, acids in corresponding protein, so that there was absolutely no doubt about the continuity of nucleotides in a gene represented by a DNA segment. This was proved to be true not only in prokaryotes but also in eukaryotes. In view of this, it was a big surprise when, in 1977, it was proved for the first time that in some mammals, birds and amphibians, a gene may not be represented by a continuous sequence of nucleotides but may be interrupted by some intervening sequences which are not represented in mRNA transcribed from the gene and utilized for the synthesis of proteins. Such genes with intervening sequences were called split genes (Fig.).

The discovery of split genes was made in 1977 by several groups in a variety of materials, which included the following:

(i) Two research groups separately headed by Phillip A. Sharp and Richard J. Roberts studied genes of adenovirus 2.

(ii) Research groups of D.S. Hogness, I.B. David and N. Davidson studied genes for 28S rRNA in *Drosophila.*

(iii) Research groups of P. Chambon, P. Leader and R.A. Flavell studied β globin genes, ovalbumin genes and tRNA genes.

In all these cases the genes were found to be interrupted

by intervening sequences. The credit for discovery of split genes, however, goes to Phillip Sharp and Richard Roberts, who won the 1993 Nobel Prize for Medicine for their work on split genes. They analysed the hybrids of late mRNA of adenovirus 2 with the adenovirus genomic DNA. When these mRNA-DNA hybrids were examined under electron microscope, the adjoining sequences of mRNA were found to hybridized with discontinuous stretches of genomic DNA of adenovirus 2.

The intervening DNA sequences were observed as loops and the phenomenon was later described as R-looping (see Fig.). Chambon's group compiled sequences of the boundaries of introns from a large number of protein coding eukaryotic genes (not ribosomal RNA or tRNA genes), which revealed the presence of consensus sequences at the intron-exon junctions. Of these GT was always found at the 5' side of the intron (left splice junction) and AG at the 3' side (right splice junction). This became popularly known as GT-AG or Chambon's rule (rarely, AT-AC occur at intron-exon junctions). Some of the diseases (*e.g.* thalassemians) are caused by mutations, which created or abolished these splice junctions.

A detailed study was conducted on *ovalbumin gene* found in chicken. This ovalbumin gene is responsible for synthesis of a protein 'ovalbumin' consisting of 386 amino acids and synthesized only by highly specialized tubular gland cells of the oviduct at the time when the hen is laying eggs. The expression of this *ovalbumin gene* is controlled by some female sex hormones. Pierre Chambon and his colleagues synthesized artificial ovalbumin gene in order to study its regulation. Such an artificial gene could be synthesized by using ovalbumin mRNA which could give rise to cDNA *(complementary DNA)* with the help of enzyme reverse transciptase. This cDNA was inserted into a plasmid and cloned in *E. coli* for its multiplication. When this cDNA was compared with corresponding genomic

DNA, it was discovered (through DNA hybridization) that the genomic DNA had additional intervening sequences.

Another important technique utilized for the study of split genes was the use of restriction enzymes which have the property of cleaving DNA at unique sites. More than 100 such restriction endonuclease enzymes are now available. In view of this, restriction endonucleases could be used to find out the presence or absence of a certain sequence in a particular gene. For example, when *EcoR1* and *HindIII* enzymes were used with cDNA for ovalbumin, it was found that no cleavage occurred suggesting that the sequences of six base pairs each recognized by these two enzymes were absent.

On the basis of this it was expected that if the DNA extracted from oviduct or another tissue was cleaved by utilizing these two enzymes, the *ovalbumin gene* will not be broken and the cleavage will occur at other places thus making it possible to isolate *ovalbumin gene* from the living cells. This DNA segment representing *ovalbumin gene* was expected to be separated with the help of hybridization with cDNA artificially synthesized. When such hybridization was done, it was found that cDNA

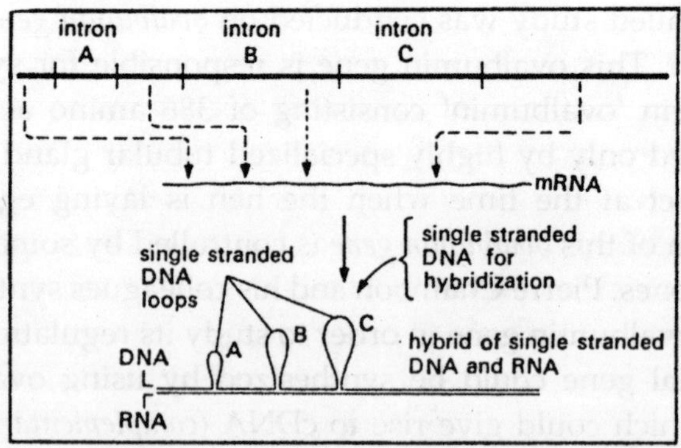

The upper figure shows a DNA sequence representing an interrupted gene with there introns (A, B, C), and the synthesis of its mRNA; the bottom figure shows the result of hybridization of mRNA with single stranded DNA obtained after denaturation of native DNA (note the loops formed by three intron regions).

could hybridize with different fragments of DNA rather than with a single fragment. Hybridization between the single stranded DNA having the gene for ovalbumin and its mRNA also showed formation of distinct loops at specific sites as observed in electron microscope (Fig.). This DNA forming the loops is obviously missing in mRNA. This kind of results eventually led to the discovery of split genes in 1977. Subsequently it could be proved that split genes are present atleast in two more cases *i.e.* gene for β-globin (a component of haemoglobin molecule) in rabbit and mouse and immunoglobulin gene (antibody gene).

By the end of 1977, it was fairly clear that the presence of split genes in case of higher organisms was a phenomenon of general occurrence. We may discuss the problem of split genes by taking a hypothetical example of a gene represented by a DNA sequence which has three pieces, called exon 1, exon 2 and exon 3 (Fig.). These three exons are separated by two long intervening sequences called introns A and B. The terms exon and intron were used by W. Gilbert for the first time and are being followed ever since. Figure shows, that in the transcription of DNA, the ultimate product *i.e.* mRNA had only those sequences which corresponded to exons, the sequences representing introns being entirely absent.

The mechanism of production of mRNA from DNA sequences having intervening sequences called introns was a matter of discussion during 1978 and 1979. Later it was discovered that both exons and introns are first transcribed and that this primary transcript is then modified. The sequences corresponding to introns are removed and the sequences corresponding to exons are joined together, in correct order to give rise to mRNA. A generalisation has also been made that the order of exons on DNA is the same as the order in which they are found in processed mRNA. There does not seem any strong reason why it should always be true and it is possible that the order of exons would sometimes be different.

It has also been shown that all classes of genes can be interrupted. These interrupted genes may include (i) nuclear genes for proteins, (ii) nuclear genes coding for rRNA, (iii) nuclear genes coding for tRNA, (iv) mitochondrial genes in yeast, (v) Chloroplast genes in a wide variety of plants, (vi) genes in archaebacteria and (vii) genes in bacteriophages of *E. coli* : they were initially believed to be entirely absent in eubacterial genomes, but during 1990s, introns were discovered even in eubacteria.

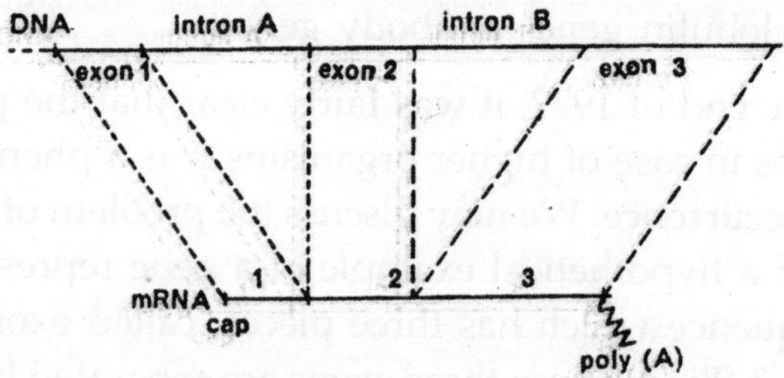

Diagrammatic representation of a hypothetical split gene having three exons (exon 1, exon 2, exon 3) and two introns (intron A, intron B) and its relationship with mRNA. There are no sequences in mRNA corresponding to those in two introns (redrawn from Science. Vol. 204, 1979).

R-loop Mapping and Restriction Mapping of Interrupted Genes: R-loop mapping is a technique in which mRNA is hybridized with double stranded DNA under conditions in which RNA-DNA hybrid is more stable than the DNA duplex. This allows displacement of one strand of DNA in the region, where it can pair with RNA. When DNA segment belonging to an interrupted gene is used in such an experiment for hybridization with mRNA, the intervening sequence or intron region loops out as double stranded DNA and the exon region loops out as single stranded DNA as shown in Figure.

The double stranded DNA loop will appear at the junction of two single stranded loops produced due to displacement by mRNA. These loops can be observed under the electron microscope as has been done for mouse minor β -globin gene showing single intron. This technique is comparable to the hybridization of mRNA with single stranded DNA.

An interrupted gene can also be studied by comparing the restriction map of the interrupted gene with that of its corresponding cDNA prepared from its mRNA through reverse transcription. When this is done, there may be two situations: (i) If the intervening sequence or intron region does not have a recognition site for an enzyme, there is only a change in the length of corresponding segment (since the two restriction sites in the regions on either side of an intron are now separated by a longer distance) "but the number of segments remains the same (Fig.). (ii) If the intron region has a restriction site for the enzyme employed, an extra cut will be made in this region. This extra cut will produce two fragments corresponding to one segment in the cDNA (Fig.).

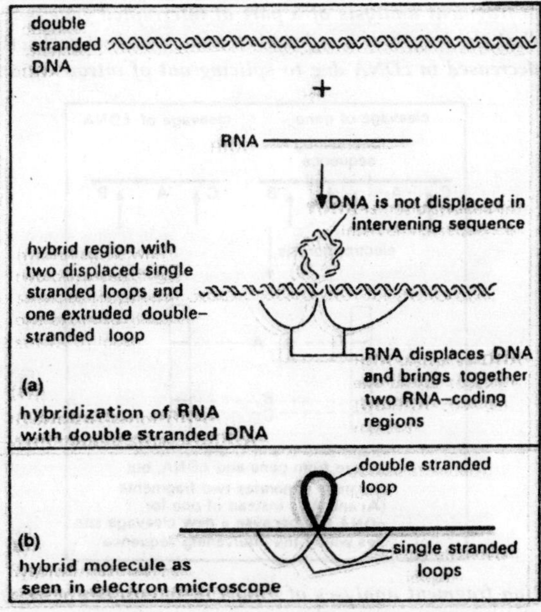

R-looping, where mRNA is hybridized with double stranded DNA (note the double stranded thick loop showing intron region and single stranded thin loops representing the exons).

When restriction maps are prepared for cDNA and the gene using a number of enzymes, we may find that the two maps may be similar at the two ends but extra sites may be

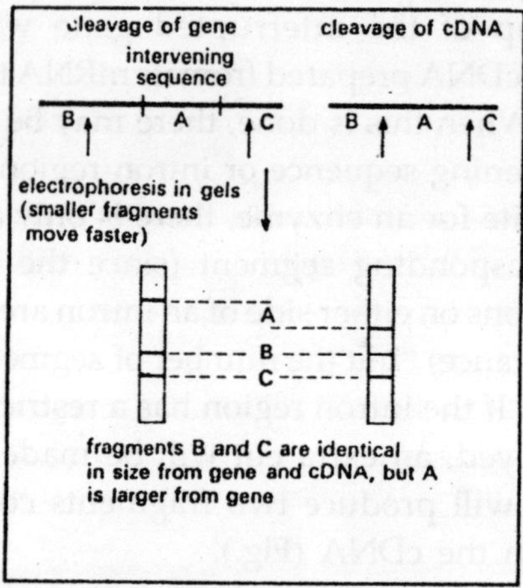

Restriction fragment analysis of a part of interrupted gene and its cDNA, showing that the length of a segment A (which had no restriction site) is decreased in cDNA due to splicing out of intron region.

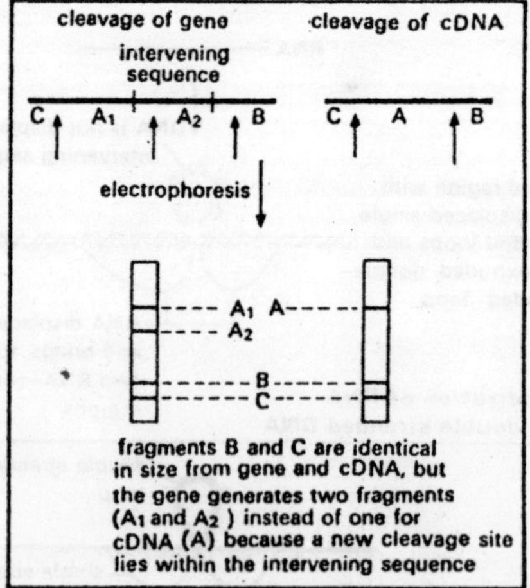

Restriction fragment analyses of a part of interrupted gene (having a restriction site in the intron region A, dividing it into two regions A1 and A2) and its cDNA showing decrease in the number of fragments in cDNA from four to three.

present in the intron regions of the gene, which are absent in cDNA or mRNA. The resolving power of restriction mapping allows detection of gene segments upto 20-30bp long. Restriction maps of cDNA and genomic DNA of mouse β -globin gene are shown in Figure. It may be seen that there are two intron regions in the gene; one is much larger, which can be visualized in R-loops under the electron microscope, but the other is smaller, that can not be visualized under the electron microscope. This demonstrates the higher resolving power of restriction mapping.

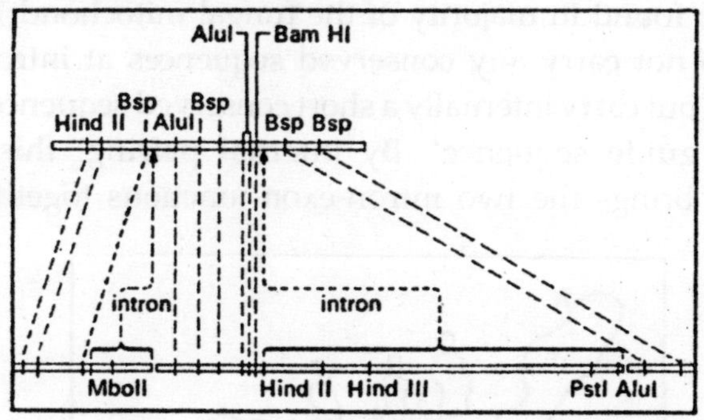

Fig. A comparison of the restriction maps of cDNA and genomic DNA for mouse β -globin gene, showing that the gene has two introns not present in cDNA. Only one of them shows up as a loop during hybridization of mRNA with denatured single stranded DNA (since the second intron is very small).

Structure of Chicken Ovalbumin Split Gene : The structure of ovalbumin gene in chicken has been studied in considerable detail and it has been found that in the split gene of ovalbumin there are eight exons and seven introns (Figs.).

The size of ovalbumin gene has been shown to be 7,700 base pairs although the mature mRNA for ovalbumin is only 1,872 nucleotides. It has also been shown, that entire ovalbumin gene having 7,700 base pairs is first transcribed to a precursor RNA to which a cap consisting of 7mG is added at the 5' end and poly A tail is added at 3' end. After the addition of cap and tail, the introns are excised, perhaps in two instalments, five introns being excised in the first step and the remaining two in second step. The exons produced due to splicing are then joined with the help of enzyme ligase to produce the mature mRNA of *ovalbumin gene*. The different steps involved in the processing of precursor RNA into mRNA for *ovalbumin gene* are shown in the Figure.

Split Genes in Fungal Mitochondria : Split genes are also found in mitochondria. Introns of these split genes in fungal mitochondria are classified into group I and group II introns on the basis of their internal organization, (i) Group I introns, which are found in majority of the fungal mitochondrial split genes, do not carry any conserved sequences at intron-exon junctions, but carry internally a short conserved sequence called 'internal guide sequence'. By internal pairing, this guide sequence brings the two intron-exon junctions together and

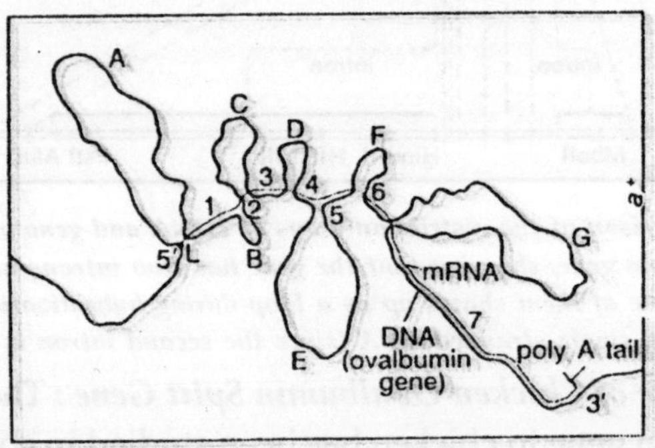

The map of electron micrograph prepared after hybridizing mRNA (thin line) of the ovalbumin gene with its single stranded DNA (thick line). The loops A-G represent 7 different introns and segments L and 1-7 represent 8 exons.

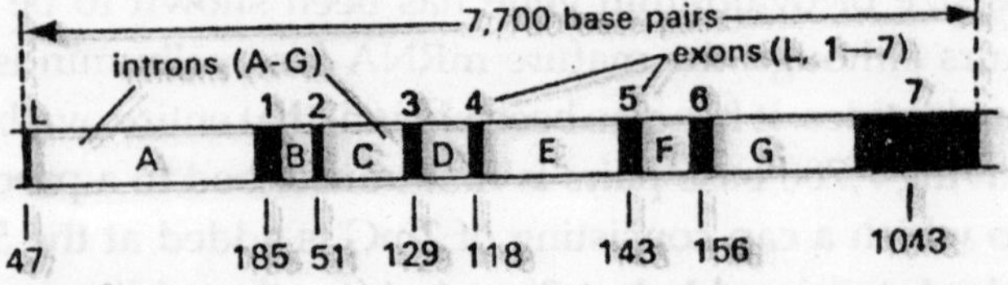

Schematic representation of ovalbumin gene showing 7 introns (A–G) and 8 exons (L, 1–7); number of base pairs in each exon are also given.

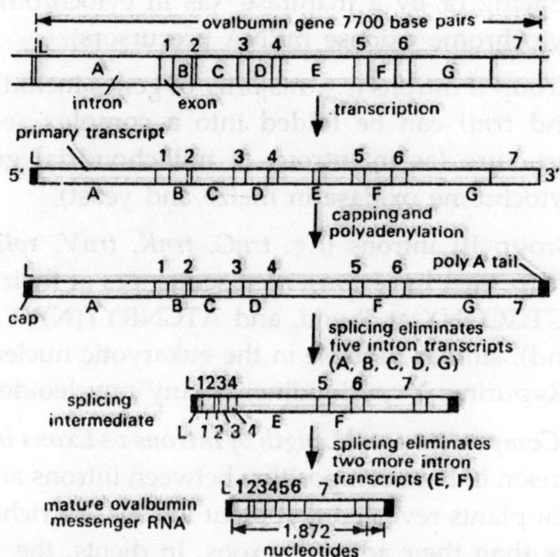

Steps involved in the production of ovalbumin mRNA (1872 nucleotides long) from ovalbumin gene (7700 base pairs long); introns (A-G) and exons (L, 1-7).

help in splicing out the introns. These group I introns are also found in the nuclear genes coding for rRNA in *Tetrahymena* (a ciliate) and *Physarum* (a slime mold). Features similar to those of group I introns are also found in introns of phage T4 genes, (ii) Group II introns resemble nuclear genes and have consensus sequences (GT and APy) and a branch sequence that resembles the TACTAAC box. These introns are excised as lariats.

Split Genes in Chloroplasts : Split genes for ribosomal RNA (rRNA), transfer RNAs and some proteins have also been reported in the Chloroplast genomes of several plants including *Chlamydomonas* and *Nicotiana.* Introns found in chloroplast genes can be classified into three groups on the basis of intron boundary sequences.

(i) Group I introns (*e.g.* in *trnL)* can be folded in a secondary structure similar to self splicing rRNA precursor of *Tetrahymena.* These can be removed either by self splicing or by a maturase' (as in cytochrome b and cytochrome oxidase mRNA precursors),

(ii) Group II introns (*e.g.* majority of genes including *trnA* and *trnl)* can be folded into a complex secondary structure (as in introns of mitochondrial genes for cytochrome oxidase in maize and yeast),

(iii) Group III introns (*e.g. trnG, trnK, trnV, rpl2, rps12, rps16,* etc.) have conserved sequences at their borders (GTGCGNY at 5' end, and ATCNRYY(N)YYAY at 3' end), similar to those in the eukaryotic nuclear genes (R=purine: Y=pyrimidine; N=any nuycleoide).

Base Composition and Length of Introns vs Exons in Plants: A comparison of base composition between introns and exons of genes in plants revealed that plant introns are richer in AT base pairs than their adjacent exons. In dicots, the AT base pairs range from 59-83% with a mean of 74% as against 55% AT base pairs in their adjacent exons. Similarly in monocots, it ranges from 31-77% with a mean of 59% as against 44% AT in their adjacent exons. In addition, plant introns are short, the average size being 80-139 nucleotides (two third of all introns studied are less than 150 nucleotides).

In both these features (base composition and length), plant introns differ from vertebrate and yeast introns, although they resemble introns in animals, like fruitfly *(Drosophila)* and the extensively studied nematode, *Coenorhabditis elegans*. These key features of plant introns provide them with specificity during pre-mRNA splicing of these introns. This splicing is not so sensitive to the environment (temperature, etc.) as that of budding yeast, fruitfly and mammalian cells. In other words, the plant introns are spliced accurately over a broad temperature range, while animal introns are sensitive to environment fluctuations.

'Intron Enhancement' / Gene Expression in Transgenic Cells: It has been shown, that addition of specific intron sequences to gene constructs used for the production of transgenic plants, leads to high level of expression of the associated genes. This phenomenon is described as 'intron enhancement' and the extent of enhancement in expression depends on:

(i) plant system (monocot or dicot),

(ii) expression system (transient or transgenic),

(iii) the intron and the promoter used and

(iv) the site of insertion.

The most commonly used intron for enhancement of the expression of transgene is maize *adh1* intron 1, which increases the expression of :

(i) CAT mRNA from an *adh1* promoter, 16-fold and

(ii) NPT II mRNA expression from CaMV35S promoter, 15-fold. There are many other examples of intron enhancement achieved in transgenic crops by using the above *adh1* intron. (CAT, luciferase and NPT II are products of reporter genes used for the production of transgenic plants).

Intron of One Gene may Contain Exon of Another Gene : Sometimes more than one mRNA sequences may be derived from the same DNA segment, due to tissue specific expressions in different tissues. In this process, part of an intron for one gene is used as exon for another gene. Following four examples illustrate this phenomenon:

(i) Amylase gene in mouse has two promoters, one of them expressed in liver and the other expressed in salivary gland. Consequently the 5' ends of precursor mRNAs differ. This leads to the presence of different coding sequences as exon 1 in the two cases (*exon S* in salivary gland and *exon L* in the liver). The derivation of two mRNAs from the same gene encoding two amylases is shown in Figure. It may be noted that exon

L is actually derived from a part of the intron of the amylase gene for salivary glands.

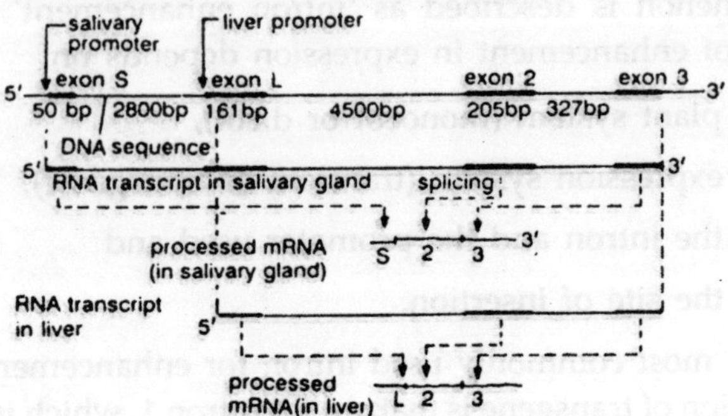

In mouse, derivation of two different amylase mRNAs from same DNA sequence, due to the presence of two promoters and differential RNA splicing.

(ii) During expression of immunoglobulin genes, substitution of one exon by another occurs at the end of the transcript, resulting in proteins having different amino acid sequences at the C-terminal ends.

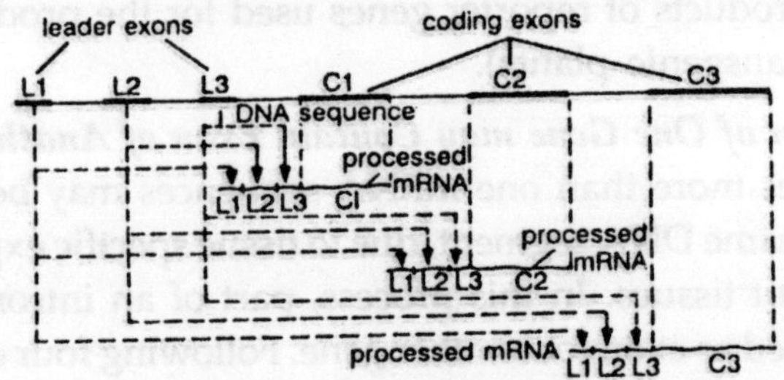

During adenovirus late expression, derivation of different mRNAs from same DNA sequence, due to cleavage, so that same leader sequence (three exons) is joined with different exon sequences.

(iii) During adenovirus late expression, the same transcript gives several mature mRNAs, each having the same leader sequence (not to be translated into protein product) derived from all the three exons (exons 1, 2 and 3) and a coding sequence derived from only one of the three available coding exons A, B and C (Fig.).

(iv) In troponin T gene for rat muscle, at the 3′ end there are five exons, of which only four each take part in forming mature mRNAs for two different forms of troponin T. The two forms differ, depending upon which of the two alternative exons (α or β) is used to construct mRNA, the other exons (W, X, Z) being common (Fig.).

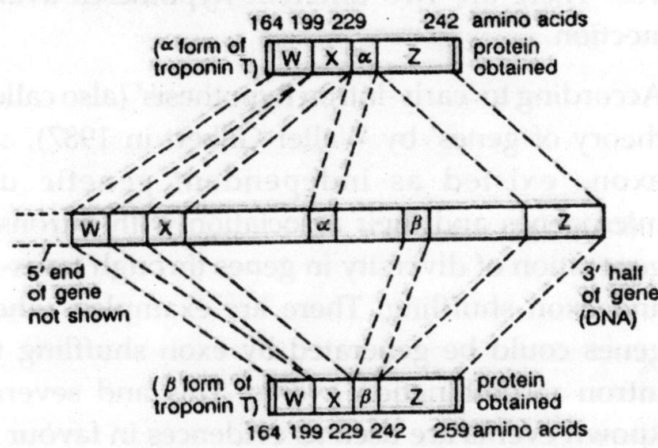

Derivation of α and β forms of troponin T from same DNA sequence, due to differential splicing patterns in rat muscle.

Exon Sequences are Conserved, but Intron Sequences Vary: Through the use of recombinant DNA technology, it could be shown that the exon sequences of more than one genes (for some related functions) in the same species or in different species, often have similar DNA sequences. However, the corresponding intron sequences differ substantially. This was demonstrated for two β globin genes of mouse. This property has been utilized in selecting exon sequences of heterologous genes for isolation of related genes in different species.

Introns with Coding Sequences : In fungal mitochondria, for some genes, it was shown that introns may code for RNA maturases or endonucleases which take part in RNA processing, and DNA recombination respectively.

Two Hypotheses for Origin of Introns 'Early Intron' and 'Late Intron' : Although introns are generally absent in protein-coding genes of bacteria, archaebacteria and some lower eukaryotes, class II type of self splicing introns have recently been reported even in bacteria. Such a distribution of introns has been used for a study of the role of introns in the evolution of genes or alternatively for a study of the origin of introns themselves. There are two different hypotheses available in this connection.

(i) According to 'early-intron hypothesis' (also called 'exon theory of genes' by Walter Gilbert in 1987), ancestral exons existed as independent genetic units or microgenes and their association with introns heloed generation of diversity in genes through trans-splicing and exon shuffling. There are examples, where new genes could be generated by exon shuffling through intron recombination events. This and several other known events are used as evidences in favour of early intron hypothesis,

(ii) According to 'late-intron hypothesis' (also called the 'insertional theory of intron origin') intron evolution was independent of the evolution of protein-coding genes. It is assumed in this hypothesis, that the self-splicing introns invaded the genome of eukaryotic ancestor, as endosymbionts and evolved into mitochondria and chloroplasts containing class II introns. It is postulated that invasion took place at the mRNA level, which was followed by reverse transcription and recombination with DNA, producing split genes. These introns gave other types of introns later. This explains, why introns are absent in bacteria, archaebacteria and lower eukaryotes. The recent report of introns in bacteria may be the initial stage of this process, of the evolution of introns. Efforts are being made to combine the two above hypotheses using strong features of both to formulate a new intermediate hypothesis.

Overlapping Genes

We will discuss the genetic code, which has been shown to be non-overlapping and triplet. In view of this if the length of DNA is known, then the maximum length of protein, which can be synthesized, can be worked out in terms of number of amino acid residues. For instance, a DNA segment of 1200 base pairs can give rise to a protein having not more than 400 amino acids. Moreover, if number of amino acids in proteins synthesized, and number of nucleotide base pairs in DNA are known, they can be compared and one would expect that the number of base pairs should be at least 3 times the number of amino acids in the protein.

Rather if there are segments of DNA which are not coding for protein, then the length of DNA segment will be more than three times the number of amino acid residues in protein. Such comparisons have been made in bacteriophage ϕ X174 which contains a single stranded DNA approximately 5,400 nucleotides in length. Complete sequence of these 5,400 nucleotides was also worked out by Fredrick Sanger and his colleagues at the MRC Laboratory of Microbiology, Cambridge, U.K. (Fredrick Sanger for this and other work got the second Nobel Prize in the year 1980).

The genome of ϕ X174 consists of nine cistrons. Amino acid sequences of major proteins coded by the genome of ϕ X 174 have also been worked out and their molecular weights are known. From the information about proteins coded, an estimate could be made of the number of nucleotides required. This estimate of number of nucleotides exceeds 6,000 which is much higher than the actual number of nucleotides present *i.e.*, 5,400.

Therefore, it was difficult to explain how these proteins

could be synthesized from a DNA segment which is not long enough to code for the required number of amino acids. On detailed study of the system, it was discovered that sequences in the same segment could be utilized by two different cistrons coding for different proteins. Such overlapping of cistrons will be theoretically possible if the two cistrons have two functions at different times and their nucleotide sequences are translated in two different reading frames.

In 1976 Barrell and his co-workers discovered that in ϕ X174, having nine cistrons (A, B, C, D, E, J, F, G, H), cistron E is present between D and J and that the cistron E overlaps cistron D. This could be done by working out the nucleotide sequences of DNA segment representing gene D, E, and J. It could be shown that amber mutations in cistron E lie within the cistron D and, these amber mutations do not influence the translation of cistron D into its protein. Similarly some nonsense mutations for cistron D lie in cistron E suggesting that the cistrons D and E overlap in the DNA sequences and that the cistrons D and E are translated in two different reading frames so that amber codon in mRNA of one cistron will not be read as termination codon during the translation of mRNA of the other cistron.

Among genes D, E, and J in ϕ X174, cistron D codes for a simple protein, cistron E codes for a protein for lysis and cistron J codes for a particle component. It was shown that enough space is not available where these three genes are located, thus confirming that there was overlapping between D and E cistrons. Overlapping has also been suggested to be present in regions of genes A and B and mutants were located in overlapping zone. Overlapping regions were also found to be present at beginnings of cistrons A, C, and J, where in each the initiation codon of a cistron overlaps with the termination codon of its corresponding previous cistron. A DNA segment showing overlapping of D and E is shown in Figure.

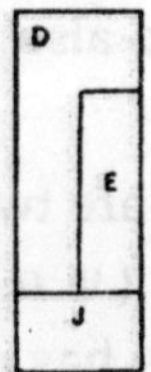

Overlapping of E and D genes in phage ϕ X 174.

Overlapping sequences have also been discovered in tryptophan mRNA of *E. coli* and several other cases. In view of wide occurrence of overlapping genes, it seems that phenomenon of overlapping genes, is an economic device to make better use of genetic material, through packing of more genetic information in less DNA. This phenomenon may be of general occurrence. Therefore, it may be necessary to modify our concept of coding potential (number of amino acids, which a DNA segment can code for) of DNA genomes, because a DNA can potentially code for many more amino acids in different reading frames than the number of amino acids expected to be coded, if no overlapping was present.

Pseudogenes

(U snRNA Series, Globin Pseudogenes and Alu Family)

In multicellular organisms, a wide variety of DNA sequences are found, which are of no apparent use. Some of these sequences are defective copies of functional genes and are, therefore, called pseudogenes. These pseudogenes have been reported in human beings, mouse and *Drosophila*. The most popular examples of these pseudogenes include the following:

(i) Human α -globin and β -globin pseudogenes (ψ), found in each of the two globin gene clusters. Complete nucleotide sequence of alpha globin psudogene is now known and it has been shown that both these genes are non-translatable, since they may have mutations in initiation codon and also frame-shift mutations along their length.

(ii) In mouse also there are two alpha globin pseudogenes (ψ), one of them ($\psi\ \alpha$ 3) is different from other pseudogenes since it has no introns which are present in functional α - globin genes as well as in other pseudogenes. The pseudogene $\psi\ \alpha$ 3 is, therefore, believed to be the reverse transcript of mRNA, obtained due to reverse transcriptase enzyme.

(iii) U snRNA series of pseudogenes in human beings includes U_1, U_2 and U_3 snRNA pseudogenes. The corresponding functional U_1, U_2 and U_3 snRNA genes may be involved in splicing of nucleic acid sequences.

(iv) In *Drosophila*, histone pseudogenes have been discovered,

(v) There are other sequences in human genome which are popularly called Alu family of dispersed repetitive elements, of which there are about 300,000 copies in a haploid human genome. Each sequence is 300 bases long and they are capable of being transcribed by RNA polymerase III (meant for 5SRNA and tRNA), rather than RNA polymerase II (meant for mRNA).

There is a number of characteristic features of these pseudogenes which throw light on their possible origin. These characteristics include the following:

(i) Most pseudogenes outnumber their genes and are therefore repetitive sequences.

(ii) Most pseudogenes are flanked by short direct repeats, 6-21 bases long, suggesting their subsequent insertion. Such pseudogenes are believed to have originated by one of the following methods.

First possibility is that the mRNA or snRNA itself becomes incorporated into retrovirus RNA, followed by the incorporation of retrovirus into the genome.

Second possibility is that reverse transcript of mRNA or snRNA was synthesized as cDNA, which subsequently became integrated. It is this second alternative which is considered to be the more probable mechanism of the origin of pseudogenes. In this hypothesis RNA takes up a configuration favourable for synthesis of DNA at its 3' end copying a part of this RNA as a template to form cDNA. The insertion of this cDNA due to staggered cleavage of genomic DNA at a specific site is shown in Figure.

Cryptic Genes

Cryptic genes are defined as phenotypically silent DNA sequences that are not expressed during the, normal life cycle of an organism. They may be activated in a few individuals of a population, due to events like mutation, recombination, transposition or other genetic mechanisms (or due to specific altered environmental conditions), and this leads to their detection. They, however, differ from other silent DNA sequences, collectively referred to as 'selfish DNA', which

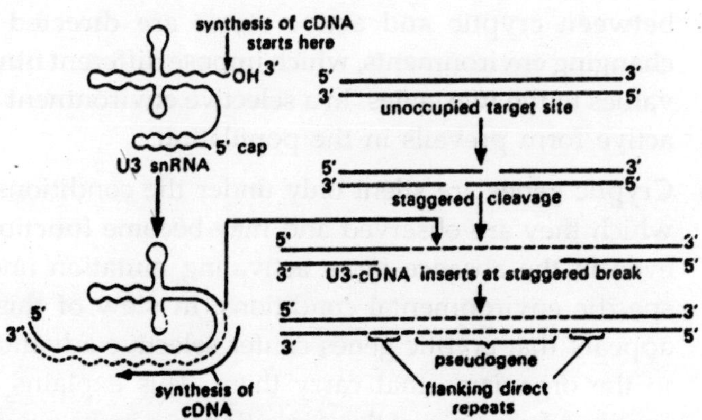

A model for origin of pseudogenes, in which cDNA (complementary DNA) is synthesized on snRNA (small nuclear RNA) and is inserted in DNA, which undergoes staggered cleavage through restriction endonuclease activity.

spread in the genome by making additional copies of themselves, without conferring any phenotypic advantage. They also differ from pseudogenes, which are homologous copies (often derived from mRNA due to reverse transcription) of active genes, but carry many mutations, so that their reversion to active form is not possible.

There are many examples of cryptic genes including the following:

(i) Genes for β-glucoside utilization in *E.coli* and *Solmonella.* Four cryptic genetic systems *(bgl, cel, asc, arb*T*)* involved in the β-glucoside utilization have been characterized in *E.coli.* They mutate spontaneously to a phenotype detectable as papillae on the indicator plates when supplemented with specific β-glucoside.

(ii) Gene for citrate utilization in *E.coli.*

(iii) A gene for alcohol dehydrogenase *(Adh* IV) in yeast.

These cryptic genes are characterized by the following two features:

(i) The cryptic state is a transient state and oscillations between cryptic and active states are directed by changing environments, which impose different fitness values to the two states. In a selective environment the active form prevails in the population.

(ii) Cryptic genes are silent only under the conditions in which they are observed and may become functional even in the absence of an activating mutation under specific environmental conditions. In view of this, it appears that cryptic genes confer selective advantage to the organisms that carry them. This explains the continued presence of these cryptic genes in the genome of an organism.

Promiscuous DNA

(Movement of DNA Between Mitochondria, Chloroplasts and Nucleus)

During 1982-84, several evidences suggesting transfer of DNA from chloroplasts to mitochondria and from both these organelles to the nuclear genome, have been reported. For instance, it was suggested that apparent translocation of DNA sequences from chloroplasts to mitochondria has taken place in corn. This conclusion was based on the discovery that a 12 kilobase segment of mitochondrial DNA (mtDNA) was homologous to a section of chloroplast DNA (cpDNA) of same organism, corn.

During the same period, a report from Texas, U.S.A., described the same phenomenon in yeast, where a segment of mtDNA showed similarity with a segment in nuclear DNA, suggesting flow of DNA from mitochondria to nucleus. A similar report of transfer of mtDNA to nucleus in sea urchin also came from California Institute of Technology. In 1984, more reports of promiscuous DNA came from different places in materials like mungbean, spinach, corn and peas, suggesting that movement of DNA sequences among organelles and between organelles and nucleus must be very common and not exceptional The movement of DNA among mitochondria, chloroplasts and nucleus are shown in Figure.

The evolutionary implications of promiscuous DNA are far reaching and it is speculated that chloroplasts and mitochondria, before invading eukaryotic cell were free living primitive prokaryotes. The endosymboint hypothesis suggests that in course of time these entrapped organisms lost their independence by losing genetic information to nuclear genome. Thus the promiscuity of DNA gave further support to endosymboint hypothesis, but exchange of DNA between mitochondria and chloroplasts is surprising, since they are independent of each other. This may be partly explained on the basis of great variation and bigger size of mtDNA, relative to cpDNA which is tightly constrained in size and sequence even between species. While mtDNA can tolerate foreign DNA and other disturbances, chloroplasts are sensitive to any such variation or modification. This is confirmed by evidence produced at Carnegie Institute of Washington, Stanford, U.S.A., where it was shown that Chloroplast sequences from a variety of plants are scattered throughout mitochondrial genomes, and some of these transfers are believed to be rather recent.

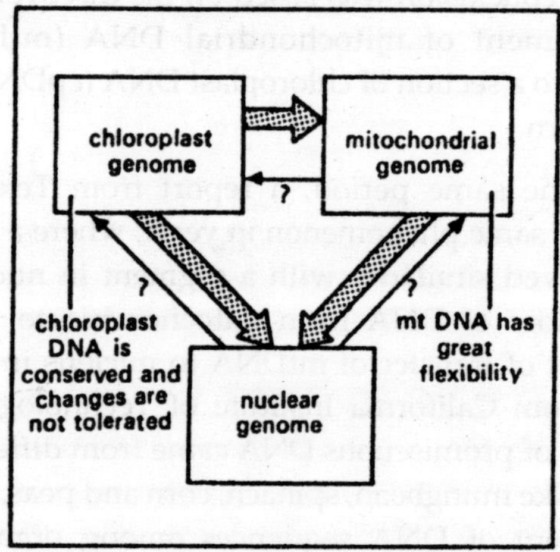

Exchange of genetic material between nuclear genome, Chloroplast genome and mitochondrial genome. Thick arrows indicate transfer of promiscuous DNA for which evidence is available.

The mechanism of this transfer of so called promiscuous DNA is not known but two possibilities are considered, (i) Some kind of a vector or transposon or transducing phage may be involved. (ii) The two organelles might have undergone fusion at some stage. This second mechanism involving fusion is preferred on the basis of electron microscopic observations.

5

Packaging of DNA

In any living system, the length of the DNA always greatly exceeds the dimensions of the compartment (*e.g.* nucleus) , which contains it. The DNA (or RNA, as in some viruses) must, therefore, be compressed very tightly to fit into the space available. The density of DNA is about 10 mg/ml in bacteria, about 100 mg/ml in eukaryotic nucleus, and more than 500 mg/ml in T4 head. The overall compression of the DNA has also been described as packaging ratio (the length of DNA divided by the length of the unit that contains it). This packaging ratio in eukaryotes reaches 1000-2000 in interphase nuclei and upto 10,000 in metaphase nuclei.

The patterns of this packaging and changes in these patterns during replication and transcription (gene expression) have been the subject of detailed study during the last more than three decades. These patterns of packaging are facilitated by associated proteins, which in eukaryotes, could be histones or nonhistones (RNA polymerase is also a nonhistone). More stable proteins in chromosomes are histones and are classified into five types called H1, H2A, H2B, H3 and H4. Another histone protein H5 (a variant of H1) has also been often described, although it may not be involved in packaging. It was thus known that chromatin consists of DNA, five different but stable proteins and a variety of nonhistone proteins.

It will be seen that the progress in the study of chromatin structure in early years, did not make much headway due to limitation of techniques. Whole mounts of chromosomes, whether within the cell or isolated, could be examined under light or electron microscope. This gave information about gross morphology of chromosomes including single stranded or multistranded nature or about the thickness of the ultimate single strands or fibres. When ultrathin sections of chromosomes were examined under electron microscope, it gave such an incomplete picture of chromatin structure that no satisfactory models could be prepared. Hundreds of speculatory models based on inadequate available data were presented and rejected before they could become known to students of cytology or genetics. However, in 1970's due to the use of improved techniques, nucleosome model for chromatin structure became available, which has been confirmed over and again during the last more than two decades. Some of the details regarding the techniques used for discovery of the nucleosome subunit, the structural details of this subunit and the higher levels of organization of these nucleosome subunits will be briefly discussed in this chapter.

Techniques Leading to Nucleosome Discovery : Adequate information for constructing suitable models for chromatin structure started coming after 1960 mainly through development of newer techniques. Techniques from biophysics, genetics and electron microscopy were developed in this connection. These techniques will be discussed in brief outlining the kind of information each produced.

X-ray Diffraction and Electron Microscopy : While studying X-ray diffraction patterns of chromatin, M.F. Wilkins and V.

Luzzati in the year 1960 observed patterns which suggested repeating units in chromatin. Similar repeats were inferred from X-ray diffraction of a mixture of DNA and histones also. Electron microscopy was also used for a study of chromatin structure. In 1974, A.L. Olins and D.E. Olins reported results of electron microscopy of chromatin obtained from interphase nuclei lysed in water.

Under electron microscope, chromatin fibre appeared as arrays of spherical particles, about 10nm (1nm = 10^{-9} m = 10Å) in diameter, connected by filaments, which were about 2 nm in diameter. These particles were called ν (nu) bodies. Later in 1975, P.M. Outdet and others used the term nucleosome for these particles. Electron microscopy was also done on chromatin subjected to nuclease digestion, and the nucleosome structure was confirmed. The structure of DNA suggests, that DNA alone can not form repeats at intervals of 10nm, which meant that the association, of histones must have conferred the repeating organization to chromatin. Association of histones is also suggested *firstly*, from the knowledge that histones are positively charged and will associate with negatively charged groups in DNA, and *secondly* from the knowledge that histones perform their function in chromatin by binding to DNA. Therefore, it suggested that repeating units in chromatin result due to association of DNA with histones and that these repeats occur at intervals of 10 nm (1 Å = 10^{-7} mm = 10^{-10} m).

Chromatin Reconstitution Experiments : X-ray diffraction patterns of native chromatin could be compared with those of chromatin, reconstituted from DNA and histones to find out the degree of successful reconstitution. It was shown by T. Kornberg (T. Kornberg is different from A. Kornberg of DNA replication fame), that 90% reconstitution was achieved, when DNA was mixed with unfractionated histone preparation. However, when purified histones were used, reconstitution failed suggesting that some denaturation of histones takes place during purification, which destroys the compactly folded native conformation of chromatin.

Nuclease Digestion and Regular Distribution of Histones: Nucleuses are a class of enzymes which will cause fragmentation of DNA at those sites which are not protected by proteins like histones. Therefore, it was expected that through a study of DNA fragments generated by nuclease treatment, location of the sites of naked DNA in chromatin will be possible and information will thus be available about the nature of association of histones with DNA in the chromatin. This technique of nuclease digestion was utilized in 1973 by Hewish and Burgoyne for fragmentation of chromatin of rat liver cells. The fragments were separated by gel electrophoresis. It was found that fragments of different sizes were produced which formed distinct bands representing sizes which were multiples (two, three, four times, etc.) of size of the smallest unit. This information came from further detailed study of bands (Figs.). On the other hand, when DNA alone and not the chromatin, was fragmented with same nuclease enzyme, uniform smear rather than bands were observed on gel. Appearance of uniform smear rather than bands meant that fragments obtained due to nuclease digestion of naked DNA, gave a gradient of size, rather than being multiples of the size of smallest unit as

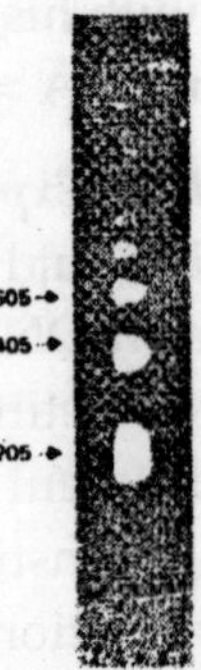

Multimeric series of DNA bands after nuclease digestion of chromatin followed by gel electrophoresis.

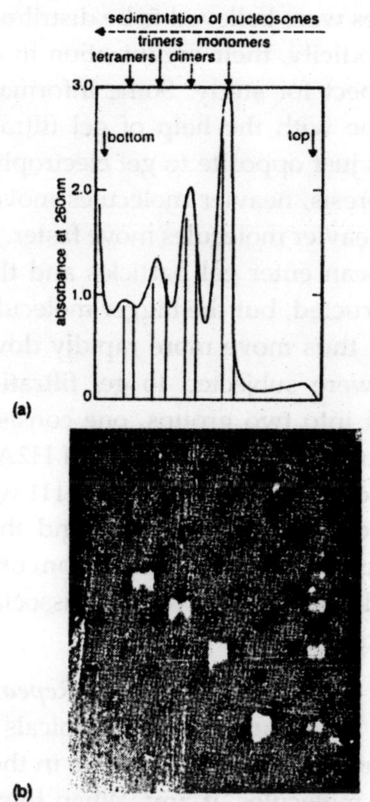

(a) Results of sedimentation of nucleosome preparation on sucrose gradient showing separation of monomers, dimers, trimers and tetramers. (b) Gel electrophoresis of DNA purified from fractions in (a) above suggesting that each multimer had a multiple of unit DNA length, (redrawn from Lewin's 'Genes VI').

shown through distinct bands in nuclease digestion of chromatin. These studies of rat liver chromatin clearly demonstrated that rat nuclease cleaved chromatin at sites spaced at regular intervals along the DNA, while naked DNA was cleaved at random. Therefore, it was concluded that histones confer a regular pattern of protection against nuclease and, therefore, proteins should be distributed along the DNA in a regular periodicity.

Since, histones were believed to be distributed along DNA in a regular periodicity, their organization in chromatin, was an important aspect for study. Some information about this organization came with the help of gel filtration technique. This technique is just opposite to gel electrophoresis, because in gel electrophoresis, heavier molecules move slowly, while in gel filtration, heavier molecules move faster. In gel filtration, small molecules can enter gel particles and their movement, therefore, is obstructed, but the bigger molecules pass around the particles and thus move more rapidly down the column. When histones were subjected to gel filtration, these were clearly separated into two groups, one consisting of H1, H3 and H4 and the other group consisting of H2A and H2B. This was considered to be surprising, because H1 was known to be double the size of the other proteins, and therefore, should have moved faster and separated itself from others. Therefore, it was speculated that H3 and H4 could associate into a dimer to have a size equal to that of H1.

Distribution of Histones within A Repeat Unit Through Cross Linking : There are certain chemicals which help in chemical cross linking and thus facilitate in the study of cross linking between molecules, if any. When this technique was applied to histones in solution, it was discovered that H3 and H4 were present as a tetramer (or a pair of dimers) *i.e.* $(H_3)_2.(H4)_2$ and not as a simple dimer (H3.H4). This tetramer nature was subsequently confirmed in chromatin also. The relationship of this tetramer with other histones and DNA was also studied. It was shown that chromosomes of most organisms had equal number of molecules of histones H2A, H2B, H3 and H4. It was also shown that 25 nucleotides of DNA were present, per histone molecule. Since H3 and H4 are present as tetramers, $(H3)_2.(H4)_2$, each repeating unit may have one such tetramer (described as kernel) and two molecules each of H2A and H2B (because in chromatin H2A = H2B = H3 = H4). Histones will thus form an octamer, $(H3)_2.(H4)_2/(H2A.H2B)_2$ associated with

200 bases in each repeating unit. One molecule of H1 is also associated with each octamer in a nucleosome unit. However, H1 was not linked with other histones, so that in core particle H1 was absent.

Properties of a Nucleosome Model.

Different constituents	Nucleosome		Core particle	
	Size	Mass (daltons)	Size	Mass (daltons)
1 DNA	200 base pairs (67 nm)	1,30,000	140 base pairs (67nm)	90,000
2. Histones (H2A, H2B, H3, H4)	Octamer (2 molecules of each)	1,08,000	Octamer (2 molecules of each)	1,09,000
3. H1 Histone	1 molecule	24,000	Absent	
Total mass		2,62,000		1,99,000

Subunit of Chromatin : The Nucleosome

Basic Structure of A Nucleosome : On the basis of X-ray diffraction, electron microscopy, nuclease digestion, and chemical cross linking, a model for nucleosome structure was proposed. Properties of this nucleosome model are presented in Table and Figure. The shape of nucleosome, arrangement of histones including H1 were also studied. The diameter of nucleosome is 11nm and its height is 6nm. The length of DNA around it is 70nm (~ 67nm DNA = approx. 200 base pairs, because 34 Å = 3.4nm = 10 base pairs), which is equivalent to 200 base pairs.

The length of DNA varies not only in different organisms (from 154 bp in a fungus to as high as 260bp in a sea urchin sperm). but also in different tissues or even in different regions within the same cell of an organism. The length of DNA, however, suggested that DNA should be coiled around nucleosome core particle. Enzyme DNAase I also led to cleavage of DNA suggesting that DNA lies on the surface of nucleosome.

It was, therefore, suggested that histones do not cover DNA, but DNA is wrapped around histone core particle. In other words, it was suggested that it will be more appropriate to describe chromatin as string on beads rather than as beads on string. Neutron scattering technique was also used to find out distances of histones and DNA from centre of nucleosome. It was shown that DNA was farther away than proteins from the centre, again suggesting that DNA is not covered by histones, but is wrapped around histone proteins. A simple model of chromatin showing DNA wrapped around histone core is shown in Figure.

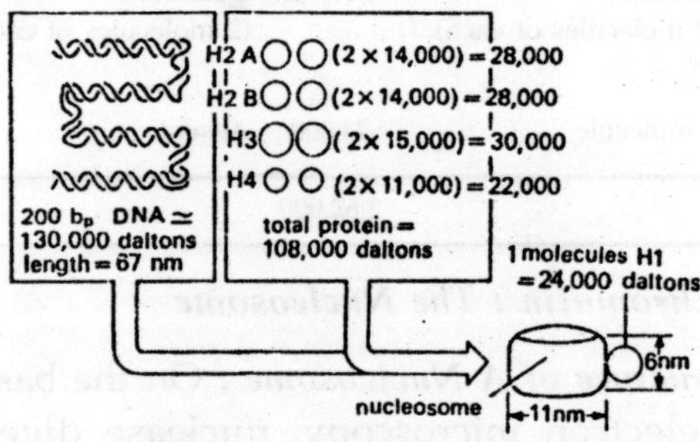

Different constituents (DNA and histone proteins) of a nucleosome and their properties.

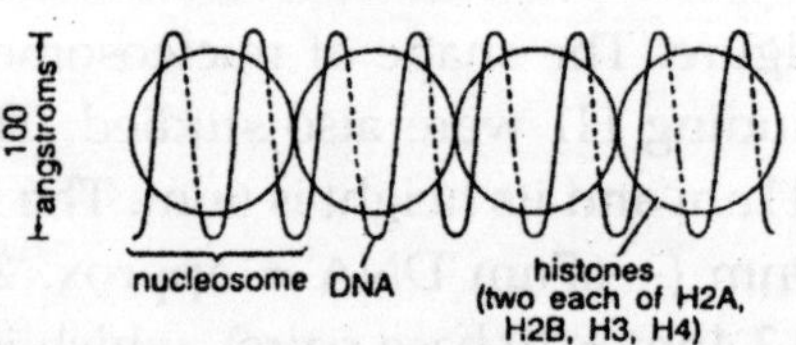

Original proposal for nucleosome repeating units, with DNA wound on a series of beads. Actual shape of histone complex or the path of DNA was not known initially (redrawn from Sci. Amer., Vol. 244, 1981).

Structural Periodicity and Super-coiling of DNA in A Nucleosome : The structural periodicity of DNA coiled around the histone octamer, was examined by treatment with enzymes DNAase I and DNAase II. Both these enzymes make single-strand nicks in DNA, so that double stranded DNA structure is not disturbed. But on denaturation following nicking, single stranded DNA fragments are released. It was shown that these cleavage sites are ~ 10 bases apart and are designated as SI to S13 (S1 is ~ 10 bases away from 5' end, S2 is ~ 20 bases away from 5' end, and so on; see Fig.). Not all sites are cut with equal frequency, so that on electrophoresis a ladder with segments of lengths, which were multiple of 10 were obtained. It was also shown that pairs of sites from S1 to S4 or S10 to S13 lie 10bp apart, while those from S4 to S10 lie 10.7bp apart.

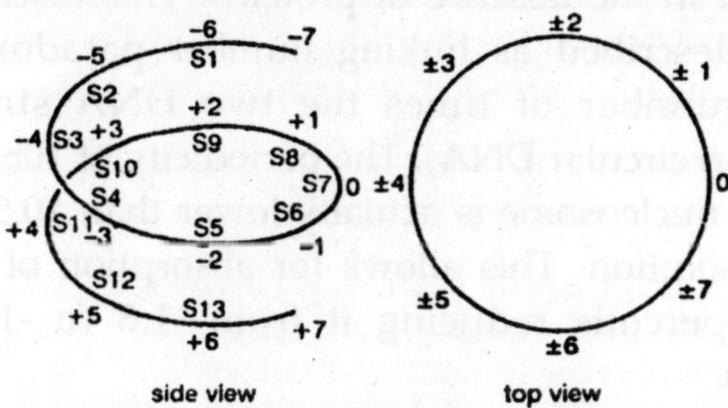

Two numbering schemes dividing core particle DNA into segments, 10bp apart; S1 to S13 are cleavage sites for DNAase I and DNAase II.

The above results differ from cleavage of DNA immobilized on a flat surface, where the cutting periodicity reflects structural periodicity (10.5bp per turn). A variation in cutting periodicity in core DNA of nucleosome (10.0bp at ends; 10.7bp in the middle) means that there is variation in structural periodicity of core DNA; in the middle, this is close to the value of DNA in solution, but at the ends it is less tightly screwed.

Much work on the nucleosome has also been conducted in virus SV40 having a circular molecule of DNA, 5200bp long (1500 nm long). The DNA is packed into a series of nucleosomes, collectively called a mini-chromosome, which is 210nm long showing a packing ratio of about 7. This minichromosome has been used for a study of supercoiling of DNA in nucleosomes. The DNA in chromatin undergoes supercoiling due to the following reasons:

(i) DNA is wrapped around the nucleosome octamer.

(ii) Nucleosomes are also coiled using 5-6 nucleosomes per turn.

Some of this supercoiling is restrained by the associated proteins (histone octamer). Theoretically, the path of DNA around one nucleosome should generate —1.8 supercoils, if no strain is exercised by proteins. Using SV40 DNA, measurements of supercoiling in DNA involved in nucleosome structure were made which gave a value of -1 turn per nucleosome as against -1.8 expected in the absence of proteins. This discrepancy is sometimes described as linking number paradox (linking number = number of times the two DNA strands are intertwined in circular DNA). The periodicity of 10bp in DNA wrapped on nucleosome is actually lower than 10.5bp found in DNA in solution. This allows for absorption of a part of negative supercoils reducing it from -1.8 to -1.0 in the nucleosomes.

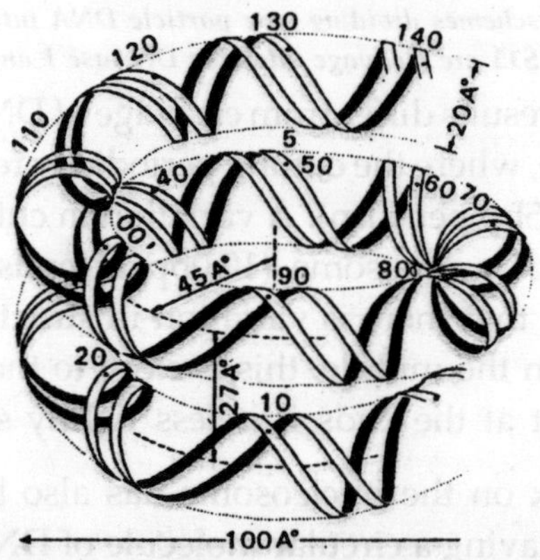

Path of DNA superhelix on a nucleosome core. There are about 80 nucleotide pairs of DNA per turn, one and three quarter turns of DNA are wrapped on histone complex (redrawn from Sci. Amer., Vol. 244, 1981).

Core Particle : When nuclease treatment was prolonged beyond the cleavage between nucleosomes, the DNA was removed from both the free ends, yielding particles containing DNA of only 146 (or sometimes 140) nucleotide pairs instead of 200 nucleotide pairs (Fig.). This reduced form of nucleosome is called core particle. Due to nuclease treatment, the length of the DNA is reduced in discrete steps as shown in Figure. Adjacent core particles, each with core DNA, are joined with each other through DNA segment called linker DNA, which is removed during prolonged digestion. The core DNA has an invariant length of 146bp, but the linker DNA can vary from 8 bp to 114bp, thus accounting for variation in DNA length of nucleosomes (154-260bp).

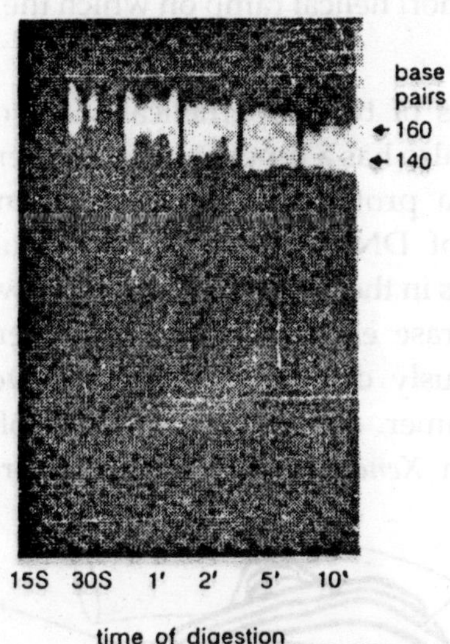

Reduction in the length of nucleosome DNA by nuclease digestion in time (15 seconds to 10 minutes).

Low Resolution (7Å) Structure of Core Particle : In the year 1977, the core particle was available in crystalline form for a study of X-ray crystallography (conducted only in 1984) to reveal the three dimensional structure. Electron microscopy and X-ray diffraction patterns also revealed that nucleosome core particle is neither spherical, nor perfectly flat, but is wedge shaped, about 11nm in diameter and 6nm high. The study of crystals suggested that DNA double helix is coiled into a large helix or superhelix, by making two turns around the histones in the middle of the particle. The two turns of the DNA double helix will almost touch, since DNA is 2nm thick and the particle is only 6nm thick. From the structure of DNA and the diameter of core particle, it has been estimated that two superhelical turns would have about 160 base pairs (80bp per turn). Since length of DNA in a core particle is only 146 pairs, it will occupy only 1.7 turns. The particle, therefore, will be thinner on one side, accounting for the wedge shape. It was also shown that histone octamer itself is also bipartite and wedge-shaped providing for a short helical ramp on which the DNA is wound (Fig.).

The closeness of two superhelical turns of DNA on the histone octamer also has a functional consequence. This allows the binding of a protein molecule simultaneously to two different sites of DNA, 80bp apart. In such a situation superhelical turns in the nucleosome may allow the association of RNA polymerase enzyme to the promoter and initiation sites simultaneously due to closeness of two turns on the nucleosome octamer. In 1993, in a study of interaction of nucleosome with *Xenopus* vitellogenin B1 promoter, it was

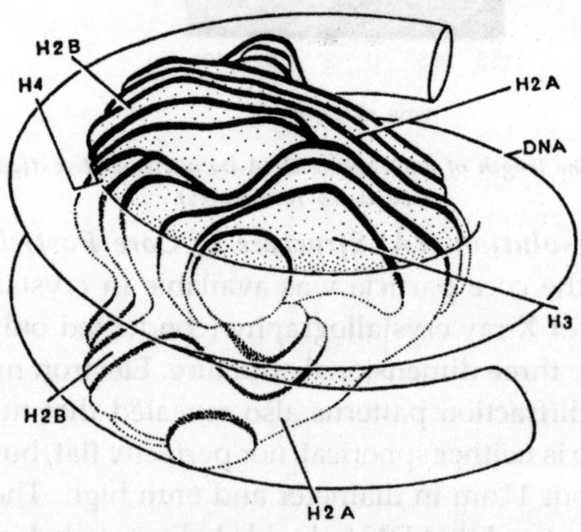

Model of nucleosome core with DNA superhelix wound on histone octamer, which has, a more or less, continuous helical ramp, on which DNA (146 base pairs) is wound (redrawn from Sci. Amer., Vol. 244, 1981).

shown that the presence of nucleosome stimulates transcription by oestrogen. It is believed, that by DNA looping, the oestrogen receptor binding site is brought closer to the site of transcription complex formation.

High Resolution (2.8 Å) Structure of Core Particle : In the year 1997, higher resolution (2.8Å) of nucleosome structure was obtained, showing how the nucleosome core is assembled (Fig.). It was shown that each histone consists of :

(i) a structured three-helix domain called histone fold and

(ii) two unstructured tails.

The histone fold helps to form heterodimers (H2A-H2B; H3-H4) each through a hand-shake like arrangement and these dimers together form the histone octamer. This histone fold has also been observed in a number of transcriptional co-activators and repressors, providing an evolutionary link between chromosome structure and the transcriptional machinery. It was shown that 121bp of the total 146bp of DNA are organized by the histone-fold domains. Following are some of the characteristic features of this organization:

(i) At the sites of DNA's entry into and exit from the nucleosome, the DNA is bound by extensions of the H3 histone fold;

(ii) each heterodimer binds about 30bp of DNA, with contacts at l0bp intervals;

(iii) the phosphate backbone of the DNA makes many contacts with the atoms of the main chains of the histone proteins;

(iv) arginine side chains peretrate all the 14 minor grooves (in a sequence of 146bp, there will be 14 minor grooves) that face the histone proteins;

(v) the path of DNA superhelix is not uniform, but distorted through bends (generated by H3-H4 and I12A-H2B) at several positions, and the average number of base pairs per turn ranges from 10.6bp (in solution) to 10.2bp; this allows the neighbouring turns of DNA superhelix to line up forming channels for histone tails.

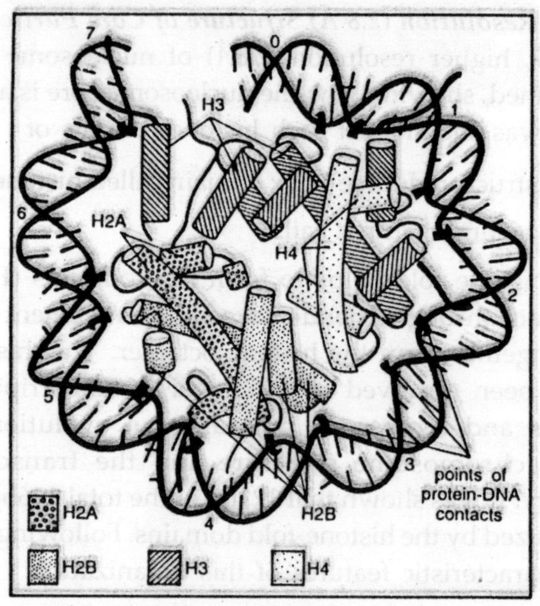

Arrangement of histones and DNA in a core particle of a nucleosome as revealed at 2.8Å resolution.

The new structure of nucleosome core also suggested that the long amino-terminal-tails of histone molecules, without assuming any secondary structure, are arranged in an orderly manner and are involved in the following nucleosome-nucleosome interactions (Fig.):

(i) tails from H3 and H2B pass through minor-groove channels, thus protruding out of DNA surface every 20bp; tails of H3s extend beyond DNA surface, coming in contact with another nucleosome;

(ii) one of the two protruding tails of H4 makes many contacts with H2A-H2B dimer of a neighbouring nucleosome;

(iii) acetylation of histone tails is associated with transcriptional activity and underacetylation is linked with repression.

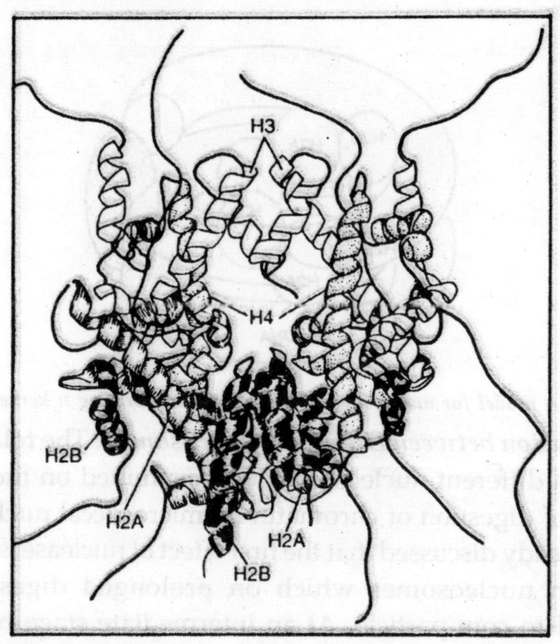

The arrangement of long amino terminal tails of different histone molecules of a core particle of a nucleosome as revealed at 2.8A° resolution.

Spatial Arrangement of Histones

Arrangement of four histones in the nucleosome was also studied through a study of specific sites in DNA and of their interaction with four histones. These sites could be assigned to specific regions of octamer and, therefore arrangement of eight histone molecules could be described.

It was shown that $(H3)_2.(H4)_2$ tetramer defines the central turn of the DNA superhelix. H2A and H2B are added as two (H2A).(H2B) dimers one on each face of the tetramer $(H3)_2.(H4)_2$ and each binding additional DNA and completing the superhelix (Fig.) $(H_3)_2$ $(H_4)_2$ tetramer makes a central 'kernel' associated with two independent dimers H2A—H2B. This explains, why H3 and H4 alone can confer nucleosome like properties on DNA but H2A and H2B can not confer this property.

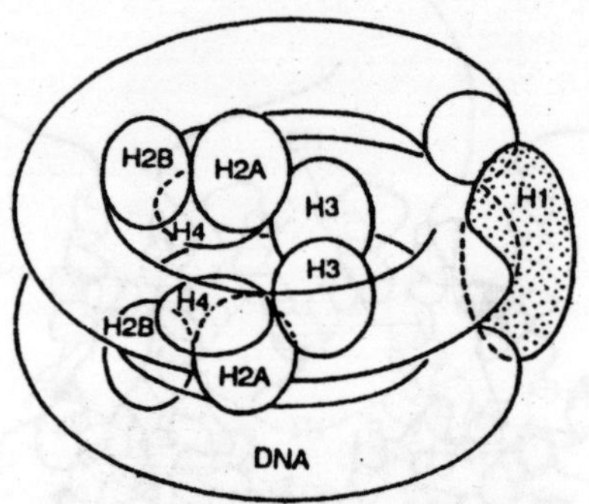

A model for nucleosome with $(H3)_2.(H4)_2$ *forming a kernel.*

Relation between Different Nucleosomes : The relationship between different nucleosomes was examined on the basis of results of digestion of chromatin by micrococcal nuclease. We have already discussed that the first effect of nuclease is cleavage between nucleosomes which on prolonged digestion, are reduced to core particle. At an intermediate stage before the core particles are produced, DNA was found to be 166 base pairs long and not 146 base pairs long as in the core particle. When this intermediate stage with 166 base pairs is further degraded to core particle, the histone protein H1 is assumed to fall off due to removal of 10 base pairs from each end of DNA.

This suggests that H1 is associated with the ends of DNA molecule. It may also be realized that 166 base pairs will make two complete turns of DNA around the nucleosome, bringing the two ends of DNA as close as possible so that a single molecule of H1 can be bound simultaneously to both the ends of DNA molecule belonging to one nucleosome subunit. In other words, HI is found on the side of a nucleosome at regions of entry and exit of DNA superhelix, as shown in Figure. Since

H1 is not necessary for coiling of DNA, it is only an accessory protein and may be involved in condensation of chromatin.

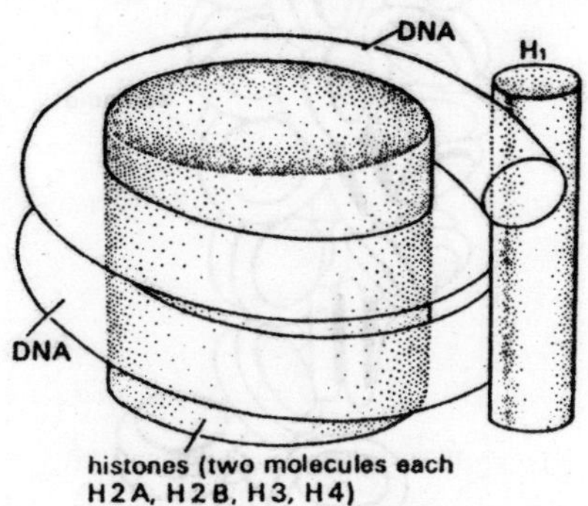

Model of full nucleosome (not the core particle) showing the role of HI histone, which is attached at the entry and exit of two turns of 166 base pairs long DNA (redrawn from Sci. Amer., Vol. 244, 1981).

Solenoid Model

It was also shown that 11nm wide fibre of nucleosomes gets coiled upon itself to form ~ 30nm wide helix with five or six nucleosomes per helix. In this helix successive nucleosome units came close together, so that their centre to centre distance was about 10nm. This 30nm structure was called a solenoid (term solenoid in physics is used for a wire coiled on a central axis). Formation of solenoid from nucleosomes can be compared with winding of a cable (DNA) on a spool (nucleosome) and then folding of wrapped spools. A suggested structure of solenoid is shown in Figure.

It was also proved that H1 protein helped in folding of 110 Å wide fibre into 300 Å wide solenoid, because when H1 was removed, this ordered folding was found to be absent and only irregular clumping of nucleosomes could be observed. It has been shown that H1 molecules aggregate by cross linking to form polymers and may thus control the formation of solenoid. It is also speculated that solenoid has to fold or coil again during condensation of chromatin.

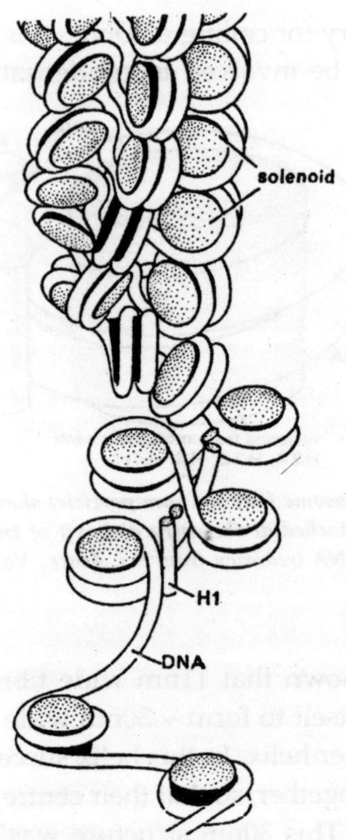

Solenoid model, with a helix having about six nucleosomes per turn, where HI molecules of adjacent nucleosomes are in contact (redrawn from Sci. Amer., Vol. 244 1981).

The above account gives patterns of coiling and packing of DNA. Since 60 nm (600 Å) long DNA (166 base pairs) is coiled in a nucleosome, only 6nm (60 Å) long, and then nucleosomes are coiled in 30nm (300 Å) wide solenoid fibres, it gives DNA a packing ratio of 1 : 50. However, in highly condensed chromosomes, the packing ratio is actually 1 : 5,000, which is 100 times greater than provided by solenoid model. This 100 times packing of the solenoid, should take place by further coiling and folding of solenoid.

Loops, Domains and Scaffolds in Chromatin : The 30nm fibres in the form of a solenoid must be folded in structures at a higher level of organization. These higher levels of organizations have been examined using metaphase and interphase chromosomes. It has been shown by DNAase digestion that DNA in the form of nucleosomes is organized in loops, each with 85kb of DNA and a length of 10–30 μm. These loops have actually been observed to emanate from a central scaffold in metaphase chromosomes.

In interphase nuclei, a nuclear matrix is found as a filamentous structure on the interior of nuclear membrane. Chromatin remains attached to this matrix through matrix attachment regions (MARs). Several MAR sequences have actually been identified. It is believed that the same sequences of DNA work as MAR in interphase nuclei and as SAR (scaffold attachment regions) in metaphase chromosomes. Both matrix and scaffold are proteinaceous in nature.

Chromatin Replication and Nucleosome Assembly : Having known the organization of DNA in the form of nucleosomes, one may like to ask the question— what happens to the nucleosomes during replication and how are new nucleosomes assembled? Cross linking experiments suggested that the old histone octamers are conserved during replication, so that when chromatin is replicated, the existing histone octamers are displaced from DNA to allow its replication. It has also been shown that the conserved octamers may reassociate with either of the two daughter duplexes. An equal number of histone octamers must be synthesized quickly, so that the nucleosomes reform more or less immediately on both the daughter DNA molecules. Although the histone octamer may be conserved, but all the conserved octamers may not reassociate with the same daughter DNA molecule. Instead they may follow a dispersed mode of distribution, in which old and new octamers are intermingled at random during association with the two daughter DNA molecules (Fig.). There is also some support for semi-conservative assembly of histone proteins. Although $(H3)_2.(H4)_2$ tetramers are stable, $(H2A.H2B)_2$ tetramers could result from heterodimers (H2A conserved and H2B newly synthesized or *vice-versa*). These two models are not mutually exclusive and both may operate in a single cell.

It has been shown that in the assembly of nucleosome, an assembly protein called nucleoplasmin (a pentamer containing a sub-unit with mass of 29,000 daltons and found in the nucleoplasm) plays an important role. Nucleoplasmin binds to histones (but not to DNA or nucleosomes) and may thus release histone octamer from the DNA in a more regulated way, so that these may not make random aggregates. It is also suggested that the nucleosomes assemble first and the DNA supercoiling follows this assembly.

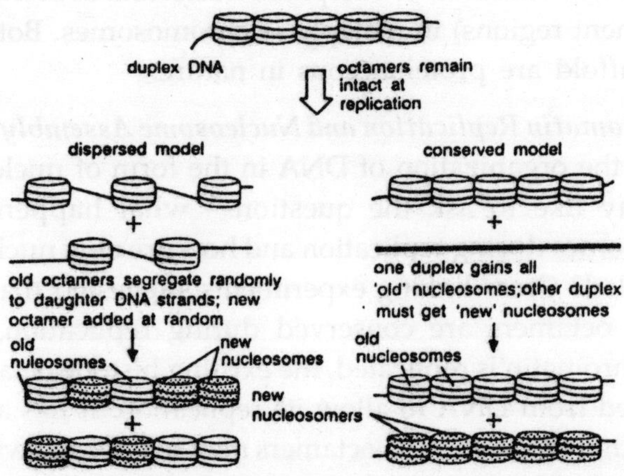

Dispersed and conserved models of nucleosome replication and assembly (dispersed model has been found to hold good).

Phasing and Modification of Nucleosomes in Active Genes : Active and inactive chromatin have been compared with respect to nucleosome organization. It is shown that nucleosome in the form of repeats are present in both cases, but in active form, nucleosomes are displaced away from the regions of DNA to be replicated or transcribed. For this purpose, the histones undergo a variety of modifications and some new non-histone proteins become associated with the chromatin.

Nucleosome Phasing : Some specificity in the distribution of DNA sequences on the nucleosome and a lack of their random distribution is called nucleosome phasing. Recent evidence suggested that nucleosomes are really phased along the length of DNA and that through their position on DNA, regulatory proteins may expose or protect certain DNA sequences. The spacing of nucleosomes is not random in statistical sense, but is rather regular and non-random. It has also been shown that spacing of nucleosomes may vary in different tissues, suggesting that the locations of nucleosomes are not constant but may vary. In the chromosomes of SV40, a DNA segment of 400 base pairs encompassing the replication region and promoters is naked, completely devoid of nucleosomes, showing phasing of nucleosomes. In several other genes at 5' end, naked regions of DNA have been described, suggesting that active genes may have to uncoil and lose their nucleosome structure temporarily, thus providing a mechanism for regulation of gene activity.

It has also been shown that the genes which are in a state of transcription are often susceptible to the enzyme DNAase suggesting that the nucleosome organization is atleast temporarily displaced from the region of transcription, presumably due to occupation of transcription sites by RNA polymerase. However, the histone octamer immediately recaptures its position as soon as the transcription is over. The susceptibility of chromatin to DNAase is attributed to lose binding of some nonhistone proteins like HMG14 and HMG17 (HMG = high-mobility group), because, once these proteins are released, the chromatin does not exhibit susceptibility. It has also been shown that some hypersensitive sites (digested with very low concentration of DNAase) are found upstream to the initiation site for transcription. The sites are free of nucleosomes and cap be 100-200bp long.

Ubiquitination, Acetylation, Methylation and Phosphorylation of Histones in the Nucleosomes : The histone proteins, which are integral parts of nucleosome undergo a variety of modifications to bring about decondensation of chromatin, to allow access of DNA replication or transcription machinery to naked DNA. These modifications include ubiquitination, acetylation, methylation and phosphorylation of some specific amino acid resides of histones. For instance, the protein ubiquitin, which is ubiquitous for its presence in all living cells, gets associated with H2A in the form of U-H2A in 10-30% nucleosomes. Usually only one of the two H2A molecules of a nucleosome is ubiquitinated. There is a tendency for U-H2A to be concentrated in the nucleosomes of the transcribed sequences.

All histone proteins present in the nucleosomes also undergo modification due to acetylation, methylation and phosphorylation. Acetylation and methylation occur on the free amino groups of lysine resides. Methylation also occurs on arginine and histidine. Similarly, phosphorylation occurs on the hydroxyl group of serine and histidine. Methylation and acetylation remove the positive charge on NH_3^+, while phosphorylation introduces a negative charge in the form of phosphate group. These modifications in histones are transient and are associated with structural changes that occur in chromatin at the time of replication and transcription.

During S-phase (DNA replication and histone synthesis) of cell cycle, histones appear to be acetylated and methylated, but the acetyl and methyl groups are removed during other parts of the cell cycle. It appears, therefore, that acetylation and methylation help in the nucleosome assembly during S-phase of the cell cycle. Acetylation has also been shown to be involved in the regulation of transcription. For instance, acetylated lysine residues in the NH_2-terminal domains of nucleosomal histones serve as landmarks for transcriptionally active chromatin within the chromosome. Hyperacetylation, also increases the accessibility of nucleosomal DNA to transcription factors. In contrast to this, deacetylation is correlated with transcriptional silencing. Histone acetyltransferases and deacetylases thus play on important role in regulating transcription. Certain deacetylases have been shown to have homology with some transcriptional regulators (e.g. Rpd3p in yeast).

Phosphorylation and dephosphorylation of H1 histone have also been observed during the cell cycle and are known to play a role in chromosome condensation during mitosis. Phosphate groups are added during mitosis and are removed soon after the division is over.

The above modifications in histone proteins also facilitate binding of certain transcription factors many of which are known to be non-histone proteins. For instance, in *Xenopus*, it was shown that placement of a nucleosome (histone octamer) on the 5S gene, blocks binding of the transcription factor TFIIIA. It was also shown that the amino terminal regions of histone molecules are either cleaved off or get heavily acetylated, which allows free access of TFIIIA to nucleosomal DNA. It is also believed that the acetylation either releases the amino terminal tails from DNA, thus making it accessible or it induces a conformational change in the nucleosome, which promotes TFIIIA binding.

As shown above, considerable information is now available about the factors involved in chromatin decondensation that is needed both for DNA replication and transcription.

6

DNA Sequences

During the last three decades, extensive survey has also been conducted for the DNA contents per genome in prokaryotes and eukaryotes. Large variation in this DNA content has been witnessed particularly in eukaryotes, where the quantity of DNA per cell is enormous, when compared with that in prokaryotes. Large proportion of this DNA in most eukaryotes has been shown to be present in the form of multiple copies of identical DNA sequences, this multiplicity ranging from a few copies to millions of copies within the same cell. These DNA sequences are together described as repetitive DNA.

The remaining DNA in the cell is found in the form of single copy DNA sequences and is therefore described as unique DNA. While the function of repetitive DNA is largely unknown, the unique DNA contains most of the genetic information in the form of structural genes controlling phenotypic traits directly or indirectly. The repetitive DNA may be clustered at one place or interspersed with unique sequences, each repeat unit ranging from a few hundred to few thousand base pairs.

These aspects of organization of DNA in the form of repetitive and unique sequences will be discussed in this chapter.

Chromosomal DNA Content and C-Value Paradox : Large variation in DNA content, among species of the same genus as well as among genera of the same family have been observed. This may be partly attributed to differences in chromosome numbers. However, species at the same ploidy level and with same chromosome number may either show little or no variation (*e.g. Hordeum, Avena,* etc.) or may exhibit many fold differences (*e.g. Lathyru.s, Vicia, Helianthus, Crepis, Allium,* etc.). This interspecific variation may be continuous or discontinuous. Intraspecific variation in DNA content has also been reported in recent years, so that the variation of DNA among plants and animals is a rule rather than an exception. This variation has sometimes also been used to explain the mechanism of evolution in specific groups.

In Table and also in Figure, contents of DNA in some plants and animals are presented, which clearly indicate that eukaryotes, in comparison to prokaryotes, have enormous DNA per cell. For instance, each human diploid cell has DNA equal to 3.5×10^{12} daltons (1 dalton = mass of one hydrogen atom), which is equal to 174cm of DNA. Since a human body may have upto as much as 100g of DNA in all cells, this DNA in the body of a human will be as long as 2.5×10^{10} km (more than 100 times the distance from the earth to the sun). If the haploid genome content in humans is 1.8×10^{12} daltons, which is equal to 2.8×10^{9} base pairs, then this genome should contain approximately as many as 3×10^{6} genes (assuming 1000 base pairs per gene coding for nearly 300 amino acids. However, an estimate of number of genes in humans has also been made on the basis of knowledge about the rate of mutation per locus. If average mutation rate per locus is known, and the mutation rate per individual is also known, then total number of genes can be calculated.

DNA Contents in Some Organisms Showing Large Variation
(1 picogram = 6.1 × 10^{11} daltons = 965 × 10^{6} base pairs).

Class of organism	Species	Number of genes	Haploid DNA (1C)		
			Picorgam	Daltons	Base pairs
1. Phages	ϕ × 174 (single strand)	11	2.8×10^{-6}	1.7×10^{6}	5,387
	Lambda (λ)	~40	46.7×10^{-6}	32×10^{6}	48,502
	T_4	~200	20.7×10^{-5}	126×10^{6}	165,000
2. Animal virus	Adenovirus	~50	21.7×10^{-6}	13×10^{6}	21,000
3. Prokaryotes	*Bacillus subtilis*	~2000	2.18×10^{-3}	1.3×10^{9}	2.1×10^{6}
	Escherichia coli	~2000	4.4×10^{-3}	2.7×10^{9}	4.2×10^{6}
4. Unicellular eukaryotes	*Saccharomyces cerevisiae*		14×10^{-3}	8.5×10^{9}	1.4×10^{7}
5. Multicellular eukaryotes	*Drosophila melanogaster*	~5000	0.18	0.11×10^{12}	0.17×10^{9}
	Zea may (corn)	< 25,000	2.6	1.6×10^{12}	2.5×10^{9}
	Xenopus laevis (toad)	< 25,000	3.2	1.9×10^{12}	3.1×10^{9}
	Mus musculus (mouse)	< 25,000	2.7	1.7×10^{12}	2.6×10^{9}
	Homo sapiens (human)	< 25,000	2.8	1.8×10^{12}	2.8×10^{9}

The frequency of deleterious mutations per locus in humans is 10^{-5} to 10^{-6} in each generation as estimated by Muller in 1967. If number of genes is 3×10^6 (3 million = 3,000,000) as outlined above, then it will yield 30 deleterious mutations in each generation at the rate of 10^{-5} mutations per locus. This will be an unbearable genetic load.

However, the actual frequency of deleterious mutations per generation per individual has been estimated to be 0.5 against an expected frequency of 30. This mutation frequency at the rate of 10^{-5} mutation per locus will give an estimate of 5×10^4 genes (50,000 genes). This suggests that actual number of genes in a human should be 50,000, and not 3 million (3,000,000) as estimated from DNA content. This anomalous situation has been described by some workers as C value paradox (the haploid DNA content in an individual is described as its C value).

There is yet another aspect of C-value paradox involving huge variation in C-values between species, whose apparent complexity does not vary correspondingly. For instance, animals usually contain 1—5pg DNA per haploid genome, but the range of this DNA in salamanders extends upto 100pg or so and in primitive angiosperms it may extend even upto 300pg. Substantial variation in the nuclear DNA content exists even between closely related species (Fig.) and can not be easily explained and hence the phrase, *C-value paradox*.

Repetitive DNA

Technique for Detecting Repetitive DNA : In the year 1964, it was discovered with surprise that much of DNA in mouse cells consisted of multiple copies of same or, very similar sequences of base pairs. It is now known that in higher organisms, this repetitive DNA generally constitutes atleast 20% and can reach upto as much as more than 90% in some cases. It is believed that these repetitive sequences do not carry

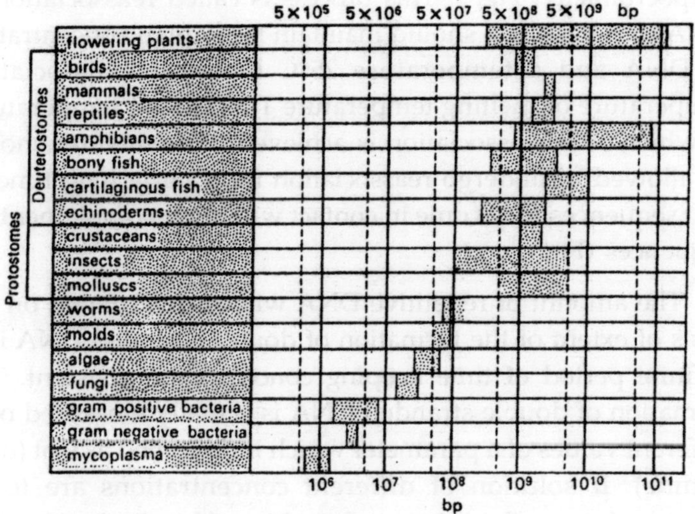

Variation in genomic DNA in a variety of animal and plants.

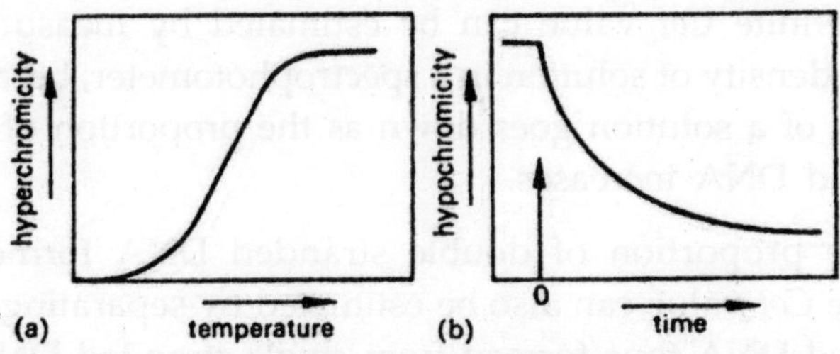

Hyperchroraicity and hypochromicity of DNA due to denaturation and reassociation respectively.

any genetic information and, therefore, do not form genes, but play some other structural or regulatory role. The presence of repetitive DNA can be detected by a simple method involving following procedure : The double stranded DNA is first denatured and converted into single stranded DNA by heating the DNA solution. This is accompanied with increase in optical density, a phenomenon described as hyperchromicity.

The solution of single stranded DNA thus obtained is then allowed to cool slowly, so that the double stranded DNA will be formed again, which accompanies decrease in optical density (hypocromicity, Fig.). This process is called reassociation of DNA for which one should maintain fairly high concentration of DNA and a temperature 25°C below the dissociation temperature or melting temperature Tm (Tm is a temperature, at which 50% reassociation is achieved). The mixture should be allowed to undergo reassociation for a fairly long time so that sequences may come in contact with their complementary sequences (Fig.).

The amount of repetitive DNA will be determined on the basis of extent of the formation of double stranded DNA in a definite period of time keeping concentration constant. The formation of double stranded DNA is actually measured over different values of a parameter which is described as Cot (*conc.* × time). If solution of different concentrations are to be compared, same *Cot* value can be achieved by altering the time allowed for reassociation. The double stranded DNA formed at a definite *Cot* value can be estimated by measuring the optical density of solution in a spectrophotometer, because the density of a solution goes down as the proportion of double stranded DNA increases.

The proportion of double stranded DNA formed at a definite *Cot* value can also be estimated by separating double stranded DNA thus formed from single stranded DNA. This can be achieved, if solution is allowed to pass through a column of hydroxyapatite crystals, which have the capacity of retaining double stranded DNA and permitting single stranded DNA to pass through (Fig.).

If a DNA sample has high proportion of repetitive DNA, reassociation will be faster, with higher degree of reassociation achieved at lower *Cot* values. Reassociation rate will also depend on genome size, because in a small genome, higher degree of repetition will be available at the same DNA concentration. Therefore, bacterial DNA with small genome size will reassociate faster, even though no repetitive DNA is found

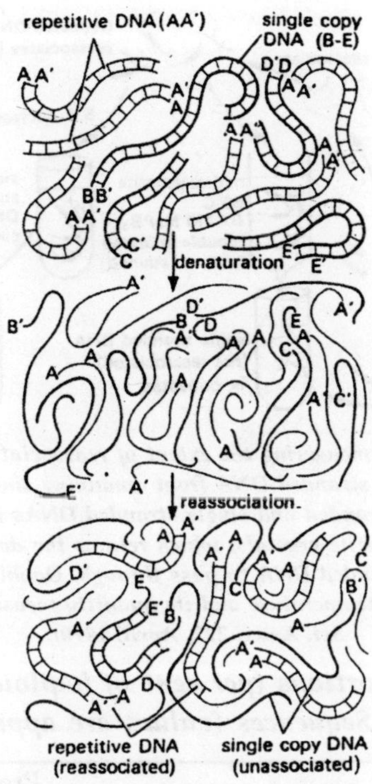

Denaturation and reassociation of DNA containing six copies of repetitive DNA (AA') shown by thin lines and one copy each of four different unique or single copy DNA's (B-B'; C-C'; D-D'; E-E') shown by thick lines. Note that more repetitive DNA has reassociated than the single copy DNA. This principle is used for determining the proportion of repetitive DNA in a genome (redrawn from Sci. Amer. 222: April, 1970).

there, so that the range of *Cot* values for complete reassociation is small. In eukaryotes with a bigger genome size, on the other hand, the range of *Cot* values is large suggesting a longer time required for reassociation. But a specific fraction of DNA reassociates faster, indicating the presence of repetitive DNA. The patterns of reassociation in a number of organisms are shown in *Cot* curves in Figures. During the last more than 25 years, hundreds of plants and animals have been analysed for the relative proportion of repetitive DNA in their genomes (Table).

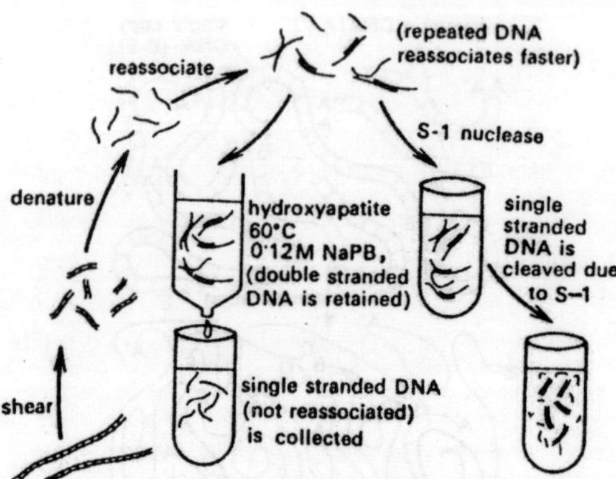

Technique used for measuring the extent of reassociation by separating reassociated double stranded DNA from remaining single stranded DNA. Mixture of double stranded and single stranded DNAs is passed through a column of hydroxyapatite crystals, which retains the double stranded DNA, and allows single stranded DNA to pass through. Double stranded DNA can be eluted by raising temperature and its quantity measured (redrawn from Sci. Amer. 222, April, 1970).

Relative Proportions (per cent of haploid genome) of Repetitive Sequences (values are approximate).

Organism		*Proportion of repetitive sequences (%)*
I Bacteria		—
II Animals	Sea urchin	38
	Snail	60
	Toad	45
	Mouse	40
	Calf	45
	Human	30
III Plants	Bread wheat	80
	Oats	83
	Maize	78
	Garden pea	75
	Barley	76
	Rye	92

Chemical Complexity vs Sequence (Kinetic) Complexity : Renaturation of DNA in solution as discussed above, depends on random collisions between the complementary strands, which follows second order kinetics, since concentrations of both (not only one of them) the complementary strands will influence the rate of reaction. The rate of reaction, when expressed through differential calculus is given by

$$\frac{dC}{dt} = -KC^2$$

(C = concentration of single stranded DNA at time *t)*

(*K* = reassociation rate constt.)

By integrating the above equation (consult a book on Integral Calculus, if necessary), between the limits of C_0 at t = 0 and C at time $t_{1/2}$ (when reaction is half complete), we get the following expression, ($C_0t_{1/2}$ = $Cot_{1/2}$, which means Cot value at which reassociation is half complete).

$$C/Co = \frac{1}{2} = \frac{1}{1+KC_0t_{1/2}}$$

$$\text{or } 1 + KC_0t_{1/2} = 2\ ;\ C_0t_{1/2} = \frac{1}{K}\ ;\ K = \frac{1}{C_0t_{1/2}}$$

In view of the above, reassociation of any particular DNA can be described by the rate constant K (in units, nucleotide moles liter^{-1} sec^{-1}) or in the form of its reciprocal *i.e.* $C_0t_{1/2}$ (in nucleotides moles × sec/liter). It is obvious that reassociation depends on concentration (Co) and time of incubation (t), i.e. *Cot* as described above. A higher $Cot_{1/2}$ means slower reaction and lower $Cot_{1/2}$ meant faster reaction. This parameter $Cot_{1/2}$ varies for different organisms and also for different DNA fractions from the same organism.

The $Cot_{1/2}$ of a reaction will be directly proportional to the DNA content, if there is no repetition of sequences within a single genome. For instance, if the bacterial genome has 0.004pg DNA (4.2 × 10^6 bp) and if Co is 12pg per unit volume, it will have 3000 copies of each DNA sequence. Contrary to this if a eukaryotic genome is 4pg, then there will be only 3 copies in 12pg i.e. concentration is 1000 times less than in bacterial genome, provided there is no repetition of sequences in the genome of 4pg. In view of this, $Cot_{½}$ will indicate the length of all the different sequences (represented only once) in a genome, which will be less than the length of total DNA in a genome when there is repetition. This will be described as kinetic complexity of the genome, and can be determined by knowing $Cot_{1/2}$ and comparing it with $Cot_{1/2}$ of a genome with known complexity (*e.g. E. coli,* which has a genome =.004pg DNA = 4.2×10^6 bp with $Cot_{1/2} = 9$). The following relationship is used for calculating the kinetic complexity.

$$\frac{\text{Cot}_{1/2}\text{ of DNA of any genome}}{\text{Cot}_{1/2}\text{ of E. coli genome (=9)}} = \frac{\text{Kinetic complexity of genome}}{4.2\times10^{6}\text{ bp}}$$

$$\text{Kinetic complexity of genome} = \frac{4.2\times10^{6}}{9} \times \text{Cot}_{1/2}\text{ of genome}$$

Figure shows reassociation curves of several DNAs, which are all kinetically pure, meaning that DNA is homogeneous

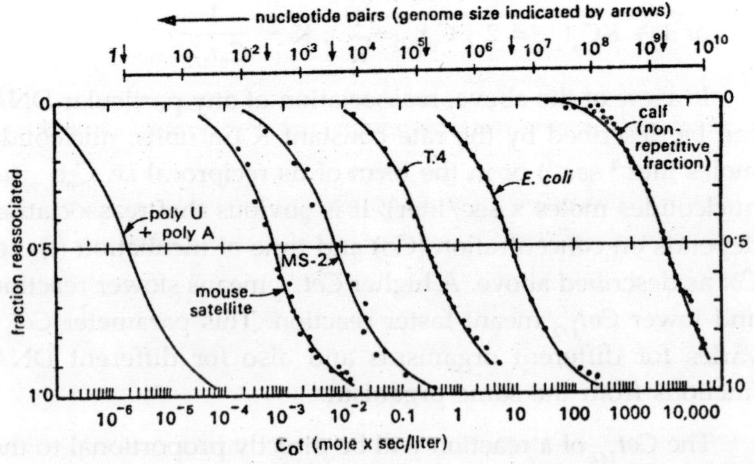

Reassociation patterns of double stranded DNA from various sources. All curves are typical S-shaped indicating homogeneity, but they are displaced to different values of Cot due to genome size.

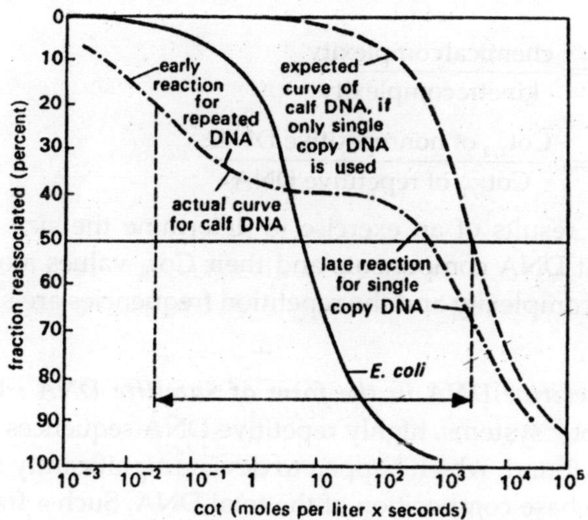

Reassociation patterns of bacterial and calf DNA. Note that the reaction on calf DNA takes place in two stages, one early representing repetitive DNA and the other representing single copy DNA. Midpoints of these two stages are indicated by dotted vertical lines, separated by a factor of 1000,000, (redrawn from Sci. Amer. 222:, April, 1970).

from the point of view of complexity. In eukaryotes, the genome contains more than one such pure components as shown in the comparison made in Figure. In this figure calf DNA has been shown to have two components, each with its characteristic $Cot_{½}$ value. However, more than two such components can also be found as shown for wheat in Figure. Proportion of each component is determined by using the formula $1 - C/C_0$ (C = concentration at $t_{1/2}$ for the corresponding component). From this proportion, if chemical complexity of the DNA component can be determined, chemical complexity of the whole genome is known. For instance, if genome size is 12×10^8 bp and the component represents 25% of the genome, then the chemical complexity of this component is 3×10^8 bp. If kinetic complexity is known from the equation given earlier ($\frac{4.2\times10^6}{9} \times Cot_{1/2}$), then repetition frequency (f) of repetitive DNA component can be determined using the following formula :

$$f = \frac{\text{chemical complexity}}{\text{kinetic complexity}}$$

$$= \frac{\text{Cot}_{1/2} \text{ of nonrepetitive DNA}}{\text{Cot}_{1/2} \text{ of repetitive DNA}}$$

The results of an exercise to determine the size of three different DNA components and their $\text{Cot}_{1/2}$ values along with kinetic complexity and the repetition frequencies are shown in Figure.

Repetitive DNA in the form of Satellite DNA : In many eukaryotic systems, highly repetitive DNA sequences of small size are found, which happen to deviate significantly from the average base composition of the total DNA. Such a fraction of DNA will have a distinct buoyant density (measured through density centrifugation), which may be higher or lower than that of the total DNA. Therefore, this fraction can be separated as a band distinct from the main band of DNA on $CsCl_2$ density ultracentrifugation and is called the satellite band (Fig.). The satellite band is found on the left of the main band if lighter and on the right if heavier than the DNA of the main band. The DNA separated as a satellite band on ultracentrifugation is therefore, called satellite DNA and has been identified in a variety of eukaryotes including mouse, rat, guinea pig, man, fruitfly, and several dicotyledonous flowering plants. There can also be more than one satellite bands in the same organism,

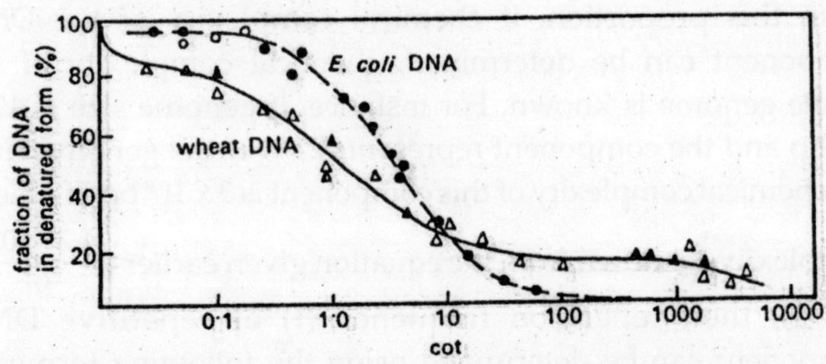

Reassociation pattern of wheat genomic DNA showing more then two homogeneous components (fast, intermediate and slow reassociation).

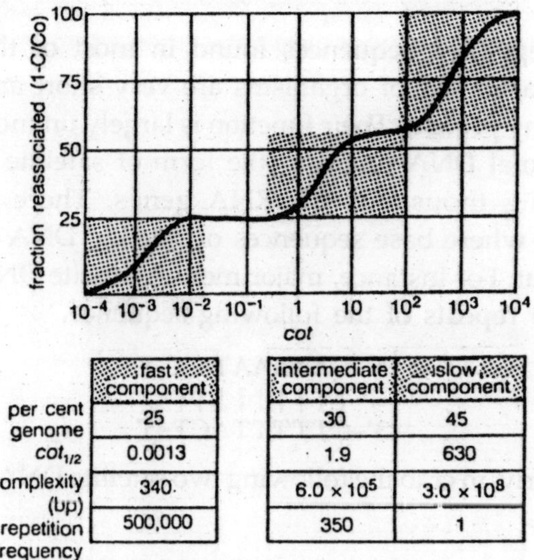

	fast component	intermediate component	slow component
per cent genome	25	30	45
$cot_{1/2}$	0.0013	1.9	630
complexity (bp)	340	6.0×10^5	3.0×10^8
repetition frequency	500,000	350	1

Reassociation kinetics of eukaryotic DNA, showing calculation of complexity and repetition frequency.

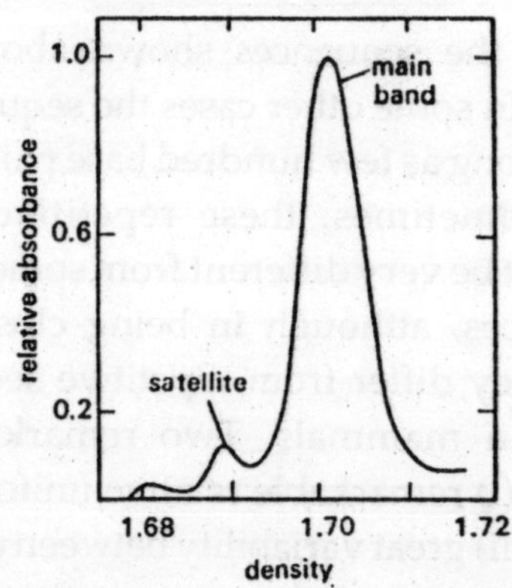

Density centrifugation of embryonic DNA of Drosophila melanogaster, showing a main band and a satellite band.

and in such a case they are designated as Sat-I, Sat-II, Sat-III, etc. (*e.g.*, in *Drosophila* and in humans). The proportion of DNA found in the form of satellite DNA may vary from 3–5% to as high as 40% in some cases. The lower limit is arbitrary, since less than 3% DNA can not be easily separated; without using special modified techniques.

The repetitive sequences found in most of the satellite DNAs in a variety of organisms are very short and can not code for any protein. Their function is largely unknown except for ribosomal DNA found in the form of satellite DNA and constituting thousands of rRNA genes. There are many examples, where base sequences of satellite DNA have been worked out. For instance, major mouse satellite DNA consists of tandem repeats of the following sequence.

5′–GAAAAATGA–3′
| | | | | | | | |
3′–CTTTTTACT–5′

Similarly, in crab the following two satellite DNA sequences are found.

5′–ATATAT–3′ 5′–ATCCATCC–3′
| | | | | | | | | | | | | |
3′–TATATA–5′ 3′–TAGGTAGG–5′

In each case, the sequences shown above are repeated millions of times. In some other cases the sequences in satellite DNA could be as long as few hundred base pairs, as in humans. In such cases, sometimes, these repetitive satellite DNA sequences may not be very different from some other dispersed repetitive sequences, although in being closely clustered in tandem arrays, they differ from repetitive sequences like *Alu* elements found in mammals. Two remarkable features of satellite DNA are (i) remarkable relative uniformity within the same species and (ii) great variability between otherwise closely related species. In view of these properties, it is difficult to assign any function to satellite DNA except in ribosomal RNA genes as mentioned above.

In-situ Hybridization of Sat DNA : The satellite DNA often lies in heterochromatic regions of chromosomes and its location can be easily demonstrated by the technique of *in-situ* or cytological hybridization. In this technique, DNA within the cell is denatured by treating the cells, that have been squashed on a cover slip. The cover slips can then be incubated in a solution of radioactively labelled satellite or any other DNA, whose location is to be determined. The sat-DNA will hybridize with denatured DNA of the cell and location can be determined by autoradiography. As shown in Figure, position of mouse sat-DNA at the chromosome ends was revealed by *in-situ* hybridization. These chromosome ends also represent the centromeres in mouse.

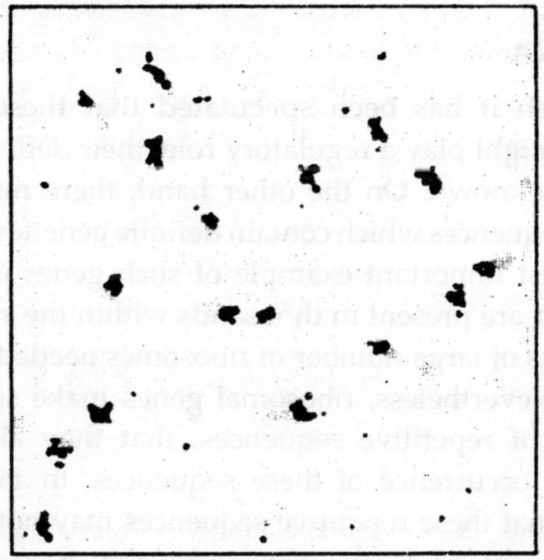

In-situ hybridization of mouse satellite DNA showing its location near centromeres.

Squash dot Hybridization : The *in-situ* hybridization technique is often used for locating the position of the repeat sequence (including sat-DNA) on a specific chromosome. However, in some cases the repeated DNA from an alien species is used for detecting the presence of chromatin material introduced from this species in a crop plant due to introgression. Sometimes, a cytogeneticist may need to screen large population of a crop species to select plants that may carry alien genetic material. Squash dot hybridization technique is designed for this purpose. In this technique one root tip from a germinated seed may be squashed onto nitrocellulose filter and the filters are hybridized with radioactively labelled repeated DNA probe derived from the alien species in question. After washing the filters in appropriate buffers, they are used for exposure of X-ray film, which on developing gives the desired information about the presence of alien genetic material. The successful, use of this technique has actually been demonstrated to detect the presence of introgressed rye chromatin in several wheat cultivars, through the use of highly repeated DNA probes derived from rye.

Selfish DNA

Although it has been speculated that these repetitive sequences might play a regulatory role, their definite function is largely unknown. On the other hand, there may be some repetitive sequences which contain definite genetic information and the most important example of such genes is ribosomal genes which are present in thousands within the same cell for the synthesis of large number of ribosomes needed for protein synthesis. Nevertheless, ribosomal genes make such a small proportion of repetitive sequences, that they alone cannot explain the occurrence of these sequences. In 1980s, it was suggested that these repetitive sequences may not confer any definite advantage to the organisms which contain them. Therefore, sometimes it is said that repetitive sequences represent the DNA which is a repository from which new functional sequences can be obtained.

There is, however, no evidence to prove or disprove this hypothesis and it may not be necessary to argue that these DNA sequences, if not of definite advantage to organism, should have been eliminated through natural selection. Actually, the cells themselves provide an environment and the repetitive sequences may be subjected to selection pressure within the cell. It can, therefore, be argued that these sequences may be maintained through non-phenotypic selection, and by a mechanism directed for the self preservation of these sequences. It is in this context that the term selfish DNA, was used for this repetitive DNA.

One of the examples of selfish genes is found in transposable genetic elements in prokaryotes. These transposable genetic elements are not indispensable and can be transposed from one place to the other or from one organism to the other. It is difficult to find reason for selection of these sequences represented by transposable genetic elements on the basis of phenotypic fitness. On the other hand, it has been argued that the transposable genetic elements are maintained due to their property of transposability which permits them to escape the pressure of selection exercised on the organism. Such transposable elements are also found in case of eukaryotes, for example Tyl element in yeast and the copia like and inverted repetitive elements of *Drosophila*.

We described selfish genes like *SD* in *Drosophila*, '*t*' in mouse, and *M* in flour beetle. The term selfish genes was used there for those genes, which are successful in propagating themselves, while being detrimental to the organisms that carry them.

7

Genetic Maps

We described methods used for preparation of genetic maps in a variety of organisms. In all these cases recombination data involving morphological, biochemical or nutritional traits were primarily utilized for preparing genetic or linkage maps. However, in using these traits, one is always handicapped due to non-availability of mutants for many genes, which therefore, can not be mapped. Linkage maps are thus often incomplete with gaps at many places. This limitation has largely been overcome in recent years due to the availability of molecular markers in the form of Restriction Fragment Length Polymorphisms (RFLPs), Random Amplified Polymorphic DNA (RAPD, pronounced as rapid), minisatellites and, microsatellites (also called SSRs = simple sequence repeats) collectively described as Variable Number of Tandem Repeats (VNTRs), DNA Amplification Fingerprinting (DAFs) and Amplified Fragment Length Polymorphisms (AFLPs), etc. These molecular markers are unlimited in number and would thus allow complete saturation of genetic maps giving a high resolution of a few centiMorgans only.

However, the genetic maps prepared through estimation of recombination frequencies (utilizing either the classical markers and/or the molecular markers) do not exactly correspond to the physical features either on the chromosomes, or on the DNA molecules, which are parts of the chromosomes. The recombination frequencies, though are function of distances between genes, but are influenced by several factors including the nature of mutants used, the position of these mutants on chromosomes, the genetic background and the environmental conditions. In view of this, attempts have been made to prepare physical maps and compare them with genetic maps, whenever possible.

At the molecular level, the mapping may sometimes involve small DNA segments, where recognition sites of many restriction enzymes may be mapped through a technique described as restriction mapping. This may later be extended to complete chromosomes through a technique called chromosome walking. This is described as physical mapping because restriction sites are separated by actual distances measured in terms of nucleotide pairs (or base pairs) and not in terms of recombination frequencies or centiMorgans. In other cases, physical positions of DNA fragments (called restriction fragments) are located through *in situ,* hybridization or deletions, so that these molecular maps are also physical in nature as described above. These molecular maps (both genetic and physical) have now been, prepared in all major crops like maize, tomato, potato, rice, barley, wheat, etc. and also in animals like *Drosophila,* mouse and humans. The technique for preparing these maps (restriction maps, RFLP/ RAPD maps or microsatellite maps) and also the results obtained, will be briefly described in this chapter.

Restriction Mapping

At the molecular level, the fine structure of a gene can be studied only through determination of nucleotide sequence of the concerned DNA segment. Sometimes this is done through isolation of DNA corresponding to a gene, which is a lengthy process. Instead we can prepare a map of the DNA by cleavage of the DNA at specific sites with the help of restriction endonucleases. These sites of cleavage can be identified and mapped to give rise to a restriction map. On a restriction map, is found a linear sequence of sites, each for a specific enzyme and the distances between them are measured as number of base pairs of DNA. This technique can not be directly used in eukaryotes, due to some inherent difficulties as will be shown later in this chapter, but has been successfully utilized for preparing *restriction maps* in prokaryotes.

In a restriction map, the nucleotide sequence of DNA between sites close to one another (300bp or less) can be determined. These regions can then be connected into a sequence of the entire gene, whose nucleotide sequence, then can be compared with amino-acid sequence of protein coded by it. This can be further extended to adjoining genes to saturate the genetic map, as is being attempted for human genome.

– ***Restriction Cleavage and Gel Electrophoresis :*** If we take a particular DNA molecule or DNA sample, digest it with a specific restriction enzyme and then subject the sample to gel electrophoresis (a technique in which DNA digest is loaded on gel slab and the DNA fragments are allowed to move under the influence of an electric current), we will notice a series of bands on the gel slab or cylinder. The position of different bands will depend on DNA fragment size, such that smaller the fragment, more rapidly it will move, and longer the fragment, more slowly will it move. It will mean that the fragment away from the loading site will be smaller and those close to the loading site will represent longer DNA fragments. The gel can be calibrated by using a mixture of DNA fragments of known lengths so that the position of bands on this standard gel can be compared with the bands in the experimental DNA

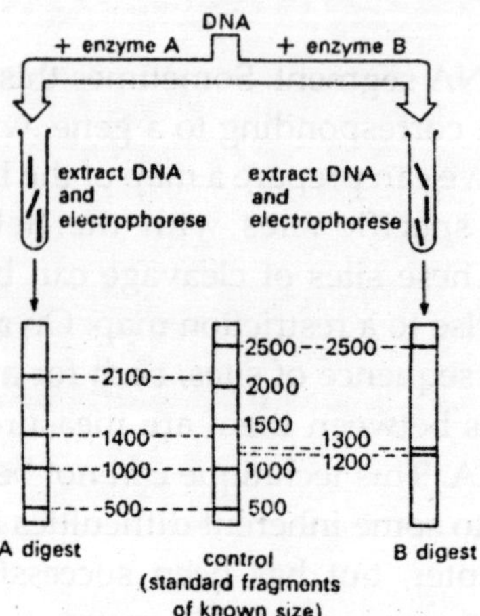

Cleavage of DNA by two restriction endonuclease A and B into fragments which can be separated by agarose gel electrophoresis.

digest and the fragment length in each band of DNA digest can thus be determined. In Figure, using a hypothetical example, we have shown the results of digestion of 5000bp long DNA molecule digested separately by two enzymes A and B. As shown in the figure, the enzyme A cleaves the DNA into four

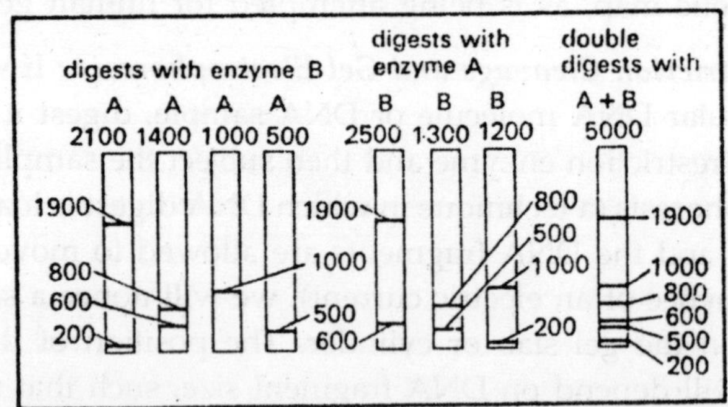

Technique of double digests to determine cleavage positions of DNA due to one enzyme with respect to those due to another enzyme (the two enzymes are A and B); the four gels on the extreme left show the result of electrophoresis after digesting with enzyme B each of the four fragments obtained after digestion with A; three central gels represent results of digesting with enzyme A each of the three fragments obtained after digestion with enzyme B and the solitary gel on the right shows the results of digesting the intact DNA with both the enzymes simultaneously.

fragments of lengths 2100, 1400, 1000 and 500bp, while the enzyme B cleaves it into three fragments of length 2500, 1300, 1200bp. These data can be used to generate a map.

Construction of a Restriction Map : The data of digestion by more than one endonucleases as discussed above can be utilized to arrange the sites of breakage in a defined order. This can be done by several methods and one of them is illustrated in Figure. This method involves double digests, where we extract each fragment produced in the individual digests with either enzyme A or enzyme B and then cleave it with the other enzyme. The original DNA sample is also digested by a mixture of both the enzymes to confirm the results of individual successive digests. In Figure 45.3, we have shown the results of such an exercise. When the fragment A-2100 is retrieved and

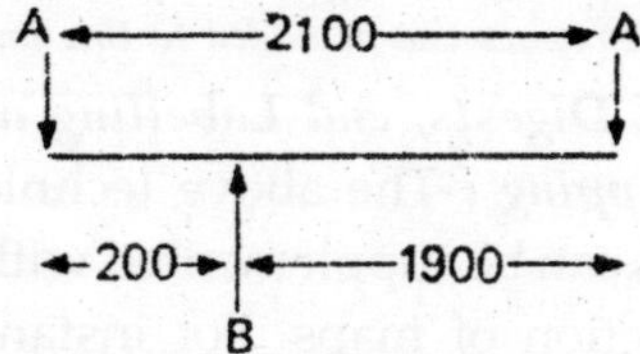

The sites of cleavage by enzymes A and B in one of the four fragments, 2100 bp long obtained due to cleavage by enzyme A, the fragment has a site for enzyme B at 200bp length from one end.

digested with enzyme B, it is cut into two fragments of 1900 bp and 200 bp. Fragment B-2500 (obtained from digestion with enzyme B) is similarly cut by enzyme A into fragments of 1900bp and 600bp, suggesting overlap of A-2100 and B-2500 (Fig.), in the region of the fragment of 1900bp which is obtained by one cut with enzyme A and the other by enzyme B. It may be seen in Figure that in reciprocal digests (A followed by B and B followed by A) same fragments are available. Using this information, we can find the overlapping regions in A and B digests and find out the sites of cleavage by A and B. This will then allow us to prepare the restriction map as shown in Figure.

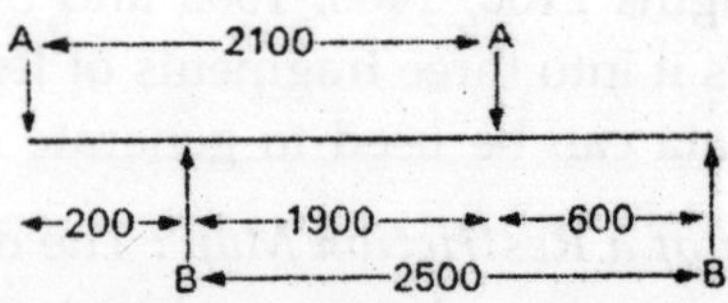

Reconstruction of restriction map showing restriction sites of enzymes A and B in two overlapping fragments (2100bp fragment obtained due to enzyme A and 2500bp fragment obtained due to B).

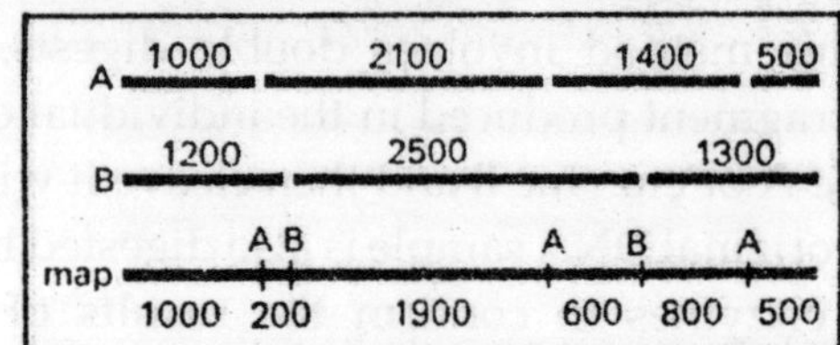

A restriction map, in the form of linear arrangement of cleavage sites, the upper two figures show sites due to individual enzymes and the bottom figure shows arrangement of cleavage sites of both enzymes (see text for details), sites of more enzymes can be added to this map by same technique.

Use of Partial Digests, end Labelling and Hybridization in Restriction Mapping : The above technique of individual and double digests can be supplemented with other techniques for actual construction of maps. For instance, by permitting incomplete digestion, fragments longer than those obtained by complete digestion may be obtained. These will be called partial digests. In partial digests with enzyme A, instead of getting A-1000, A-2100, A-1400, A-500, we may get A-3100, A-1400, and A-500, which will suggest that A-1000 and A-2100 lie in adjacent regions. In another partial digest if we get A-1000, A-3500 and A-500, this will suggest that A-2100 and A-1400 are adjoining, which will mean that A-1000 and A-1400 are found on the two sides of A-2100. This technique will thus allow the arrangement of fragment in a linear order.

Another useful technique involves labelling of the ends (end labelling) of DNA molecule before digestion, so that the fragments containing these ends can be identified due to labelling, even after digestion. In above example if A-1000 and A-500 are radioactively labelled, these will be present at the two ends of the restriction map. Some of the features of restriction map can be confirmed by nucleic acid hybridization. For instance, if A-2100 and B-2500 have an overlapping region of 1900bp, they should hybridize with each other, which will confirm overlapping.

The above technique will help in accurate completion of a restriction map, but will require that we have a complete set of restriction fragments which make the entire DNA region being mapped. Once the restriction map is ready, it can be compared with the genetic map. Although large changes (due to deletions, duplications, inversions and interchanges) on the genetic map can be easily located on restriction map, but point mutations can not be easily mapped on the restriction map, since the restriction sites often do not change due to mutations. In such situations one may like to determine the nucleotide sequences of individual fragments and compare them in the normal and the mutant individuals. Since sequencing of nucleotides is more laborious and can not be easily undertaken for the whole genome, another technique of mapping a variety of molecular genetic markers including restriction fragment length polymorphisms (RFLPs) has been utilized during 1980s and 1990s to generate molecular genetic linkage maps in a variety of animal and plant systems. This technique and the different types of molecular marks will be briefly discussed.

Molecular Genetic Maps

In order to prepare a genetic map in a species, two diverse parents belonging to this species are used to produce a mapping population either by selfing to produce F_2, or by backcrossing to produce BC_1. F_1 individuals can also be used to produce a mapping population in the form of doubled haploids or DHs (using pollen culture from F_1, pollen) or recombinant inbread lines or RILs (using repeated selfing for several generations). The advantage of using DHs and RILs is that they can be easily maintained due to homozygosity, so that DNA samples can be repeatedly obtained. F_2 individuals, on the other hand, are heterozygous and can not be maintained and used repeatedly for DNA extraction.

A variety of molecular markers have been used for the preparation of molecular genetic maps. These markers are first screened for detecting polymorphism between the parents, that are used for the production of the above mapping population (F_2's, BC_1, DHs, RILs, etc.). The markers which detect polymorphism between parents, are then used for recording data on the two parents, and also on individuals from the mapping population. The data obtained on parents and the mapping population is then used for preparing a genetic map, through computer software like MAPMAKER. The different molecular markers that have been used for the preparation of molecular genetic maps are described in the following text.

Restriction Fragment Length Polymorphisms (RFLPs) : The genomic DNA (DNA extracted from any organism), when subjected to digestion with a restriction endonuclease and electrophoresed, may give rise to a continuous array of different sizes of fragments, which may be difficult to identify in eukaryotes. Therefore specific cloned sequences of DNA can be used to identify variation at the DNA level. These variations are monitored as changes in the length of defined individual DNA fragments produced by digestion of DNA sample with restriction endonucleases. Therefore, these variations are termed as 'Restriction Fragment Length Polymorphisms (RFLPs)'. These polymorphic markers have been used for genetic mapping in human beings, mice, fruit-fly and in a number of crop plants.

An RFLP can be demonstrated using the following experimental procedure:

(i) Extract and purify DNA from several individuals which may differ in some respects among themselves.

(ii) Digest DNA samples with a restriction endonuclease.

(iii) Subject each DNA digest to gel electrophoresis separately, but on the same gel slab.

(iv) The DNA fragments of different lengths resulting from digestion can be separated by gel electrophoresis, but the differences in the distribution patterns of fragments in several DNA samples due to digestion can not be detected directly. This is because the number of fragments is large and the range in size is rather continuous, such that they form a continuous smear on the gel. Electrophoresis is therefore, followed by the following steps.

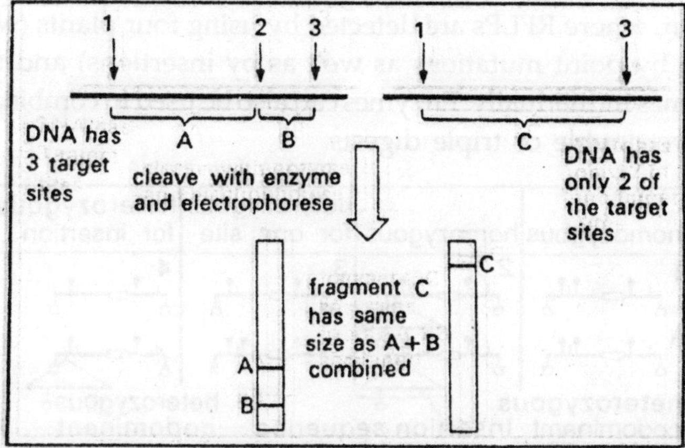

Detection of a change in DNA as revealed by the appearance or disappearance of a restriction site (an additional restriction site is present in the DNA segment shown on the left, which is absent in the segment on the right).

(v) Fragments are transferred from gel to solid support such as nitrocellulose filters—Southern blotting.

(vi) The filters with small fragments will be hybridized with radioactively labelled DNA probes (non-radioactive probes labelled with biotin or digoxigenin are also often used), so that the presence of individual fragments, fully or partially homologous to the probe, can be detected by autoradiography after hybridization. Since DNA clones having repetitive DNA may hybridize with many fragments, they will give a smear on autoradiography. Therefore, unique DNA sequences are generally used as probes for detecting RFLPs.

If two or more than two individuals differ in a restriction site affecting the number and length of DNA fragments homologous to the probe, the bands may differ in number and/or will appear at different locations in their corresponding autoradiograms, which will be called an RFLP at the phenotypic level. Figure shows how the presence of an additional site will produce two fragments and its absence will produce only one fragment. In Figure, results of a hypothetical experiment are shown, where RFLPs are detected by using four plants (which differ by point mutations as well as by insertions) and three enzymes individually. Enzymes can also be used in combination to give double or triple digests.

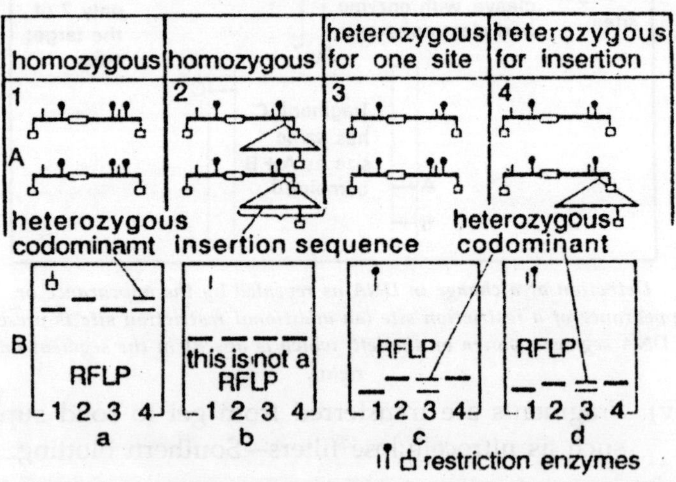

The results of digestion of genomic DNAs of four hypothetical diploid individuals (1-4 shown in A : two homozygous—1 & 2 and two heterozygous—3 & 4, for restriction sites and an insertion) with three enzymes (individually and in combination), followed by hybridization with a specific probe and autoradiography. Note the polymorphisms in three of the four cases (a-d shown in B) due to different enzymes.

Once a large number of RFLPs are available in a species, the parents, F_1 and F_2's (or BC_1 or DHs or RILs) can be used to study their inheritance and linkage relationship and genetic linkage maps can be prepared. Such linkage maps have been prepared by using more than one thousand RFLPs each in maize and tomato. The RFLPs have also been assigned to specific chromosomes using aneuploids involving specific chromosomes in several crops. Details of these RFLP maps in several crops are given in Table. Efforts are also being made to study linkage of RFLPs with morphological and economic traits so that they can be used for practical plant breeding.

Random Amplified Polymorphic DNA (RAPD) using PCR: Polymerase chain reaction (PCR) was described which has revolutionized molecular biology research. If artificially synthesized primers with arbitrary sequences, ~20 nucleotides long (random primers) are used for PCR amplification, DNA segments to be amplified will be selected at random which will thus provide a truly random sample of DNA markers, which are described as random amplified polymorphic DNA (RAPD = pronounced as 'rapid'). An example of RAPD markers demonstrating genetic diversity in maize genotypes is shown in Figure. RAPDs have several advantages over RFLPs and can be used as their substitute in several studies. The main advantages include the following:

(i) since same primers with arbitrary sequences can be used for different species, no species specific probes are needed for different species as required in RFLPs. For instance, primers that provide polymorphic markers in maize, also provided polymorphic markers in soybean, *Arabidopsis*, humans, *Neurospora*, etc.;

(ii) collection of data using RAPDs proceeds much more quickly than those using RFLPs, since there are fewer steps involved (Table). It has been shown that generation of RAPD data is five times quicker than RFLPs.

A list of some Crops, where RFLP Maps have been Prepared (Mbp=106 base pairs),

Crop	Gametic chromosome no. (n)	Number of RFLP loci	Map length (cM)	Genome size (Mbp)	Resolution Mbp/cM
1. Maize (*Zea mays*)	10	> 1000	2200	2500	1.14
2. Tomato (*Lycopersicon esculentun*)	12	> 1000	1500	1000	0.67
3. Potato (*Solatium tuberosum*)	12	350	690	2000	2.90
4. Rice (*Oryza sativa*)	12	2275	1520	430	0.28
5. Wheat (*Triticum aestivum*)	21	> 1000	6300	16000	2.54
6. Rye (*Secale cereale*)	7	7500	> 1000	7600	7.60
7. Barley (*Hordeum vulgare*)	7	> 500	1400	4900	3.50

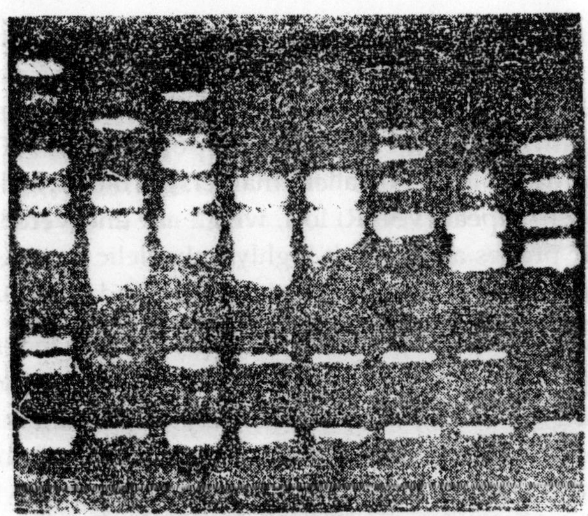

Random amplified polimorphic DNAs (RAPDs) in Maize

RAPD technique may also be combined with RFLPs to increase the efficiency of molecular markers for genetic analysis. A RAPD marker known to be polymorphic may be labelled and used as a probe for RFLP analysis, thus eliminating the need for recombinant DNA cloning of probes in bacteria.

Different steps involved in RFLP and RAPD analyses.

RELP	*RAPD*
1. Extraction of DNA	1. Extraction of DNA
2. DNA digestion with restriction enzyme(s)	2. Amplification of DNA by PCR using random primers
3. Separation of DNA fragments on agarose gel	3. Separation of amplified DNA on agarose gel
4. Transfer of DNA fragments—Southern transfers	4. Visualization of markers on the gel and photography of the gel
5. Hybridization of Southern transfers with probe(s)	
6. Detection of hybridized probe using autoradiography or using a non-radioactive system	

Minisatellites and Microsatellites (VNTRs) : It has also been emphasized that polyallelic markers will be very useful for mapping both the simple Mendelian traits as well as polygenic traits in segregating populations. The most widespread of these polyallelic markers are the variable number of tandem repeat (VNTR) loci, which are uncovered by locus specific probes and exhibit highly polyallelic fragment length variation. VNTRs include minisatellites and microsatellites.

Minisatellites have been extensively utilized for DNA fingerprinting but have not been very popular for genetic mapping, since they are not uniformly distributed throughout the genome. An interesting and common example of VNTRs involves the number of tandem repeats associated with rRNA genes in rDNA concentrated at nucleolar organizing regions (NORs) of specific chromosomes of an organism. Like rRNA genes, most minisatellite loci are concentrated in proterminal regions of chromosomes, and thus may not be able to provide the desired density of markers. This problem has largely been overcome in microsatellites or SSR loci.

Microsatellites or Simple Sequence Repeats : (SSRs) represent variation in the repeat number of sequences, 1-6bp long (minisatellite repeat can be as long as 1 kbp), such as (AC)n or (AAC)n, etc. Being smaller in size, SSRs are more common and amenable to PCR analysis (Fig.), Recently, microsatellite maps have been prepared in mouse, humans, barley, wheat, etc. Similar efforts are also being made in several other crop plants.

Sequence Tagged sites (STS) and Expressed Sequence Tags (ESTs): In using RFLPs described above, the probes used and the genomic sequences exhibiting polymorphism are unknown and therefore, the information on RFLPs generated by one worker can not be directly used by another worker elsewhere, unless he first gets samples of the relevant probes from the earlier worker. This made the use of RFLP maps difficult. In order to make them more useful and understandable, the RFLP markers were sequenced in many cases, so that the sequence information for each RFLP marker is made available to other interested workers. Further, these sequences can be used as primers for PCR reactions for speeding up the generation of data. In this manner, in several cases RFLP markers on a genetic map were characterized by specific sequences, which marked the sites which were then described as sequence tagged sites (STS). STS primer sequences are now available in many cases and are being used to convert RFLP maps into STS maps, which are globally understandable.

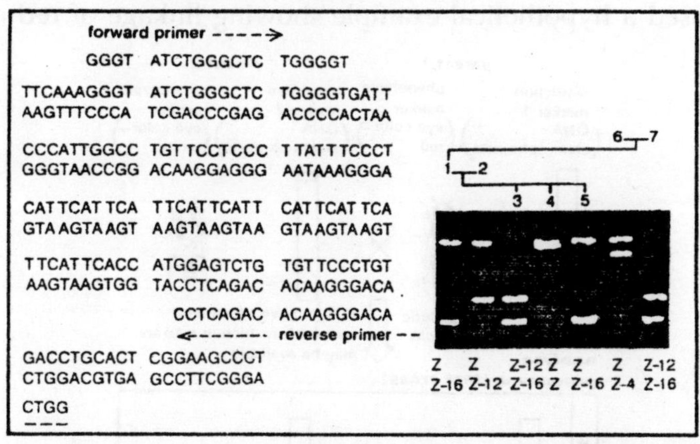

Generation and visualization of simple sequence repeats (SSR) length polymorphism.

Another problem with RFLP and RAPD maps is that the DNA sequences that are used for mapping have an unknown function. These may or may not belong to functional genes and therefore many of these sequences may not express. In order to overcome this problem, mRNA has been used for the synthesis of complementary DNA (cDNA), so that this cDNA is now used for the preparation of molecular genetic maps. Such maps will have markers which will all represent expressed sequences. The sites on such molecular genetic maps are therefore described as expressed sequence tags (ESTs). EST maps in humans have been prepared using mRNA isolated from brain, since majority of genes are expressed in the brain of human beings. Similar EST maps have been attempted in a number of animal and plant species.

Linkage and Recombination between Molecular and Phenotypic Markers : In order to study linkage relations between a molecular marker and a phenotypic marker, we may examine the RFLP/PCR patterns and the phenotypes of two parents that differ morphologically as well as for molecular markers. By making crosses between two such parents, we can estimate the recombination frequencies between molecular markers and phenotypic markers in F_2 generation. In Figure, we used a hypothetical example showing linkage of red and

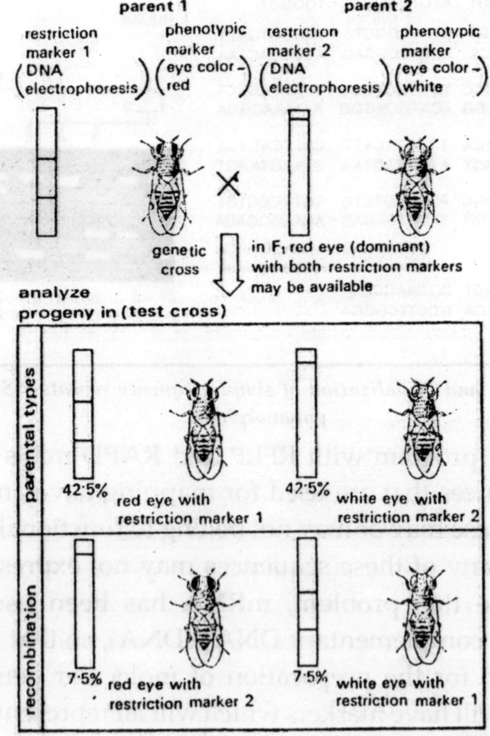

A diagrammatic representation showing linkage and recombination between a RFLP (a molecular genetic marker in the form of restriction pattern) and a phenotypic marker (eye colour).

white eyes with different molecular markers in *Drosophila*. It can be seen that as a result of recombination, in F_2 generation, 15% progeny shows association of red eye with a molecular marker associated with white eye in a parent and *vice-versa*. Therefore, it is concluded that restriction marker is 15 map units away from the marker eye colour, although it has no causal relation with eye colour. Such a linkage between molecular marker and phenotypic marker allows identification of genetic loci at the molecular level. A tight linkage of such a molecular marker with a disease may also be used for diagnosis of the disease and for isolation of its gene. In many of our crop plants, these molecular markers are also recommended (i) as tools for identification of cultivars, and (ii) for selection of desirable plants in segregating populations. In view of this, RFLPs and other molecular markers will be increasingly utilized in future.

Molecular Physical Maps

The distances between markers in a genetic map represented in the units of centiMorgans (cMs) are not proportionate to physical distances, which will be represented either in the units of megabase pairs (10^6bp) or else as proportion of a whole chromosome arm (*e.g.* 0.1 representing a distance equal to 1/10th of the arm). A number of techniques have been used for the preparation of these physical maps. Some of these techniques are described in the following text.

Physical Mapping using Chromosome Deletions : If same molecular marker is used with genomic DNA samples extracted from a normal individual and a deletion stock, the absence of marker or its homologous sequences in the deletion stock will suggest that the marker is located within the deleted segment. This technique has actually been used in humans for physically locating the molecular markers closely linked to the tapeto-choroidal dystrophy or TCD disease (an eye disease) on the X-chromosome. Similarly, deletion stocks were also used for physical mapping of a number of molecular markers in bread wheat.

Physical Mapping using in Situ Hybridization (ISH) : Molecular markers could also be mapped physically using the corresponding probes for *in-situ* hybridization (ISH) on the chromosomes. The probes are redioactively labelled and used for hybridization with DNA sequences on well spread metaphase chromosomes. Physical location of ribosomal DNA on chromosomes has been done in many species using this technique. In wheat, exact physical location of rDNA on chromosomes 1B and 6B could be achieved. This method is generally suitable for physical mapping of repetitive DNA sequences. DNA sequences obtained in the form of yeast artificial chromosomes (YACs) have also been used for hybridization on *Drosophila* giant salivary gland chromosomes, so that physical maps of *Drosophila* genome are also available.

Chromosome Walking for Physical Maps : When a probe is used for the identification of a gene sequence in a genomic library, the probe may hybridize with a number of clones, each carrying a part of a large gene fragmented during preparation of genomic library. If we obtain partial digests (by digesting the DNA only partially) from the genome, different genomes (from large number of cells) may give fragments which have overlapping sequences, because sites cleaved in different genomes of the same organism, will differ being random. Since none of these fragments may have its entire sequence represented in the probe, overlapping sequences may be used to construct the original genomic sequence. Identification of fragments with an overlapping sequence may be a key to the reconstruction or characterization of large chromosome regions. This is achieved by the technique popularly called chromosome walking.

The technique of chromosome walking involves the following steps :

(i) from the genomic library select a clone of interest (identified by a probe) and subclone a small fragment from one end of the clone (there is a technique available to subclone a fragment from the end);

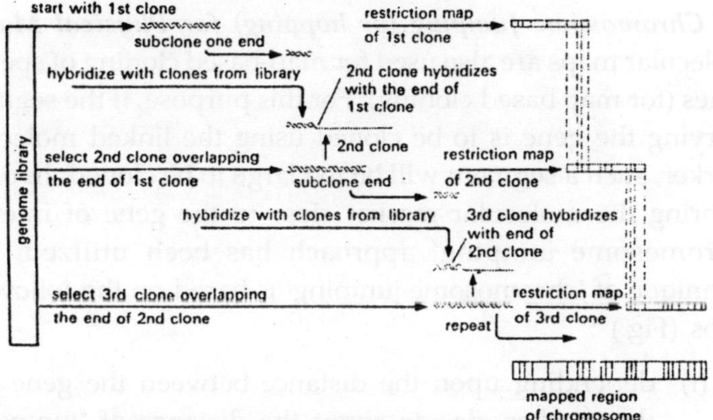

The technique of chromosome walking through successive hybridization between overlapping genomic clones.

(ii) the subcloned fragment of the selected clone may be hybridized with other clones in the library and a second clone hybridizing with the subclone of the first clone is identified due to presence of overlapping region;

(iii) the end of the second clone is then subcloned and used for hybridization with other clones to identify a third clone having overlapping region with the subcloned end of the second clone;

(iv) third clone identified as above is also subcloned and hybridized with clones in the same manner and the procedure may be continued;

(v) restriction map of each selected clone may be prepared and compared to know the regions of overlapping as shown in Figure, so that identification of new overlapping restriction sites will amount to walking along the chromosome or along a long chromosome segment.

Regions of chromosome approaching 1000kb have been mapped following the above technique. Restriction maps of entire chromosomes can be prepared in this manner following the technique of chromosome walking.

Chromosome Jumping (or hopping) for Physical Maps : Molecular maps are also used for map-based cloning of specific genes (for map based cloning). For this purpose, if the segment carrying the gene is to be cloned using the linked molecular marker, such a segment will be too large to be cloned. In order to bring the molecular marker close to the gene of interest, 'chromosome jumping' approach has been utilized. The technique of 'chromosome jumping' is based on the following steps (Fig.) :

(i) depending upon the distance between the gene and the marker, decide about the distance of 'jumps' or 'hopsize' (*e.g.* 100kb or 200kb);

(ii) genomic DNA molecules in the range of selected size (say 80kb-130kb in case of 100kb 'hopsize' or 160–240kb for 'hopsize' of 200kb) are selected through pulsed-field gel electrophoresis;

(iii) for circularization of DNA segments, ligation between two ends of each long linear DNA molecule was allowed using T4 ligase in the presence of *supF* (*supF*$^-$ is a suppresser tRNA gene that will allow selection of clones representing junction fragments by cloning them in an amber mutated phage *supF*$^+$);

(iv) DNA circles obtained in (iii) above are digested with *EcoRI;*

(v) the vector λ Ch3 Δ Alac, an amber-mutated phage vector (*sup* F$^-$) is also cut with *EcoRI* and used for cloning small DNA fragments representing the junctions of the circularized genomic DNA molecules and carrying *supF*$^+$;

(vi) the cloned DNA fragments obtained in

(v) above represent the jumping library, which can be plated on a bacterial host and screened through the technique of plaque hybridization.

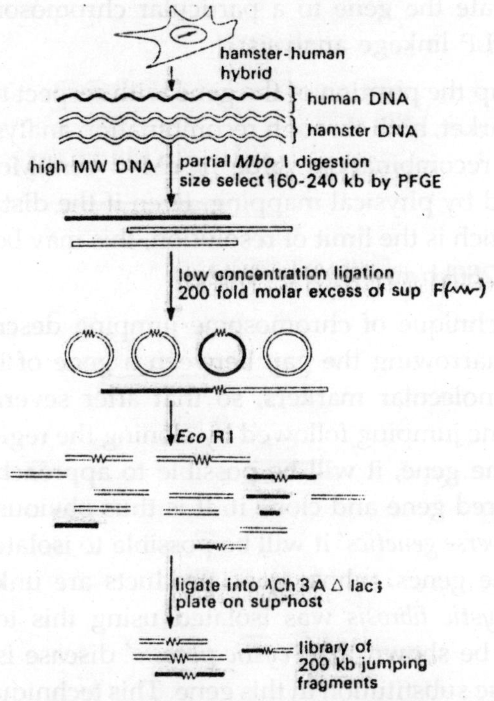

Steps involved in chromosome jumping

Molecular Maps, Map-based Cloning and Reverse Genetics: In most cases, we identify the genes by their protein products, which help in the isolation and cloning of a specific gene. However, for many human disorders which are each controlled by a single gene (*e.g.* huntington disease, cystic fibrosis, duchenne muscular dystrophy, etc.), the gene product is not known despite intensive investigations. Such genes are cloned by a process called 'reverse genetics', in which the gene is identified primarily by its map position. Once the gene is cloned, it can be utilized for getting information about the encoded protein, the gene mutation and the metabolic defect that may be associated with the disease. Since the information about the genetics is coming in the reverse order, this is popularly described as *reverse genetics.* The reverse genetics approach involves the following two steps :

(i) locate the gene to a particular chromosome through RFLP linkage analysis;

(ii) map the position of the gene with respect to molecular market, both through recombination analysis (in terms of recombination value in cM = centiMorgan units) and by physical mapping. Even if the distance is lcM, which is the limit of resolution, this may be equivalent to 280kb to 7,600kb (Table).

The technique of chromosome jumping described above will help narrowing the gap between a gene of interest and available molecular markers, so that after several cycles of chromosome jumping followed by cloning the regions that are closer to the gene, it will be possible to approach very close to the desired gene and clone it. It is thus obvious that in the field of *'reverse genetics'* it will be possible to isolate, clone and characterize genes, whose gene products are unknown. The gene for *cystic fibrosis* was isolated using this technique. It could also be shown that *'cystic fibrosis'* disease is caused by a single base substitution in this gene. This technique of *'reverse genetics'* has become a very active area of research particularly in human genetics in the coming years. Paul Berg (Noble Laureate), in 1991 redefined 'reverse genetics' (see author's book *'Genetics,* for more details).

8

Regulation of Gene Expression

***A Variety of Mechanisms in Eukaryotes (Including Cell Receptors and Cell Signalling)* :** During the last two decades, considerable information has become available on the regulation of gene expression in eukaryotes, although in the 1960s only bacteria, were considered suitable for such studies. Due to inherent difficulties in working with multicellular eukaryotes, initial studies in these systems were restricted to the following:

(i) formulation of theoretical models for the regulation of gene expression and

(ii) conducting experiments to identify whether histones or non-hislone chromosomal proteins are relatively more important for the regulation of the activity of specific genes in time and space (*i.e.* at specific developmental stage and in specific tissue).

More recently (particularly after 1980), regulation of specific genes has been examined in greater detail giving valuable information. Specific DNA sequences and proteins (*e.g.* transcription factors) involved in regulation of specific genes have been identified. These specific genes express either in specialized cells or respond to specific stimuli and are variously described as luxury genes or smart genes as opposed to house keeping genes, which are constitutive in their expression. A large number of transcription factors have been identified which take part in the formation of a transcription complex, which activates the enzyme, RNA polymerase and associated proteins to initiate transcription.

Specific regulatory DNA sequences have also been identified (sometimes through the use of transgenic plants), which respond to stimuli and provide sites for binding of transcription factors. For transmission of stimuli (external or internal) to the target DNA sites, specific signal transduction pathways (involving intermediate receptor molecules) have been worked out in detail in many cases. Examples are also available now, where regulation of gene expression in eukaryotes is exercised at the post-transcriptional level involving RNA processing or translation. A remarkable phenomenon of regulation through genome reorganization or gene rearrangement has also been discovered in (i) human immune system involving antibody production, (ii) yeast mating types involving production of pheromones-receptors and in (iii) trypanosome's variable surface antigen-glycoprotein (VSG) production. Another important area of research, which is related with regulation of gene expression in eukaryotes is cell signalling. The cells do possess structures and mechanisms, which permit them to perceive stimuli and transmit the signal to the interior of the cell.

Consequently, appropriate genes are expressed at appropriate time and in appropriate tissue. The cells possess receptor molecules on their surface or inside the cell, which help in receiving the signals, and also possess signal transduction pathways which transduce the signals to the genes, whose expression needs to be regulated. This area of research has been actively persued during 1980's and 1990's and spectacular success has been achieved. All these aspects will be briefly described in this chapter. Limited space in this book does not allow detailed treatment of the subject, which has expanded rapidly in recent years.

Regulation at Transcription Level

As in prokaryotes, regulation of gene expression is exercised in majority of cases at the level of transcription, although examples of post-transcriptional control are also known. In view of this, major part of this chapter is devoted to regulation at transcriptional level. This regulation may involve both activation and repression, because genes may have to be switched on in some cases and switched off in other cases. DNA sequences involved in regulation, were also identified through the use of DNA foot printing as well as through the use of gene constructs for production of transgenic plants. All these aspects will be briefly described in this section.

Activation of Transcription

Activation of transcription as a means of regulation of gene expression has been studied during the last more than 20 years, using techniques, which have undergone a revolution through the development of recombinant DNA technology during 1980s. In a chronological order, initially (1969-73), these studies involved formulation of theoretical models based on the occurrence of large proportion of repetitive DNA in eukaryotes. Later (1973-75), chromatin reconstitution experiments demonstrated a positive control exercised by non-histone chromosomal proteins and still later (particularly in 1980s and 1990s), the role of transcription factors was examined in detail in large number of cases.

Britten-Davidson Model for Unit of Transcription: Although various models for regulation of protein synthesis in eukaryotes have been proposed, a popular model, known as Britten-Davidson model or the gene-battery model was one of the first models which was proposed by two scientists, Britten and Davidson in 1969. It was subsequently elaborated by them in] 1973. The model assumes the presence of four classes of sequences:

(i) *Producer Genes:* A producer gene is comparable to a structural gene; of prokaryotic operon.

(ii) *Receptor Site:* A receptor site is comparable to operator gene in bacterial operon and one such receptor site is assumed to be present adjacent to each producer gene or a set of such producer genes.

(iii) *Integrator Gene:* It is comparable to regulator gene and is responsible for synthesis of an activator RNA that may or may not give rise to proteins before it activates the receptor site (Fig.).

(iv) *Sensor Site:* A sensor site regulates the activity of integrator gene, which can be transcribed only when the sensor site is activated. The sensor sites are recognized by agents which like hormones and proteins change the pattern of gene expression. For instance, hormone-protein complex or a transcription factor may bind to a sensor site and cause the transcription of integrator.

The inter-relationship of the four classes of sequences given above are illustrated in Figure. In this model, the genes (producer gene and integrator gene) are those sequences which are involved in RNA synthesis. On the other hand, sites (receptor and sensor sites) are those sequences which help only in recognition without taking part in RNA synthesis. It is also proposed that receptor sites and integrator genes may be repeated a number of times, so as to control the activity of a large number of genes in the same cell. Repetition of receptor ensures that same activator recognizes all of them and several enzymes of one pathway are simultaneously synthesized.

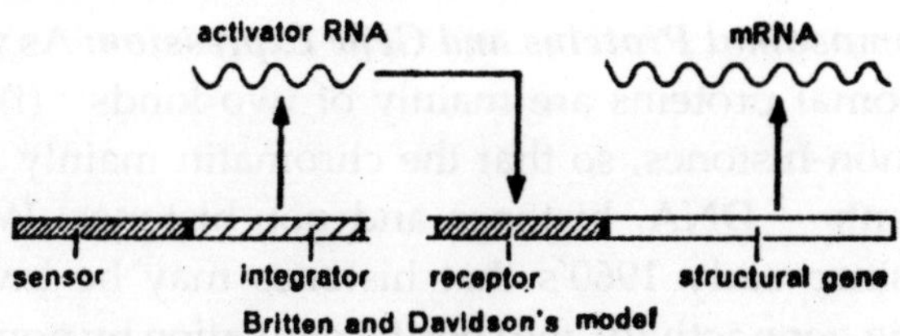

Different components in Britten-Davidson's model.

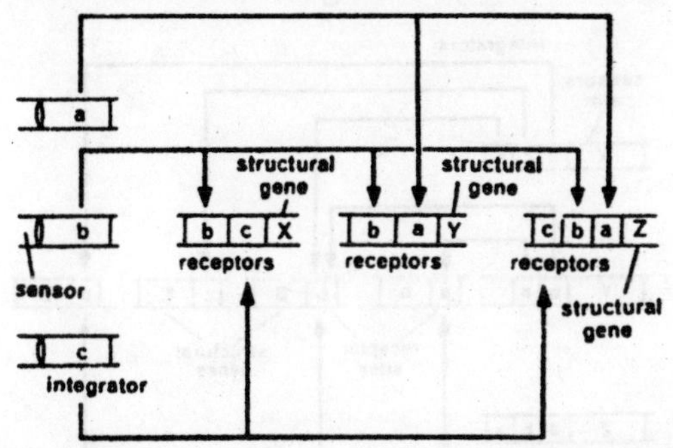

Redundancy of receptor sites (as proposed in Britten-Davidson's model).

It is also possible that the transcription of same gene be needed in different developmental stages. This can be achieved by multiplicity of receptor sites and integrator genes. Each producer gene may have several receptor sites, each responding to one activator (Fig.) so that a single activator though can recognize several genes, but different activators may activate the same gene at different times. An integrator gene may also fall in cluster with same sensor site (Fig.).

Gene Battery: In Britten-Davidson model, a set of structural genes controlled by one sensor site is termed a battery. Sometimes, when major changes are needed, it is necessary to activate several sets of genes. If one sensor site is associated with several integrators, it may cause transcription of all integrators at the same time thus causing transcription of several producer genes through receptor sites. The repetition of integrator and receptor sequences is consistent with the knowledge that sufficient repeated DNA occurs in eukaryotes.

Chromosomal Proteins and Gene Expression: As we know, chromosomal proteins are mainly of two kinds : (i) histones and (ii) non-histones, so that the chromatin mainly has three components— DNA, histones and non-histones. While it is known since early 1960's that histones may be involved in repressing gene activity, the specific regulation by non-histones was demonstrated only during 1970's.

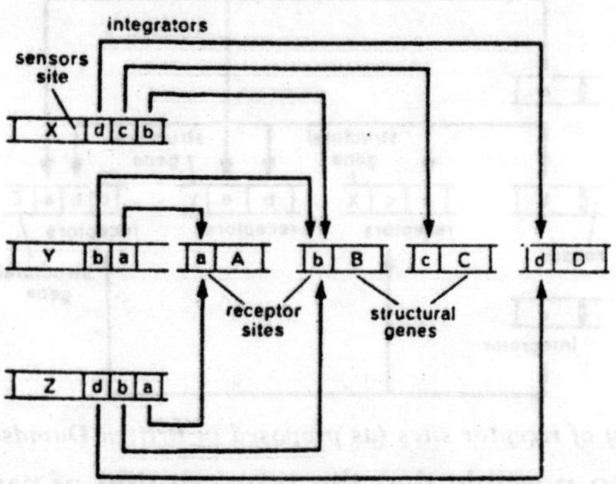

Redundancy of integrator genes (as proposed in Britten-Davidson's model).

The most important experiments in this connection were pioneered by Gilmour and Paul (1973). They isolated the chromatin from different tissues separately and then dissociated it into DNA, histones and non-histones. This was followed by chromatin reconstitution using either the three components derived from chromatin of the same tissue or by combining non-histones of one tissue with the DNA and histones of another tissue (Fig.). It could be demonstrated in such experiments that the mRNA which is synthesized *in vitro* from reconstituted chromatin, mainly depended on the source of non-histone proteins (for a review consult O'Malley *et al.*, 1977).

There are atleast three specific proteins, whose synthesis in eukaryotes has been studied using chromatin reconstitution experiments. These are :

(i) the globin synthesized in reticulocytes,

(ii) the ovalbumin synthesized in chick oviduct due to hormonal stimulation and

(iii) the histones synthesized during '5' phase of cell cycle.

In all these three cases, similar observations were available which are as follows :

(i) Same quantity and quality of mRNA was synthesized on native chromatin as well as on chromatin reconstituted from DNA, histones and non-histones derived from same tissue.

(ii) When heterologous chromatin was reconstituted by using non-histones from one tissue and DNA and histones from another tissue, mRNA synthesized was similar to that synthesized on the native chromatin to which non-histones belonged (Fig.).

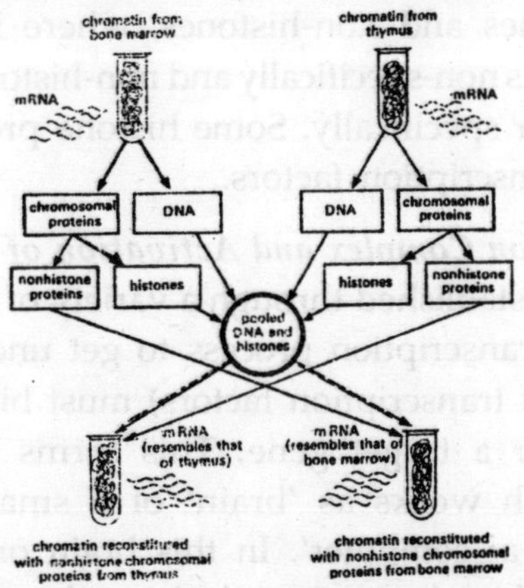

Chromatin reconstitution experiment demonstrating the positive role of non-histone proteins.

Above observations suggested a positive regulatory role of non-histones in eukaryotes. This role of non-histones is also suggested by the following additional information.

(i) Non-histones are present in increased quantity in tissues active in RNA synthesis, while quantity of histones is constant,

(ii) Non-histones exhibit 2-3 times more diversity than histones,

(iii) Non-histones have tissue specificity and DNA binding specificity,

(iv) Certain non-histones are known to induce RNA synthesis *in vitro*.

(v) Synthesis of specific non-histones is associated with induction of gene activity.

(vi) A large number of transcription factors are actually non-histone proteins.

Above observations though suggest a specific role of non-histones, but in the absence of histones, non-histones have been shown not to induce RNA synthesis. This suggests that regulation of protein synthesis actually involves an interaction between histones and non-histones, where histones inhibit protein synthesis non-specifically and non-histones induce RNA synthesis rather specifically. Some histone proteins may also function as transcription factors.

Transcription Complex and Activation of Smart Genes: It has now been established through a variety of studies, that for the elaborate transcription process to get underway, several proteins (called transcription factors) must bind to the DNA sequences near a target gene. This forms a transcription complex, which works as 'brain' of a smart gene and is comparable to a 'computer'. In this brain or computer, the signals are combined to make a decision about whether or not to switch on the smart gene. Only by the right combination of transcription factors or signals, the transcriptional enzyme (RNA polymerase) is activated, and produces RNA. In contrast to this computer like complex control of transcription in eukaryotes, the control in prokaryotes is simple and can be compared to mere switches.

In eukaryotes, a number of DNA binding proteins and

their corresponding binding regions on DNA have been identified and isolated in a variety of organisms. While in prokaryotes, there is a maximum of three protein binding sites near the transcription start point, in eukaryotes, these sites may be atleast five and may be as many as 20 for some smart genes (not for house keeping genes). For instance, in sea urchin, for the regulation of a gene producing actin, as many as 20 DNA regions near this gene have been identified, which are recognized by regulatory proteins. Atleast five of these 20 sites are used for activation of actin gene.

Seven transcription factors (TFIID, TFIIA, TFIIB, TFIIE, TFIIF, TFIIJ, TFIIH), which are essential for transcription of all genes including house keeping genes and smart genes. However, for transcription of smart genes, additional transcription factors must form a complex to activate the obligatory proteins coupled to the RNA polymerase II enzyme that transcribes the gene into mRNA. Only when all components (including the obligatory proteins and the right combination of transcription factors forming a complex) are present, the transcription of a smart gene will take pace.

Several hundred different transcription factors have been found to date, two third of these falling into four recognizable groups—helix-turn-helix, helix-loop-helix, zinc finger and leucine zipper. The remaining one third may be unique to their types. Each transcription factor has separate domains for (i) recognition and binding a specific DNA sequence and for (ii) activation of obligatory proteins. These are called DNA binding domain and activation domain.

Some of the earliest studied transcription factors include GAL4 and GCN4 in yeast, which activate genes involved in metabolism of galactose and those involved in general control of nitrogen metabolism (amino acid biosynthesis) respectively. The details of these two transcription factors are shown in Table.

GAL4 and GCN4 Showing DNA Binding and Activation Domains

Protein	*Total length*	*Domains**	
		DNA binding	*Activation*
GAL4	881	1-74	147-196 & 768-881
GCN4	281	1-60	107-125

A number of other transcription factors are listed in Table.

Some Transcription Factors and then Properties

Transcription factor	*DNA binding domain*	*Specific function*
GR	Zn finger	Glucocorticoid receptor
Spl	Zn finger	General
CTF/NF-1	—	General
C-Jun	Leucine zipper	Binds AP-1 sites (for dimerisation)
Fos	Leucine zipper (for dimerisation)	—
C/EBP	Leucine zipper (for dimerisation)	Hepatocytes and fat cells
AP2	—	Confers TPA and cAMP inducibility
CREB	Leucine zipper	Confers cAMP inducibility and El A transactivation
OCT-1	Homeodomain	Binds octamer; general
OCT-2	Homeodomain	Binds octamer; lymphoid
SRF	—	Serum
NF-κB	—	Lymphoid
HSTF	—	Heat shock

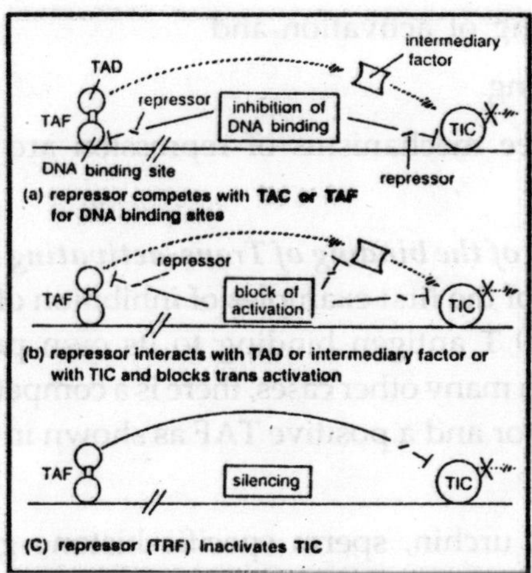

Three categories of transcriptional repression (TAP = trans-activation factor; TIC = transcription initiation complex; TRF = trans-repressing factor).

The mechanism involved in binding of these transcription factors to DNA and their involvement in protein-protein interactions. It should be recognized that the discovery of regulatory DNA sequences and transcription factors was done independent of each other in many cases, so that in all cases it is not known, which transcription factor regulates the activity of which specific genes. However, certain transcription factors are known to be found in some specialized cells and not in others, so that some idea about the genes regulated by them can be inferred. In other cases, where genes have been isolated, their regulatory sequences could also be identified. Similarities are available among the different regulatory DNA sequences as well as among amino acid sequences of different transcription factors, suggesting mechanisms of DNA-protein interactions.

Repression of Transcription

In recent years, many examples of negative control of transcription have also become available in eukaryotes. These examples can be divided into three main classes :

(i) inhibition of DNA binding,

(ii) blocking of activation and

(iii) silencing.

These three mechanisms of repression are illustrated in Figure.

Inhibition of the binding of Trans-activating Factors (TAF) to DNA: One of the first examples of inhibition of TAF binding involved SV40 T antigen binding to its own promoter as in prokaryotes. In many other cases, there is a competition between a negative factor and a positive TAF as shown in the following examples :

(i) In sea urchin, sperm specific histone gene *H2B-1* is repressed in embryos, due to interference by a negative factor with a positive CCAAT binding factor.

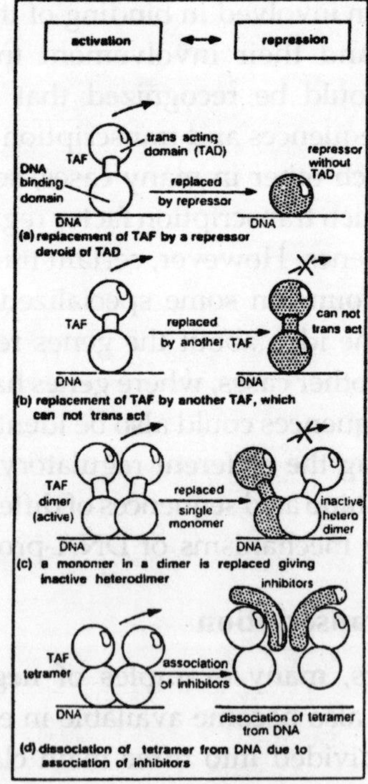

Four ways in which DNA binding of TAF can be inhibited.

(ii) In humans, fetal γ -globin gene is repressed in adults due to binding of a factor NF-E, which inhibits binding of positive factor CP1 to CCAAT box.

(iii) Several genes are repressed, when GC-box factor competes with Spl for binding to GC-rich sequences. During such an inhibition, a positive factor for one gene may also function as negative factor for another gene due to similar DNA sequences. This has been shown in several developmental genes in *Drosophila* and also for glucocorticoid receptor molecule,

(iv) In still other cases, a component (*e.g.* Fos) of a heterodimer (*e.g.* Jun-Fos) can be replaced to give inactive complex (*e.g.* C-Jun-Jun B), which competes with heterodimers; or an inhibitor (*e.g.* IkB) may cause a TAF dissociate from DNA and inactivate it (Fig.). Heat shock protein (hsp90) also associates with glucocorticoid receptor and renders it incapable of binding DNA.

Blocking of Activation: In several examples, TAF can bind DNA, but its activation domain is not allowed to function (Fig.). Following alternatives are possible :

(i) Negative factor may bind near the binding sites meant for positive factors. For instance, in *c-myc* gene, the negative factor myc-PRF binds next to positive factor myc-CFl, and for polyoma virus, repressor PEA2 binds near PEAl.

(ii) Negative factor may bind the positive factor, rendering its trans-activation domain ineffective as done by binding of GAL-80 to TAF GAL4 in yeast; similar is the situation with binding of AP-1 with P67SRF needed for activation of *C-fos* gene (AP-1 is usually a TAF, but here it works as a repressor).

(iii) In signal transduction, sometimes an intermediary

protein is involved, which is blocked by another trans-activating domain, which may be present in high concentration (*e.g.* receptors of several steroid hormones). Ela (an adenovirus protein immortalizing primary cells) also blocks activation of several genes by binding intermediary proteins involved in activation due to TAFs (AP-1, GAL4, VP16), which bind to SV40 enhancers.

Silencing of Transcription: After the discovery of enhancer elements involved in enhancing transcription in 1985, elements were discovered in yeast, which mediated repression and therefore were called silencers. Like an enhancer, a silencer also functions irrespective of its position (upto 2.5kb away) and orientation relative to the gene, whose expression it controls. The role of silencers can be illustrated using the example of

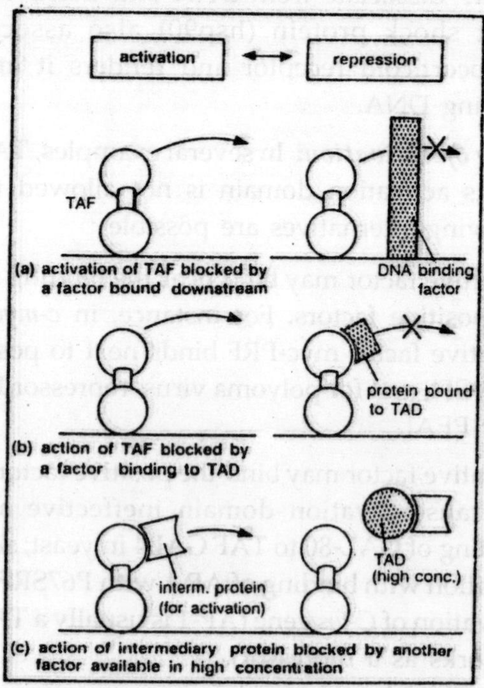

Three ways in which activation (not binding on DNA) by TAF is blocked.

mating type loci in yeast, where a mating type active locus *MAT* is associated with two silent loci, one *HMR* on the right side and the other *HML* on the left side. These silent Icci can be 'α' mating type or 'a' mating type and can replace *MAT* to bring about a change in mating type. The loci *HMR* and *HML* are each repressed by a silencer located - 1 kb upstream in each case and described as E and I silencers or as *ER* (or *HMR-E)* and *EL* (or *HML-E)*. Besides, E and I silencers, there are several transacting factors involved in silencing. For instance, *HMR-E* has been shown to have binding sites for a transcription factor RAP-1 and another protein Abf-lp needed at replication origin. Both proteins are needed for repression. Similarly α_2 protein coded by one of the genes at *HML* locus binds to *HML-E* and represses mating type gene(s). There are also four *SIR* genes *(SIR1, SIR2, SIR3, SIR4)* involved in silencing. Each of these four genes have been cloned and their products are believed to take part in silencing by interacting with histones to modify chromatin structure.

Silencer sequences have also been reported for human immunodeficiency virus (HIV) and called negative response element (NRE), and also for chicken lysozyme gene. The silencer factor (a protein) either locks the transcription initiation complex and makes it unavailable for activating factors or it disorganizes the transcription initiation complex (Fig.).

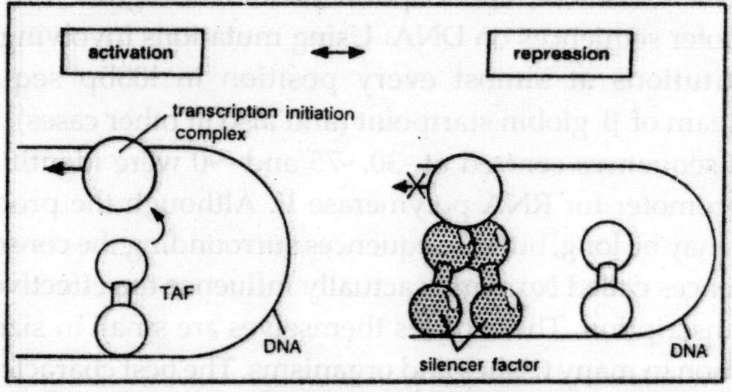

Mechanism of the action of silencer factors, rendering the transcription initiation complex ineffective.

It should be realized that repression of gene activity is not mediated by a single mechanism and one or more of the above three mechanisms may be involved in the same case or even additional mechanisms may be involved. It has been shown that the steroid receptor superfamily utilizes all the three mechanisms for repression and is also involved in activation of transcription although of different genes. More information in this connection will become available in coming years.

Specific DNA Sequences Controlling Transcription : A large number of promoter DNA sequences have been identified with a variety of genes in eukaryotic genomes. These regulatory DNA sequences may be present either upstream (before the start site) or downstream (after the start site), and used for binding of a number of proteins, most of them being transcription factors. Some of these sequences are present in all genes, but there are others, which are specific and are therefore used for regulation of gene expression.

Upstream DNA sequences controlling transcription by RNA polymerase II in eukaryotes (TATA box, CAAT box and GC box). Eukaryotes have three RNA polymerases (I, II, III). RNA polymerase I synthesizes rRNA, RNA polymerase II transcribes protein coding genes and RNA polymerase III synthesizes tRNA and 5SRNA. These three RNA polymerases, with the help of a number transcription factors, recognize different promoter sequences on DNA. Using mutations involving base substitutions at almost every position in 100bp sequence upstream of β globin startpoint (and also in other cases), three short sequences centred at -30, -75 and -90 were identified in the promoter for RNA polymerase II. Although the promoter itself may be long, but the sequences surrounding the consensus sequences called *boxes* may actually influence the effectiveness of transcription. These boxes themselves are small in size and common in many tissues and organisms. The best characterized of these modules called boxes are TATA box, CAAT box and GC box (Fig.).

TATA box has a resemblance with Pribnow box of prokaryotes and was identified, in the laboratory of Dr. Hogness, by Dr. Goldberg during his Ph.D. work. It is invariably AT rich. Therefore it is often referred to as Goldberg-Hogness box or TATA box. It has been shown that the presence of TATA box is essential for transcription. This has been proved with the help of both *in vitro* (in culture) and *in vivo* (living system) experiments, where deletions or mutations in TATA box resulted in reduced or complete halt of transcription. However, this is the least effective component of the promoter. It has also been shown that other sequences both downstream from TATA box (between TATA box and initiation site) and upstream from TATA box upto a distance of 200 bases may be important in controlling transcription or in other words, the gene activity.

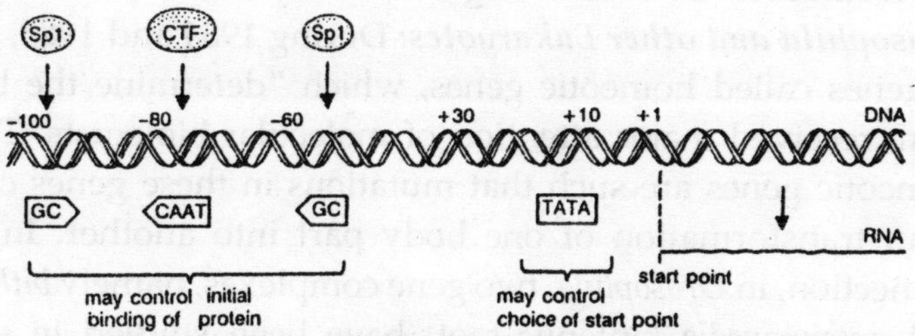

Different consensus sequences (boxes) in the promoter region for thymidine kinase gene (orientation of sequence is shown by arrows; Spl and CTF are transcription factors binding to GC and CAAT boxes).

Further upstream from the transcription initiation site, at position between -70 and -80, is also found a consensus sequence, which has the pattern, GG (C or T) CAATCT. It is called CAAT box, which is the most effective component of the eukaryotic promoters. The GC box, which may occur between -50 and -60 or between -90 and -100 contains the sequence GGGCGG and may be found in multiple copies in either orientation.

Downstream Promoter for RNA Polymerase III : Promoter sequences for transcription have often been located only upstream the starting point. However, the genes coding for

5SRNA in *Xenopus laevis* have been shown to have promoters downstream the initiation site and within the transcription unit between +55 and + 80 base pairs from the startpoint of the gene. There are also other cases of RNA polymerase III transcription, where promoters are located downstream. For instance in VA1 (adenovirus gene), the promoter lies between +9 and +72 base pairs, and in $tRNA_f^{met}$ gene of *Xenopus laevis*, the promoter lies in two parts, one lying between +8 and +30 and the other lying between +51 and +72. Any deletions in these internal sites have led to failure of initiation although deletions in other regions of the gene did not cause any failure of initiation. Some other tRNA genes in eukaryotes, also have similar promoter regions.

Homeotic Box and Regulation of Gene Activity in Drosophila and other Eukaryotes: During 1984 and 1985, a set of genes called homeotic genes, which "determine the body plan, received major attention of molecular biologists. These homeotic genes are such that mutations in these genes cause clean transformation of one body part into another. In this connection, in *Drosophila*, two gene complexes, namely *bithorax* and *antennapedia* (antenna-foot) have been studied in some detail and have now been cloned. In each of these two complexes, there are several genes, and not all of them are homeotic. It has been demonstrated that in almost all these genes, homeotic or non-homeotic, a nucleotide sequence (180 base pairs long) is always found near the 3' end of the gene. This sequence seems to be highly conserved and has now been discovered in other organisms like *Xenopus* (toad), mice and humans (in humans, atleast five such homeotic genes having this 180 base pair long sequence are known). This sequence (180bp long) has been named Homeo box by Walter Gehring of the University of Basel in Switzerland. Some of these genes will be described. In *Drosophila*, a number of genes of *antennapedia* complex include *'Antp'* (antennapedia), *'ftz' (fushi tarazu) 'engrailed'* and *'deformed'*. Similarly genes of *bithorax* complex include *ubx* or *ultrabithorax* and *infrabdominal-2*. Among these genes *'ftz'* genes are interesting and are characterized as segmentation genes that follow the 'pair rule', since every alternate segment is missing thus reducing the number of segments to half (*ftz* homozygote, which dies after early larval stage; *fushi tarazu* is a Japanese word meaning not enough segments). In homozygous individuals for mutation in *'engrailed'* gene also, the alternate segments are defective thus following 'pair rule'. Similarly, *'Antp'* gene in its mutant form, in homozygous condition, causes the conversion of antenna of adult flies into legs. These genes can prove useful in the study of genetic control of developmental pattern.

In some of the above genes, 'homeo box' has been sequenced as in *'fushi tarazu' (ftz)*. It has been shown, that although the genes having homeo boxes may widely differ in their phenotypic effect to the extent that some of them are not really homeotic, but the 'homeo box' sequence seems to be similar and highly conserved. Therefore, it is believed that their role in regulation of gene activity should be similar. The 60 amino acids coded by 'homeo box' seem to give rise to a domain within the protein products of each of the homeotic genes that constitute the complexes like bithorax and antennapedia. It has also been shown that the 'homeo boxes' of *Drosophila* genes are homologous to yeast *MAT* $\alpha 1$ and *MAT* $\alpha 2$ genes, which regulate sets of genes determining the mating type of yeast cells (see later in this chapter for more details). The protein products of *MAT* $\alpha 2$ gene of yeast and that of *'engrailed'* gene of *Drosophila* have been shown to bind specifically to a specific DNA sequence thus regulating the activity of either those genes to which the homoeo box belongs or even other unrelated genes.

Response Elements for a Variety of Stimuli: In signal transduction pathways, smart genes are expressed eventually due to the presence of unique modules in DNA. These modules represent short consensus sequences, usually < 200 bp upstream of the gene to be expressed. These are called response elements (RE) and may be present (either in the promoter or in the enhancer) in single or multiple copies. The response elements are used by certain proteins (*e.g.*, receptor protein or transcription factor), the binding region of which extends for a short distance on either side of RE. Some examples of response elements representing these proteins are given in Table.

Some Response Elements on DNA, which are Recognized in Response to Specific Stimuli in Eukaryotes.

Agent of stimulus	*Factor & size (daltons)*	*Module & consensus*	*DNA bound*
1. Heat shock	HSTF (93,000)	HSE (CNNGAANNTCCNNG)	27bp
2. Glucocorticoid	Receptor (94,000)	GRE (TGGTACAAAGTTCT)	20bp
3. Cadmium	? (?)	MRE (CGNCCCGGNCNC)	?
4. Phorbolester (TPA)	API (39,000)	TRE (TGAC TCA)	22bp
5. Serum	SRF (52,000)	SRE (CCATATTA-GG)	20pb
6. Light —	—	LRE	—

Ubiquitin Protein and Regulation of Heat Shock Genes: A massive increase in the transcription of DNA sequences included in heat shock genes has been observed due to increase in temperature. Therefore, heat shock genes offer an ideal system for the study of regulation of gene activity. These genes occur in wide range of organisms ranging from bacteria to humans. It has been shown that a protein named ubiquitin plays a key role in this regulation.

The promoters of eukaryotic heat shock genes possess a response element called heat shock element (HSE) (Table) located upstream of these genes. The HSE is recognized by a 'heat shock transcription factor (HSTF)', which is active (perhaps through phosphorylation) only during heat shock and binds to a site (70bp) carrying HSE (15bp). About 20 genes carrying same HSE are switched on simultaneously, producing heat shock proteins. Both HSE and HSTF have been conserved during evolution. For instance, in yeast and fruit fly similar HSTF proteins and similar HSE (DNA sequences) have been found.

It is suggested that perhaps the heat shock transcription factor (HSTF) itself remains ubiquitinated under normal condition and that failure of ubiquitination renders this factor free to be available to attach to a sequence (HSE) meant to initiate transcription of heat shock genes. The role of ubiquitin in modification of nucleosome structure through its association with histone proteins has also been suggested.

Transgenic Plants to Study Regulatory Sequences : The response elements have often been identified through the use of gene constructs, consisting of sequences flanking the gene in question and a reporter gene (*e.g. cat* gene for chloramphenicol acetyltransferase or *gus* gene for β glucuronidase). These gene constructs are used for getting transgenic plants where the response of reporter gene under the control of RE is assessed.

Transgenic plants have particularly been found useful in the study of development, since with their help, gene expression can be studied in time and space *i.e.* in different environmental conditions causing induction or suppression of gene expression. Transgenic plants thus facilitated the recognition of regulatory sequences involved in these differential expressions of gene activity. For this purpose, sequences upstream to a number of structural genes were cloned either with their own structural genes or with other unrelated structural genes (*e.g.* reporter genes) to study the specific regulatory role of these sequences in time and space. A number of structural genes, whose regulatory sequences were examined in this manner are listed in Table, and the details of some of these are described in the following text.

Gene for Small Subunit of Ribulose Bisphosphate Carboxylase (rbcS): Dozens of genes in plants are affected by light and many of these regulated by phytochrome. The expression of phytochrome gene is depressed and that of RUBISCO (ribulose, 1,5 bisphosphate carboxylase/ oxygenase) gene is stimulated by red light. RUBISCO is involved in CO_2 fixation and is the most abundant enzyme on the earth. It consists of two subunits, each encoded by a separate gene, the smaller subunit rbcS encoded by a nuclear gene and the larger subunit rbcL encoded by a chloroplast gene. Although, rbcS from different plants are fairly similar, small differences, that exist, can be used to distinguish mRNA of a transgene from that of the, host plant using specific probes. The pea *rbcS* gene was transferred to petunia and tobacco plants, and soybean *rbcS* gene was transferred to petunia. The gene expressed itself in transgenic plants, *rbcS* subunit from the introduced gene sometimes even combining with *rbcL* subunit of the host plant to yield the enzyme. 'Deletion mutants' having sequences deleted upstream of the transgene *rbcS* led to the identification of a sequence upstream (-166 to -149 bp) of TATA box, which was light responsive and was, therefore, conveniently called light responsive element (LRE). This LRE sequence could also regulate genes other than *rbcS* gene, so that the activity of these unrelated foreign genes became light, organ and age specific in a manner similar to *rbcS* gene (high expression in leaves, low expression in stem and no expression in roots). LRE was shown to have three regions, boxI, boxII and boxIII and it was shown that box II region also had a negative regulatory element. This LRE also seems to be repetitive (with two repeats) and each repeat unit can mediate light responsiveness and organ specificity. However, it has also been shown that both repeats of LRE are required for full expression of the *rbcS* gene in youngest leaves of tobacco plants.

It has also been shown using transgenic plants, that a different sequence upstream to *rbcS* is responsible for age responsiveness. For example, it was shown that the region from -410 to -166 bp is essential for high level of expression in immature leaves, but is not important for light responsive expression in mature leaves, where the region -166 to -149 bp was most important.

Gene for Chlorophyll a/b Binding Protein (Cab): Examples of the transfer of *Cab* gene from pea, wheat and *Nicotiana plumbaginifolia* are listed in Table. In this case it was shown that a 247 bp sequence for pea *CabABm* (-347 to -100) and a 268 bp sequence (-357 to -89) ,for wheat *Cab-l,* when separately fused with heterologous TATA sequence fused to neomycin phosphotransferase (NPTII) coding sequences, conferred on this coding sequence a light responsive expression. This regulatory sequence when present induced expression of NPTII by red light and the expression could be reversed by a flash of far-red light as observed in phytochrome mediated expression. In another experiment, negative elements suppressing expression were also identified, when *Cab-E* gene of *Nicotiana plumbaginifolia* was studied in a similar manner in transgenic plants.

Transgenes that are Inducible (by a stimulus) and/or Exhibit Tissue Specific Expression in Transgenic Plants.

Gene product	*Donor plant*	*Transgenic host*	*Inducer stimulus*	*Specific host tissue*
1. rbc S	Pea	Tobacco, petunia	Light	Green tissues
rbc S	*N, plumbaginifolia*	*N. Plumbaginifolia,* petunia, tobacco	Light	Green tissues
rbc S	Soybean	Petunia	Light	Green tissues
2. Cab	Pea, wheat	Tobacco	Light	Green tissues
Cab	*N. plumbaginifolia*	Tobacco	Light	Green tissues
Cab	Wheat	Tobacco	Orcadian clock	Green tissues

Contd..

Gene product	*Donor plant*	*Transgenic host*	*Inducer stimulus*	*Specific host tissue*
3. ST-LSI	Potato	Tobacco	Light	Green tissues
4. Chalcone synthase	*Antirrhinum majus*	Tobacco	Light	Flower
5. Hsp	Maize	Petunia	Heat	All tissues
	Soybean	Tobacco	Heat	All tissues
6. Protease inhibitor	Potato	Tobacco	Wounding	All tissues
7. Patatin	Potato	Potato	Tuber growth	Tuber
8. Leghaemoglobin	Soybean	*Lotus*	Rhizobium	Nodule
9. Nodulin	Soybean	*Lotus*	Rhizobium	Nodule
10. β–phaseolin	French bean	Tobacco	Rhizobium	Seed
11. β–conglycin	Soybean	Petunia, tobacco	Seed development	Seed
12. Lectin	Bean	Tobacco	Seed development	Seed
	Soybean	Tobacco	Seed development	Seed+root
13. Glutenin	Wheat	Tobacco	Seed development	Seed
14. Hordein	Barley	Tobacco	Seed development	Seed
15. Legumin	Pea	*N. plumbaginifolia*	Seed development	Seed
16. EPSP synthase	Petunia	Tobacco	Herbicide	Flower pollen
17. Ribonuclease	1. *Aspergillus oryzae*	Tobacco	Anther development	Anthers
	2. *Bacillus amyloliquefaciens*	Tobacco	Anther development	Anthers
18. Ribonuclease inhibitor	*Brassica amyloliquefaciens*	Tobacco	Anther development	Phloem cells

Heat Shock Genes: Thermal stress induces the transcription of a small set of heat shock genes in diverse organisms, and also reduces the expression of many other vital genes. The upstream region of heat shock genes regulating their expression is highly conserved, so that when a gene construct carrying NPTII reporter gene, fused with upstream region of heat shock gene *hsp70* from *Drosophila,* was introduced into tobacco, induction of NPTII expression due to heat shock was comparable to that of endogenous plant heat shock genes (inducible in all organs except pollen). Similarly, in maize a homologue of above gene, *hsp70* with 1.1 kilo bases of upstream sequence, when introduced in transgenic petunia, exhibited heat inducibility.

Alcohol Dehydrogenase Gene (ADH) : Alcohol dehydrogenase gene expresses under anaerobic conditions involving oxygen deprivation. Using transgenic plants, it could be shown that regulatory sequences upstream to *ADH* gene from maize lead to suppression (down regulation) of gene activity under aerobic conditions, this activity being normal and uninducible under anaerobic conditions. The inability of the *ADH* upstream regulatory sequences to induce gene expression became evident when these sequences failed to induce chloramphenicol acetytransferase (CAT) expression under aerobic as well as anaerobic conditions in transgenic tobacco plants. Other enhancer sequences (from octopine synthase or CaMV35S gene), when transferred at a position upstream of *ADH* gene, led to induction under anaerobic conditions. Therefore, under normal conditions, *ADH* activity is inhibited under aerobic conditions due to negative control and expression is promoted under anaerobic conditions simply due to the absence of a negative control rather than the presence of positive induction.

Genes for Seed Storage Proteins in Legumes and Cereals: The genes for seed storage proteins provide an excellent example of a cell and tissue specific expression. Using transgenic plants, sequences could be identified which were responsible for tissue or cell specific expression of these genes. For instance, when wheat glutenin gene or barley hordein gene (with its regulatory sequence) was fused to chloramphenicol acetyltransferase (CAT) gene and introduced into tobacco, CAT activity was found only in the endosperm. Another reporter gene utilized for the same purpose is β-glucuronidase (GUS) gene, which also showed endosperm specific expression when fused with a wheat glutenin gene (with its regulatory sequence). The regulatory sequence thus identified was called endosperm box of the storage protein genes.

EPSP Synthase Gene : 5-enopyruvyl shikimate-3-phosphate synthase (EPSPS) is an enzyme involved in aromatic amino acid biosynthetic pathway and is the target of lyphosate based herbicides. In petunia, enzyme is synthesized in flowers at a high level and is hardly found in leaves, although in tomato it is synthesized both in flowers and leaves. Transgenic tobacco plants carrying CAT gene or GUS gene fused with deletion mutant forms of EPSPS gene were analysed and it was concluded that sequences between -1800 to -800 of petunia EPSPS gene were responsible for tissue specific expression of this gene.

Modification of DNA Sequences/transcripts and Histone Proteins : Regulation of gene expression can also be achieved through methylation and demethylation of coding DNA sequences or their transcripts. Histone proteins involved in the organization of nucleosomes can also be modified (acetylated) to allow access of transcription machinery to the DNA sequences, thus regulating gene expression.

Methylation of DNA/RNA Sequences: During 1980s, a definite role of methylation and demethylation in gene control was also established. It was shown in many genes, that methylation of specific sites in the vicinity of a gene may lead to suppression of gene activity and demethylation may lead to expression of gene. With the help of restriction enzymes like HpaII and MspI (they are called isoschizomers *i.e.* they recognize same restriction sequence, but one of them does not cleave it when methylated), degree of methylation at CpG (cytosine - guanine) rich sites could be studied to show that the sites are undermethylated in tissues, where the gene is active. Evidence of suppression of gene activity due to methylation is available from several experiments, including the following:

(i) Herpes *thymidine kinase (tk)* gene was first methylated *in vitro* and then used for transformation of mouse *tk*$^-$ cells. It was found that efficiency of transformation was low and the few transformed *tk*$^+$ cells available had partially demethylated DNA in comparison to the methylated *tk* gene used for transformation. Similar results for another gene *apart+* in mouse were also available.

(ii) In mouse retroviruses, using transfection experiments, it was shown that the inactive provirus sequences were heavily methylated, while active provirus sequences were undermethylated. It was also shown that non-methylated proviral sequences were infectious, and methylated sequences were not.

(iii) In cultured mouse embryo cells, differentiation could be artificially induced by inhibiting methylation using a base analogue 5 azacytidine (5 azaC), which is specific in inhibiting DNA methylation *in vivo.* It was also shown in these differentiated embryo cells that the DNA actually became demethylated. Effect of 5azaC was also shown in avian retrovirus, where it induced gene expression associated with demethylation at an endogenous level. In human inactivated X-chromosome in mouse-human cell hybrids also, 5azaC enhanced reactivation, presumably due to demethylation.

The above evidences suggest and support the model that

absence of methylation is necessary for gene expression in many cases. It is believed that methylation or demethylation takes place at specific critical sites, which represent a small fraction of sites available in DNA. This results in alteration of a site in such a manner that a particular regulating protein may interact with the site in a very specific manner, thus regulating gene activity.

Acetylation of Nucleosome Histones and Regulation of Gene Expression: It has been shown that acetylation of histones can destabilise nucleosomes and allow access of transcription factors to recognition elements, relieving the gene sequences from repression due to histones. Conversely, deacetylation of histones stabilizes the repressed state. Transcriptional co-activators are now known, which have histone acetyl transferase activity. CBP/p300 is one such co-activator having acetyl transferase activity, integrating transcriptional-activation signals -from diverse regulatory proteins including steroid and nuclear-hormone receptors. On addition of a ligand (like oestrogen and thyroid hormone), the steroid and nuclear hormone receptors stimulate CBP/p300 mediated histone acetylation, thus rendering the chromatin much more accessible to transcriptional machinery.

Alternative Splicing of Transcripts : Another regulatory mechanism in eukaryotes involves alternative splicing. The expression of a large number of genes has been shown to be regulated in time and space by this mechanism. In this mechanism, the splice sites selected by splicing machinery depend on the conditions demanding differential expression of genes.

Regulation at Translation Level

Information about translational control in eukaryotes is coming from experiments designed to answer following questions regarding the mechanism of operation of translational machinery.

(i) When there are more than one AUG codons in a mRNA molecule, how does the ribosome know, which one to use for initiation of translation?

(ii) What determines the frequency with which a particular mRNA message is read?

(iii) After finishing translation of an open reading frame (ORF), can a ribosome start again, if an AUG codon follows in the same mRNA within a reasonable length?

(iv) Can a mRNA remain untranslated in a cell and translated only when necessary?

It has been possible to determine the nucleotide sequence of natural mRNA, to alter these mRNAs as desired and to make synthetic mRNA using recombinant DNA methods. These techniques helped in answering some of the above questions and also yielded information regarding regulation of gene expression at the level of translation. While describing Kozak's scanning hypothesis for reading of mRNA by ribosomes in eukaryotes.

Activation and Repression of Translation using mRNA binding Proteins : We described mechanisms of repression as well as activation of translation in prokaryotes. Several proteins involved in repression and activation of translation are thus known in prokaryotes. However, in eukaryotes, no activator protein is known, and there is only one documented example of a translational repressor protein (iron-responsive element-binding protein = IRE-BP). On the other hand, regulation of translation in eukaryotes is also brought about by modification of general components (*e.g.* eIF2) of the translational machinery, a phenomenon unknown in prokaryotes. Regulation is also exercised during elongation of polypeptide chain, although its mechanism is not well understood.

Activator Proteins: In eukaryotes, till recently, no unambiguous examples were available, where a m-RNA binding protein directly activates translation. However, one possible mechanism that may be involved in the action of an activation protein is shown in Figure. In some cases, the non-coding leader sequence at the 5' end forms a secondary structure, so that the 5' end is not available for binding of 40S subunit of ribosome.

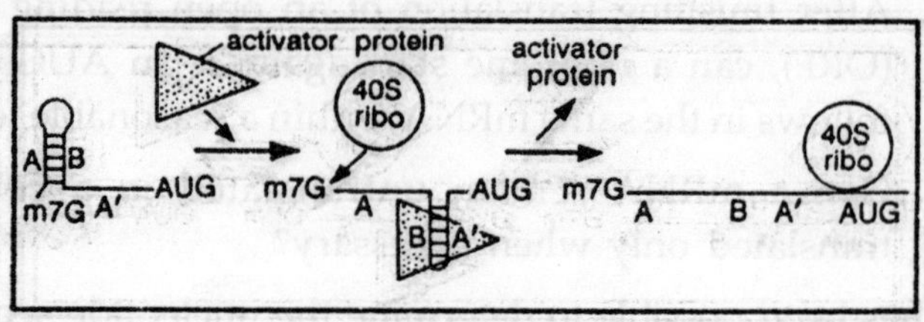

Hypothetical mechanism for activation of translation in mammalian systems. (a) In the absence of activator, 5' end acquires a secondary structure, which does not allow binding of 40S; (b) binding of activator (triangle) opens the 5' end, by making an alternative stem-loop structure; (c) the activator is displaced by processivity of scanning 40S subunit of ribosome.

The activator protein binds to mRNA and leads to the formation of a hairpin structure, which is accompanied with the exposure of 5' end of mRNA for ribosome binding. The activator protein is later displaced by the scanning 40S subunit of ribosome, till it comes across the initiating codon AUG.

For transcription, many gene specific initiation factors (transcription factors) are known. For translation also, general initiation factors are known both in prokaryotes and eukaryotes, but no gene specific or m-RNA specific translation initiation factors have yet been identified in eukaryotes (although for translation of bacterial mRNA and mt mRNA, such factors are known).

Repressor Protein, IRE-BP

(a) *Down-regulation of Ferritin Synthesis:* The synthesis of ferritin (an iron-sequestering protein) in mammalian cells depends on iron availability. Its regulation is achieved through inhibition of the translation of ferritin mRNA, due to binding of a repressor protein at the 5' end of this mRNA. This 5' end contains the iron-response element (IRE), which makes a stem loop structure. The repressor protein is similarly called iron-response element binding protein (IRE-BP). IRE-BP resembles mitochondrial aconitase and has aconitase activity, enabling it to sense iron levels. When iron is scarce, the reduced form (3Fe-4S) of the protein binds the IRE and when iron is abundant, the oxidized form (4Fe-4S) dissociates from IRE. The stem-loop structure of IRE, located +10 to +30 nucleotides from 5' end of mRNA, allows binding of IRE-BP, inhibiting the initial binding of 40S subunit of ribosome. This reduces or stops the synthesis of ferritin from its mRNA (Fig.).

(b) *Up-regulation of Transferrin Receptor (TfR) Synthesis:* The IRE-BP, while represses the synthesis of ferritin, it up-regulates the synthesis of transferrin receptor (TfR). The 3' end of TfR mRNA is prone to cleavage and IRE-BP binds there giving stability to TfR mRNA, so that it can be efficiently translated into TfR (Fig.).

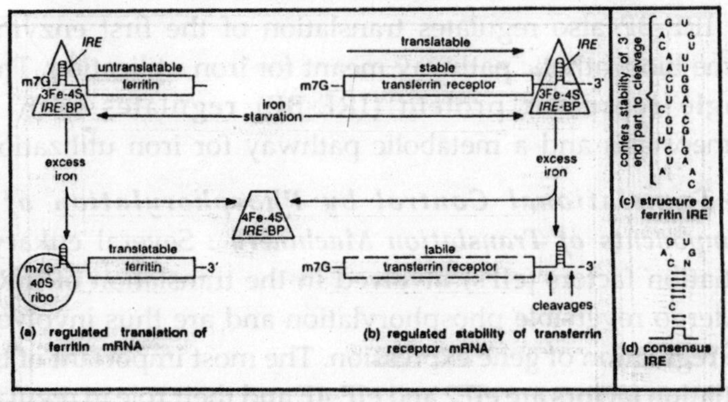

Post-transcriptional regulation of two genes involved in the uptake and detoxification of iron in mammalian cells. (a) translation of mRNA for ferritin, a protein that sequesters excess iron, is inhibited by the binding of IRE-BP to mRNA at its 5' end; (b) translation of transferrin receptor (TfR) involved in the binding of IRE-BP to TfR mRNA at its 3' end; (c) actual sequence of one IRE (TfR mRNA has five such IRE motifs at 3' end; Only one is shown here); (d) consensus IRE structure.

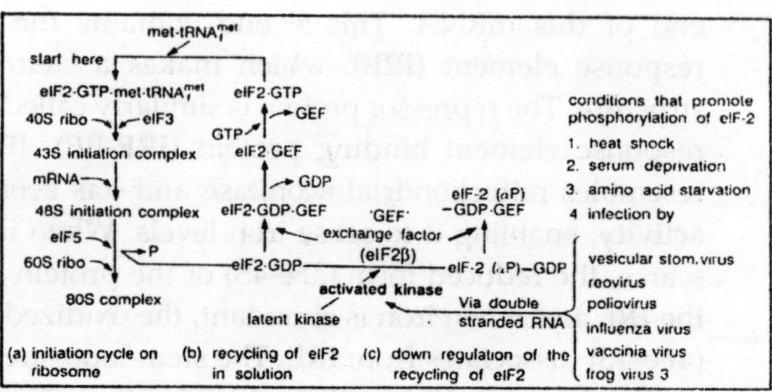

The eIF-2 cycle in mamalian cells. The eIF-GDP complex released from the 40S ribosomal subunit in step (a) is stable. The exchange of GDP for GTP in step (b) is mediated by a protein, designated GEF which is usually limiting in cells. Because GEF has a much higher affinity for eIF-2 (aP) than for unmodified eIF-2, phosphorylation of eIF-2 sequesters GEF in a stable complex (step c), thus causing the global translational capacity to decline. The modification of eIF-2 and consequent inhibition of translation are reversed by phosphatases, which are not depicted. The enzymes that catalyse phosphorylation of eIF-2 include p68-kinase (DAI) and the heme-regulated kinase HRI. The increase in phosphorylation depicted in step (c) is itself subject to down-regulation, for example, by VA-RNA of adenovirus. The other viruses marked with an asterisk also appear to encode or induce components that circumvent the virus-induced phosphorylation of eIF-2.

IRE-BP also regulates translation of the first enzyme of home biosynthetic pathway meant for iron utilization. Thus a single repressor protein (IRE-BP) regulates both iron homeostasis and a metabolic pathway for iron utilization.

Translational Control by Phosphorylation of the Components of Translation Machinery : Several eukaryotic initiation factors (eIFs) involved in the translation of mRNA, undergo reversible phosphorylation and are thus involved in the regulation of gene expression. The most important of these initiation factors are eIF2 and eIF-4E and their role in regulation of gene expression by reversible phosphorylation has been elucidated.

Reversible Phosphorylation of eIF-2α. In mammalian systems, reversible phosphorylation of eIF-2α has been known for years, although its role in regulation of translation has been proved only recently. Such translational control by eIF2 α phosphorylation is not known in plants.eIF-2 forms a complex with GTP, which binds with met- $tRNA_i^{met}$ met and carries it onto the 40S ribosome subunit, forming a 43S complex. The 43S complex engages at the 5' end of mRNA and scans down to the AUG codon, where eIF-5 joins and causes hydrolysis of GTP. The resulting eIF-2-GDP has lower affinity for 40S and is therefore released, leaving 40S met- $tRNA_i^{met}$ behind. The 60S then joins giving 80S ribosome, ready to form the first peptide bond (Fig.).

Affinity of eIF-2 for GDP is 400-fold higher than for GTP and therefore regeneration of eIF 2 GTP from eIF-2-GDP requires a helper factor, designated eIF-2 β or GEF (guanine exchange factor). (In prokaryotes the function of GEF is performed by EF-Ts). However, the regeneration of eIF-2-GTP may be prevented, if a subunit of eIF-2 undergoes phosphorylation, which is promoted by a variety of conditions including heat shock, amino acid deprivation or virus infection.

GEF has a 150-fold higher affinity for eIF-2(α P) than for unmodified eIF-2, so that formation of stable eIF-2 (α P)GDP-GFF does not allow the formation of eIF-2-GTP, causing protein synthesis (translation) to stop (Fig.). An example of eIF-2 mediated regulation of the translation of mRNA for the yeast gene *GCN4* is discussed below.

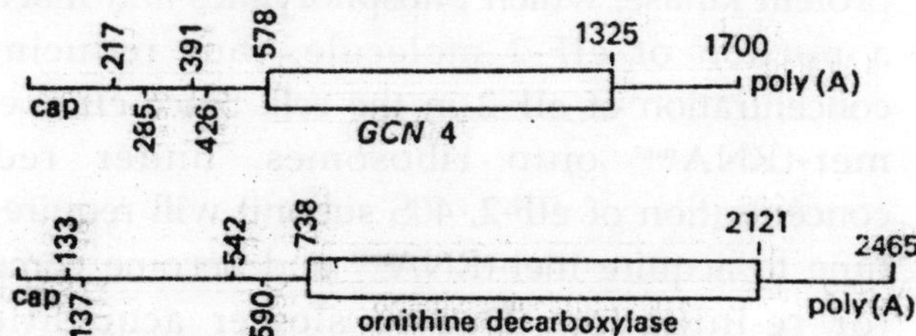

Structures of mRNAs corresponding to genes for GCN4 (general control of nitrogen metabolism) and ornithine decarboxylase in yeast, showing untranslated sequence at 5′ end having false AUG codons.

(a) *Translational Control of the Gene GCN4 in Yeast via eIF-2:* In yeast, expression of *GCN4* gene (involved in general central of nitrogen metabolism) is regulated primarily at translation level, since although under different conditions, the level of protein encoded by *GCN4* changes more than 100 fold, the level of mRNA for this gene does not change. *GCN4* gene protein (a transcription factor) regulates transcription of 30-40 genes involved in amino acid biosynthesis and its own synthesis is boosted 100 fold when yeast is starved of an essential amino acid. Its level of expression also depends on a repressor specified by *GCD1* locus, which in its turn is regulated by *GCN1, 2* and *3* loci.

The structure of mRNA of *GCN4* gene explains the mechanism of its translational regulation, on the basis of Kozak's scanning hypothesis. It has a 577 nucleotides long untranslated sequence at 5' end, which is much longer than normal and has four AUG codons upstream of the real AUG (Fig.). The presence of four AUG codons is utilized by the eukaryotic initiation factor, eIF2 in the regulation of translation of GCN4-mRNA. The initiation factor, eIF-2 forms a complex eIF-2-GTP-met-tRNA, whose concentration is regulated by amino acid concentration. Under amino acid starvation, uncharged $tRNA_i^{met}$ accumulates and activates a protein kinase, which phosphorylates and inactivates a portion of eIF-2 molecule, thus reducing the concentration of eIF-2 in the cell. Since eIF-2 escorts met-$tRNA^{imet}$ onto ribosomes, under reduced concentration of eIF-2, 40S subunit will require more time to acquire met-$tRNA^{imet}$ and become competent for re-initiation. Due to slower acquisition of competence, some 40S subunits bypass one or more AUG codons and gain access to the GCN4 start site.

On the other hand, under condition of no starvation (amino acid), eIF-2 concentration is adequate so that the 40S subunits become competent quickly and can not bypass the fourth AUG codon. Consequently, a small peptide is synthesized and released at the termination site of ORF4 (open reading frame-4). The termination at ORF4 is somehow incompatible with re-initiation at GCN4, so that under condition of no starvation, GCN4 is not expressed. Translation of GCN4-mRNA is also inhibited, if intercistronic distance between ORF1 and ORF4 is increased by inserting an unknown sequence. In this case, a longer distance allows all 40S subunits become competent before reaching ORF4, even under starvation condition. This will not leave any 40S subunit to bypass ORF4 and reinitiate translation at GCN4. (Fig.).

(b) *Translational Control of the Synthesis of Ornithine Decarboxylase:* An example analogous to *GCN4* gene in yeast, is the mRNA corresponding to ornithine decarboxylase enzyme involved in polyamine biosynthesis. It contains a 737 nucleotides long 5' untranslated region containing four AUG codons (Fig.) followed by termination codons 16, 2, 4 and 10 nucleotides later respectively. The enzyme is extremely rapidly produced in cells stimulated by hormones and growth factors, but we do not know if translational efficiency changes with the physiological state of the cell. If so, the long leader sequence and the false AUG codons may play a part.

Reversible Phosphorylation of eIF-4E and rpS6 (Ribosomal Protein S6): Besides eIF-2, eIF-4E and rpS6 have also been observed to undergo striking changes in the level of phosphorylation under specific conditions like serum deprivation in culture medium (eIF-4E) and heat shock. Under heat shock, phosphorylation of eIF-2 increases and that of rpS6 decreases. Also, while increased phosphorylation of eIF-2 leads to reduced translation, increased phosphorylation of rpS6 apparently leads to enhanced translation.

In case of eIF-4E, the principal site of phosphorylation is `serine-53' and a mutation on this site renders the factor non-functional in translation assays. This suggests that eIF-4E may undergo reversible phosphorylation each time it is used. In activated B lymphocytes also, increased phosphorylation was associated with increased translation rate.

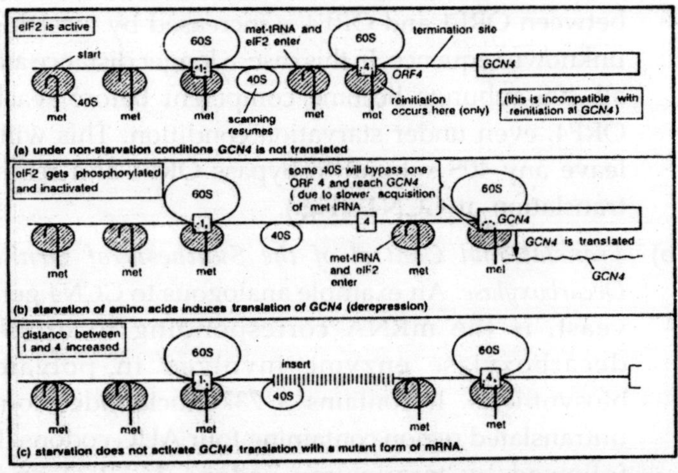

Translation control of GCN4 gene in yeast through re-initiation mechanism; only two of the four AUG codons (shown as 1 and 4) are depicted.

One of the cell division cycle *(cdc)* genes in yeast also encodes eIF-4E, which may regulate cell division by reversible phosphorylation. Phosphorylation of eIF-2, eIF-4E and rpS6 is also associated with phosphorylation of other initiation factors, elongation factors and aminoacyl tRNA synthetases. However, unequivocal evidence of the regulation of translation by reversible phosphorylation is available only in case of eIF-2.

Reversible Phosphorylation and Targeting Specific mRNA: The regulation at the translation level requires that the phosphorylated or dephosphorylated forms of the components of translational machinery, should identify a specific mRNA from the bulk mRNA population. This may be achieved in one of the following ways :

(i) a mechanism like that for *GCN4* described above (several AUG codons upstream of real start codon);

(ii) activation of kinases or phosphatases in a localized region of cytosol, where a specific mRNA is localized (compartmentalisation of the cytosol);

(iii) variation in secondary and tertiary structures among the non-coding leader sequences at the 5' ends of mRNA molecules, which modulate ribosome-mRNA interaction.

The best example of differential translation of mRNAs is the virus infected cells, where viral genes are preferentially expressed. This may be achieved by one of the following mechanisms :

(i) Cellular mRNAs are degraded or host's transcription is shut off.

(ii) Viral mRNAs are more abundant (*e.g.* reovirus).

(iii) Viral mRNAs are better designed for initiating translation, due to unstructured 5' leader sequence. (*e.g.* adenovirus).

(iv) Overall translational capacity of the host is reduced due to partial phosphorylation of eIF-2α, induced by double stranded RNA generated during the course of virus transcription and/or translation, (*e.g.* adenovirus, poliovirus, reovirus, etc.). In many viruses, factors are also generated which inhibit completion of phosphorylation of eIF-2α. An adenovirus encoded transcript, designated VA-RNA, actually binds to and blocks the activation of p68 kinase thus making unaltered eIF-2α available for efficient translation.

Masked mRNA in Eggs of Sea Urchin and Xenopus : There are also several examples of 'masked mRNA' in eggs and oocytes, which may indicate translational control. This *'masked mRNA'* is found in abundant quantity and remains untranslated until after fertilization. There are atleast two examples of this *'masked mRNA'* :

(i) the mRNA coding for small subunit of ribonucleotide reductase in clam and sea urcin eggs and

(ii) the mRNA for fibronectin in *Xenopus* unfertilized eggs, remain untranslated until about 6-7 hours after fertilization (midblastula stage). The mechanism of this unmasking is not clearly understood.

Regulation by Gene Rearrangement

Expression of Immunoglobulin Genes : Antibody molecules, called immunoglobulins, are coded by genes, which undergo DNA rearrangements during cell differentiation, thus suggesting that although a genome usually remains constant in different cells, it does undergo a change with regard to immunoglobulin genes in mammals. Immunoglobulins are produced by cells originating in bone marrow (B-cells), and later differentiating into lymphocytes. These may differentiate into several clones, producing different antibodies, but each clone producing only one antibody. For this work on the rearrangement of DNA sequences that results in a large variety of sequences for production of antibodies, S. Tonegawa of MIT (USA) was awarded the 1987 Nobel Prize for Physiology/ Medicine. This work will be briefly described in this section.

Structure of Immunoglobulin Chains: Antibody molecules (immunoglobulins, abbreviated as I_g) are tetrameric, each containing two identical 'heavy' (H) chains of about 500 amino acid residues, and two identical light (L) chains of about 250 amino acid residues (Fig.). In man, mouse, rabbit and other mammals there are seven kinds of heavy chains (alpha = α ; gamma = $\gamma_1, \gamma_{2a}, \gamma_{2b}, \gamma_3$; delta = δ ; mu = μ) and two kinds of light chains (kappa = κ and, lambda = λ). I_g molecules are named according to the type of H chain, which they contain (*e.g.* I_gG has gamma chains, I_gM has mu chains and I_gA has alpha chains).

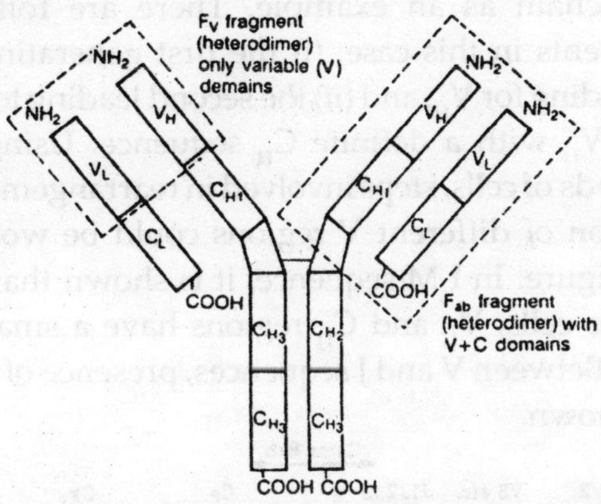

Heavy and light chains and their association to generate different V (variable) domains and C (constant) domains in immunoglobulin protein.

Each I_g chain consists of two distinct parts (i) C-terminal constant part or C-region and (ii) N-terminal extremely variable part or V-region. For each C-region (C_α, C_μ, C_λ, etc.), there can be several V sequences. Separate sets of V sequences are known for heavy chains (V_H) and light chains (V_κ = for kappa and V_λ for lambda). The $V_{H'}$ V_κ and V_λ sequences have fair degree of homology and possess only some variable regions.

I_g Gene in Germline and Differentiated Cells: By using cDNA as probe, it could be shown that C-and V-coding sequences were closer to each other in the antibody producing cells, than in the germ line. This was predicted because, same C-region was found combined with different V-regions in the same individual. Later experiments showed that C—C splicing at the DNA level is responsible for generating the coding genes for both light and heavy chains. Although, the genes for immunoglobulin molecule may be found on more than one chromosome, but cutting and rejoining of DNA takes place only within a chromosome and not across chromosomes. In addition to C and V regions, both light and heavy chains have another region called J (junction of joining sequences). In heavy chain, in addition to C, J and V regions, another short sequence D (diversity) is also found. We will discuss DNA splicing only for heavy chain as an example. There are following two rearrangements in this case; (i) the first generating the DNA sequence coding for V_H, and (ii) the second leading to association of specific V_H with a definite C_H sequence. Using probes in different kinds of cells, steps involved in rearrangements leading to association of different V regions could be worked out as shown in Figure. In I_gM sequence, it is shown that in embryo or germ line cells, V_H and C_H regions have a small region of J sequence. Between V and J sequences, presence of D sequence was also shown.

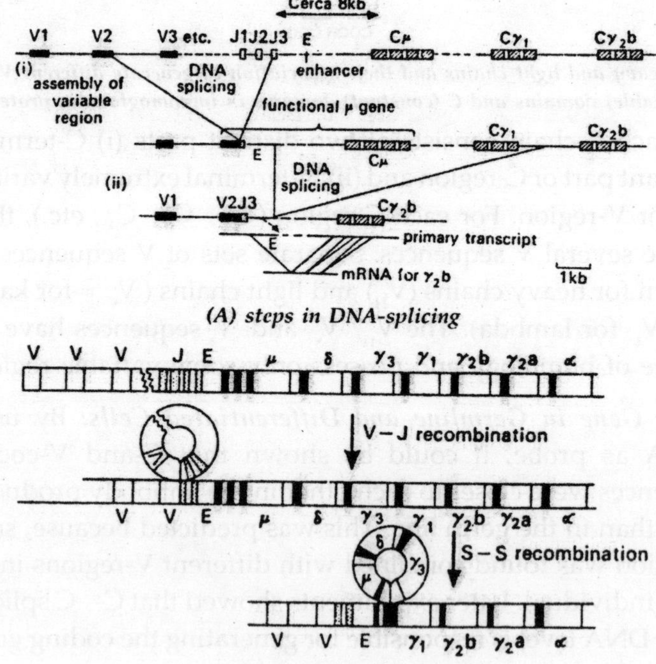

(A) steps in DNA-splicing

(B) model of heavy-chain switching

(A) structure of immunoglobulin gene and the steps involved in DNA rearrangements in different regions.

(B) a model explaining DNA rearrangements in heavy chain switching.

V-regions through DNA Splicing and Joining: Each heavy chain V_H region is encoded by a DNA sequence assembled from three gene segments a V segment, a J segment and D segment. The antibody diversity in the *Vsub* region is based on (i) inherent diversity in the sequences representing V, J and D segments and on (ii) the combinatorial joining of these segments (called combinatorial diversity). Any of the 500 V segments in the mouse heavy-chain pool can be joined to any of the 4 J segments and to any of at least 12 D segments, to encode at least 24,000 (500 x 4 x 12) different, heavy chain V_H regions. Similarly in the light chain at least 1000 V_L regions can be obtained, so that there can be as many as 24,000 x 1000=25 x 10^6 divergent antibodies. The number of possibilities are further increased by the joining mechanism itself (there may be loss or gain of nucleotides during joining). This random loss or gain at joining sites is described as junctional diversity. Two genes (*rag-1* and *rag-2; rag* = recombination activating genes) are involved in the U(D)J recombination system to produce heavy chain V_H regions.

V-C Association through DNA Splicing: Second phase of DNA splicing involves association of a V segment with C segment. It has been shown that the same V segment can attach with different C-regions, thus providing for H-chain class switching. This is achieved by successive deletions of DNA, leading to the production of IgM in primary response and I_gG and I A as secondary response. The V-D-J segment may first attach to C_μ (which is first exon in C-region) and give I_gM. Later the cell may switch and may produce another immunoglobulin. This class switch may require further deletions of C-region exons. The deletion is mediated by a repeated DNA segment located just before each C-exon and designated as S-region (switch). I M has all kinds of heavy chains; in I_gG3 segment C is missing and in I_gG_1, C_μ and $C_{4\gamma}^{3}$ are missing. Further deletions may occur in other cases as shown in Figure.

Enhancers in Regulation of Immunoglobulin Genes: Enhancers are certain DNA sequences, that increase the transcription of genes, although this is not the only characteristic of DNA segment to be called enhancer. Other essential characteristics of an enhancer are :

(i) it is essential for transcription;

(ii) it can be moved around on either side of gene or may be inverted and

(iii) it can work with genes from sources other than its own.

These enhancers have been discovered in immunoglobulins genes and have been found to work only in cells that make immunoglobulins. In this manner, enhancer is a remarkable regulatory sequence that is cell specific. The enhancer sequence is found (i) between V-D-J and C in DNA segment for heavy chain and (ii) between V-J and C in DNA segment for light chain (Fig.). The role of enhancer sequence is demonstrated by the following evidences:

(i) If intron region between V-D-J and C regions of heavy chain is removed from a cloned gene system, and then the latter is injected into mouse myeloma cells, the cloned genes lose their ability to make a heavy chain,

(ii) When cloned intron region was cleaved into three fragments, and each is separately linked with genes for T-antigen of SV40, or with (3 globin gene of mammals, only one of these fragments, presumably carrying the enhancer, stimulated the transcription.

(iii) When cloned intron region of heavy chain gene was attached to a bacterial gene, increased gene product was achieved.

To obtain correct initiation of transcription, the enhancer must interact with other control sequences which are located upstream (towards 5' end) from the beginning of V-region. Normally V-D-J (or V-J) joining the upstream sequence and enhancer are far apart, but rearrangements bring them close enough for transcription to occur. The enhancer functions even after every class switch and thus can continue to activate transcription.

Yeast Mating Type Switching : Life cycle of budding yeast, *S. cerevisiae* (Fig.) shows that there are two mating types (MATa and MATα), which fuse to give a diploid (MATa/MATα). The diploid cell sporulates and through meiosis forms four spores, two of each mating type. Each mating type is characterized by secretion of a pheromone (MATa secretes a factor and MATα secretes α factor) and a surface receptor for the pheromone of the opposite type. The diploids (MATa/MATα) produce no pheromone or surface receptor. The pheromone-receptor interaction activates a cell membrane protein called G protein consisting of three subunits ($\alpha\ \beta\gamma$) (see later in this Chapter for details on G proteins; Nobel Prize for discovery of G protein was awarded to M. Rodbell and A. G. Gilman in the year 1994). Activation needs a GTP molecule, which displaces GDP associated with $\alpha\ \beta\gamma$ complex and releases α from $\beta\gamma$ dimer. This separation of α from $\beta\gamma$

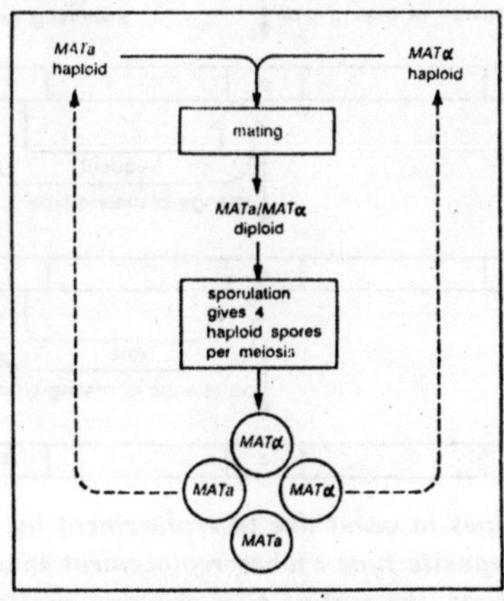

Life cycle of yeast involving mating of MATa and MATα haploids.

activates the next protein in signal transduction pathway. Therefore G protein is also called a transducin. The monomer α alone (and not $\beta\gamma$) is capable of bringing about the activation. The above information has come from a study of mutants (called STE = sterile), which are incapable of mating. At least ten such mutations (STE 1-7, 11, 12, 18) are known.

A remarkable feature in yeast is the ability of some strains (which carry dominant *HO* allele and not those which carry recessive *ho* allele) to switch their mating type from a to α and *vice-versa.* Consequently an *HO* haploid strain, irrespective of its original mating type, will soon have cells of both mating types giving diploids *(MATa/MAT*α*).* This is achieved by the presence of a *MAT (a or* α*)* locus with an active cassette (type *a* or α) associated with two silent cassettes at loci *HML(*α*)* on the left and *HMR(*a*)* on the right. Only the active cassette at *MAT* locus is expressed which can be *MATa* or *MAT*α. This can be replaced by *HML*α or *HMRa* respectively leading to change in mating type (a$\rightarrow\alpha$ *or* $\alpha\rightarrow$a). *MATa* is almost

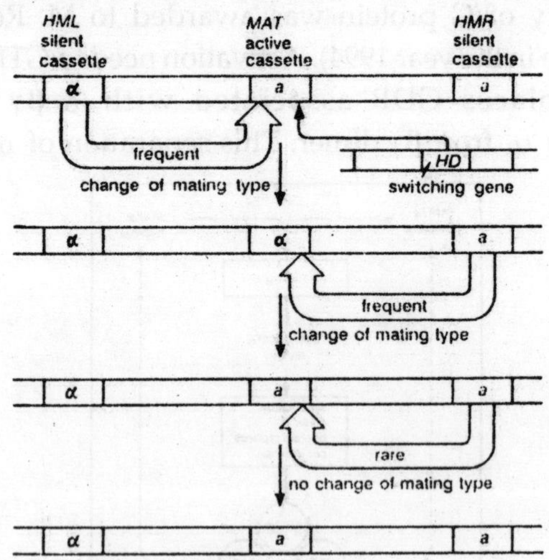

Change of mating types in yeast due to replacement by active cassette by silent cassette of opposite type : when replacement involves cassettes of same type, the mating type remains unaltered.

always replaced by, *HML* and *MAT*α by *HMR*, so that if *HML*(α) is mutated to *HMLa* or *HMR* is mutated to *HMR*α, replacement by them does not change mating type (Fig.). The *HML* and *HMR* are constitutively repressed (rendered silent) by some repressors binding to silencer sequences, *HML-E* or *EL* and *HMR-E* or *ER* (also called *E* and / silencers) located ~ 1 kb upstream of these loci. The switching is achieved by the activity of an endonuclease coded by the dominant gene *HO*. The gene *HO* is also repressed by several control systems, which need to be lifted before *HO* plays a role in switching.

Trypanosome Surface Antigen (VSG) Switching : Sleeping sickness in humans (and a related disease in cow) is caused by a unicellular parasite called trypanosome, which alternates between two hosts, tsetse fly and a mammal (Fig.). Trypanosome undergoes important biochemical changes involving diversity in the variable surface glycoprotein (VSG), a major component of surface coat, which provides antigenic reaction. Although, at any one time a trypanosome expresses only one VSG, its ability to change into any of the ~100 possible different VSGs is the secret of parasite's survival during fly-mammal cycle.

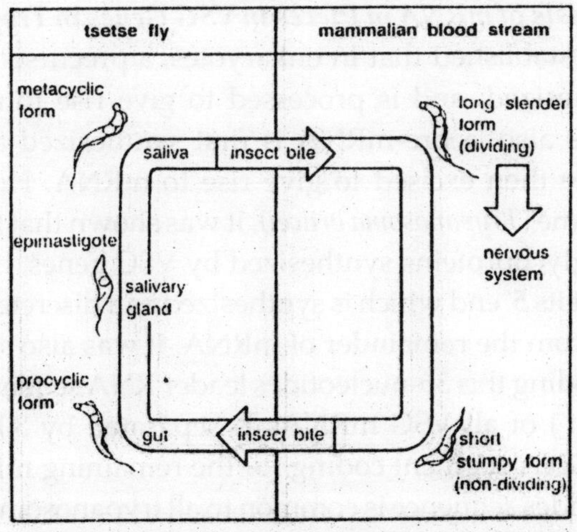

Life cycle of a trypanosome alternating between two hosts (tsetse fly and a mammal) causing sleeping sickness.

A tsetse fly, during a bite, gains the parasite, which loses its VSG but re-acquires a new VSG after, three weeks when, it differentiates in 'metacyclic form'. This form enters mammalian bloodstream during a bite, where VSG keeps on changing every 1-2 weeks, so that the immune response always lags behind the change in VSG and the parasite evades the immune surveillance, perpetuating indefinitely, till it enters the central nervous system and often causes death.

Genes (about 1000) for ~100 different VSGs are present scattered on different chromosomes in the genome of trypanosome. Diversity, therefore, depends on changing expression from one pre-existing gene to another. Each VSG is coded by a single basic copy gene, which may be telomeric or internal in location, and there may be several isogenes for same VSG or similar VSGs. The copy of the gene, which is active, is called expression linked copy (ELC) and is located on an expression site. Creation of an ELC may involve transfer of a basic copy gene to the expression site or *vice-versa.* Almost all switches in VSG type involve replacement of the ELC by a pre-existing silent copy.

Synthesis of mRNA in Pieces in VSG Genes in Trypanosome: It is well established that in eukaryotes, a precursor mRNA is first synthesized, and is processed to give rise to mRNA. In split genes also, a pre-mRNA is first synthesized and intron regions are then excised to give rise to mRNA. However, in trypanosome *(Trypanosoma brucei),* it was shown that the mRNA for some glycoproteins synthesized by VSG genes, has a small segment at its 5' end which is synthesized as a discrete transcript separate from the remainder of mRNA. It was also shown that DNA encoding this 35-nucleotides leader RNA sequence (called 'mini exon') of all VSG mRNAs is separated by 30 kilobases from the DNA segment coding for the remaining mRNA. This 35-nucleotides sequence is common to all trypanosome mRNAs. No precursor giant mRNA, which could give rise to this mRNA by post-transcriptional splicing, could be observed. It was therefore, concluded that 35-nucleotides *'mini exon'* RNA was synthesized as a part of a separate transcript and then either (i) joined to the remainder of the m-RNA through trans-splicing or (ii) acted as a primer for the transcription of the longer segment of mRNA.

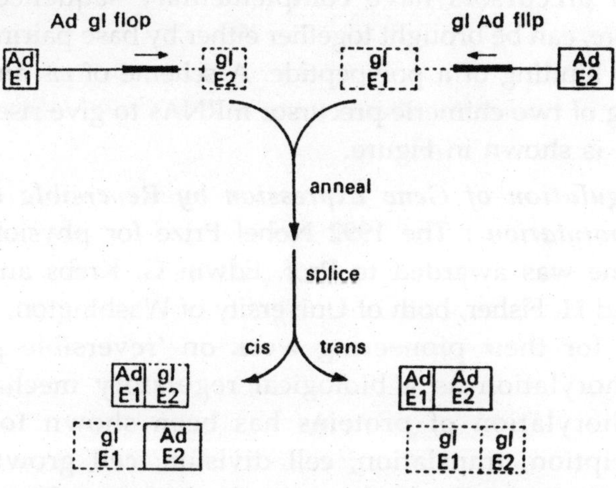

Production of mRNA through splicing and rejoining of segments from different precursor mRNA molecules.

It has also been shown that mRNA can be produced from two precursor molecules through trans-splicing followed by joining. This became possible through the availability of a technique for *in vitro* splicing systems. In a study from Massachusetts Institute of Technology, USA, two artificial mRNA precursors could be synthesized, one having L1 exon and the other having L2 exon. These two exons could be spliced together under optimum *in vitro* conditions to enable them to join together. In another study from Yale University, USA, chimeric precursors had either (i) the exon adE_1 and $g1E_2$ or (ii) the exons adE_2 and $g1E_1$. In each of these chimeric precursors, the two exons were connected by an intron and it was shown that splicing may be (i) *cis* leading to production of mRNA having exons from the same precursor or may be (ii) *trans* leading to the production of mRNA having exons from two precursors. The former is an intramolecular reaction and the latter is an intermolecular reaction and both were shown to proceed at the same rate.

In the above experiments, it was shown that the introns of two precursors have complementary sequences and, therefore, can be brought together either by base pairing or by mutual binding of a polypeptide. A scheme of *cis* and *trans* splicing of two chimeric precursor mRNAs to give rise to one mRNA is shown in Figure.

Regulation of Gene Expression by Reversible Protein Phosphorylation : The 1992 Nobel Prize for physiology or medicine was awarded to Prof. Edwin G. Krebs and Prof. Edmond H. Fisher, both of University of Washington, Seattle, U.S.A. for their pioneering work on 'reversible protein phosphorylation as a biological regulatory mechanism'. Phosphorylation of proteins has been shown to affect transcription, translation, cell division, cell growth, cell differentiation, cell transformation and many other cellular processes. Regulation of enzyme activity by reversible phosphorylation has several advantages over *de novo* synthesis of enzyme. The response is fast and the signal gets amplified. One molecule of kinase can phosphorylate several molecules of an enzyme. Activity of several non-enzymatic proteins like transcription factors is also regulated by reversible phosphorylation. It is estimated that a eukaryotic cell has about 1000 genes for different protein kinases (for phosphorylation) and perhaps same number of genes for protein phosphatases (for dephosphorylation).

The first protein (enzyme in this case), whose activity was shown to be regulated by phosphorylation was glycogen phosphorylase (involved in breakdown of glycogen in muscle and liver). This enzyme exists in two forms, phosphorylase a (not dependent on AMP for its activity) and phosphorylase b (dependent on AMP for its activity). Reversible phosphorylation leading to interconversion of these two forms has been shown.

Almost all of the enzyme is in an inactive 'b' form in resting muscle, which gets converted to the active 'a' form, when the level of hormone, epinephrine, increases in the blood (Fig.). Phosphorylation of protein occurs on many residues, but mainly serine (Ser; 90-95%), threonine (Thr; 5-10%) and tyrosine (Tyr; <10%) are phosphorylated. Phosphorylation does not always lead to increase in enzyme activity, and in some cases may actually lead to decrease in activity (*e.g.* glycogen synthase). Sometimes positive and negative regulatory phosphorylation sites are present in the same enzyme, so that same enzyme may need two kinases and two phosphatases.

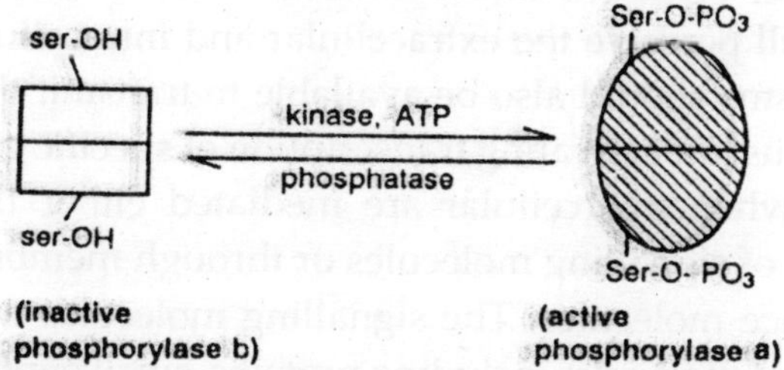

Reversible phosphorylation of phosphorylase enzyme at serine residue causing conformational changes in the enzyme.

WD-repeat Proteins and their Regulatory Role : WD-repeat proteins are made up of conserved repeating units, each unit ending with Trp-Asp (WD). They are found in all eukaryotes, but not in prokaryotes. During early 1990s, this conserved repeat unit was found for the first time in beta (β) subunit of , G-proteins, which transduce signals across the plasma membrane. Later, these WD repeats were found in more than two dozen proteins. Each repeat has a region of variable length, followed by a core of more or less constant length having GH (Gly-His) on one end, and WD (Trp-Asp) on the other end, so that the repeats can be represented as- $\{X_{6\text{-}94}\text{– (GH – }X_{23\text{–}41}\text{ – WD)}\}_n$.

All WD-repeat proteins are regulatory in nature and none is an enzyme. They regulate cell functions, such as signal transduction (*e.g.* G_β, PR55, RACK1), RNA processing (*e.g.* CstF, Makll, PRP4), gene regulation (*e.g.* COP1, $TAF_{II}80$, TUP1), vesicular traffic (*e.g.* β'COP, SEC 13) and cytoskeletal assembly/cell cycle (*e.g.* CDC4, CDC20, βTrcp). Several of these WD proteins form multiprotein complexes, sometimes interacting with other proteins through the WD-repeat region.

Cell Receptors, Cell Signalling and Regulation of Gene Expression : Since the regulation of gene expression should meet the requirements of a cell during different developmental stages and during adaptation to different environmental conditions, there should be mechanisms at the cellular level, which will perceive the extracellular and intracellular signals. Mechanisms should also be available to transmit the signal to the nucleus, for activating transcription of specific genes. These signals, when intercellular are mediated either through the secretion of signalling molecules or through membrane bound cell surface molecules. The signalling molecules may include a variety of molecules including proteins, small peptides, amino acids, nucleotides, steroids, retinoids, fatty acid derivatives and even dissolved gases like nitric oxide and carbon monoxide. In each case, the signalling involves a ligand (the signalling molecule) and a receptor. The receptor can be an *intracellular receptor* or a *cell surface receptor* (Fig.).

The signalling molecules that are secreted by cells can mediate signals in one of the following four ways (Fig.) :

(i) *paracrine signalling* involves signalling molecules that are secreted to influence only cells in the vicinity;

(ii) *synoptic signalling* involves the use of neurotransmitters, which transmit signals through specialized junctions called synapses found between nerve cells;

(iii) *endocrine signalling* involves the use of signalling molecules called hormones;

(iv) *autocrine signalling* involves the secretion of molecules that bind receptors on the same cells or cells of the same type (as in contraction of smooth muscles or aggregation of platelets).

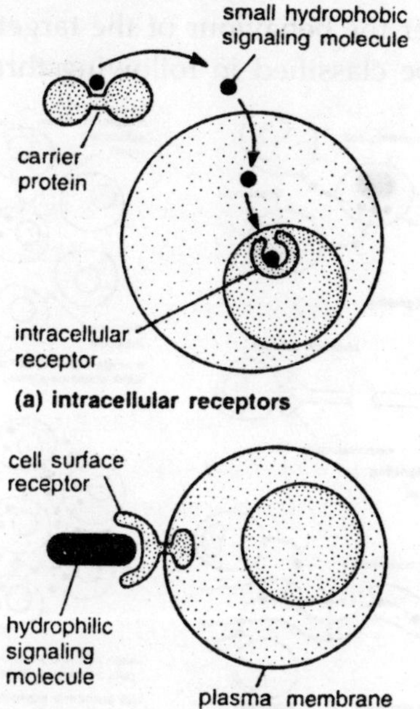

(b) cell-surface receptors

Intracellular receptors and cell surface receptors, each providing site for the binding of extracellular signalling molecule (ligand), which may be hydrophobic (can cross cell membrane) for intracellular receptors and hydrophilic (unable to cross cell membrane) for cell surface receptors (redrawn from Alberts et al., 'Molecular Biology of the Cell', 1994).

Intracellular Receptors : Steroid hormones, thyroid hormones, retinoids and vitamin D, when act as signalling molecules (ligands), diffuse across the plasma membrane of target cells and bind receptors available within these target cells. These ligands activate the intracellular receptors, which then regulate transcription of specific genes. These receptors are collectively called intracellular receptor superfamily (or *steroid-hormone receptor superfamily*, Fig.).

Cell Surface Receptors : The cell surface receptor proteins act as signal transducers. The ligand binds to a cell surface receptor with high affinity and activates the receptor, which converts the extracellular signal into one or more intracellular signals that alter the behaviour of the target cell. Cell surface receptors can be classified in following three groups (Fig.);

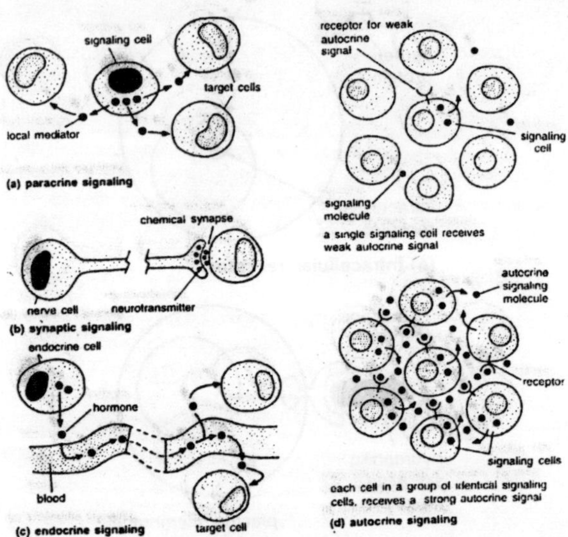

Four types of signalling mediated by secreted molecules, and differing in speed and selectivity with which the signals are delivered to their targets (redrawn from Alberts. 'Molecular Biology of the Cell', 1994).

(i) *Ion channel*: linked receptors (also called *transmitter-gated ion channels)*. These are involved in rapid signalling at synapses,

(ii) *G-protein linked Receptors:* These are involved in activating another membrane bound target protein through a third protein called G protein (a trimeric GTP-binding regulatory protein). This target protein activated through G-protein, either alters the concentration of one or more intracellular mediators or modifies the ion permeability of the cell, leading to an alteration in the behaviour of the cell. All G-protein-linked receptors belong to a large superfamily of *seven-pass transmembrane proteins,*

(iii) *Enzyme-linked Receptors:* These receptors, when activated, function directly as enzymes or are associated with enzymes (mostly kinases), thus bringing about phosphorylation of specific sets of proteins in the target cell.

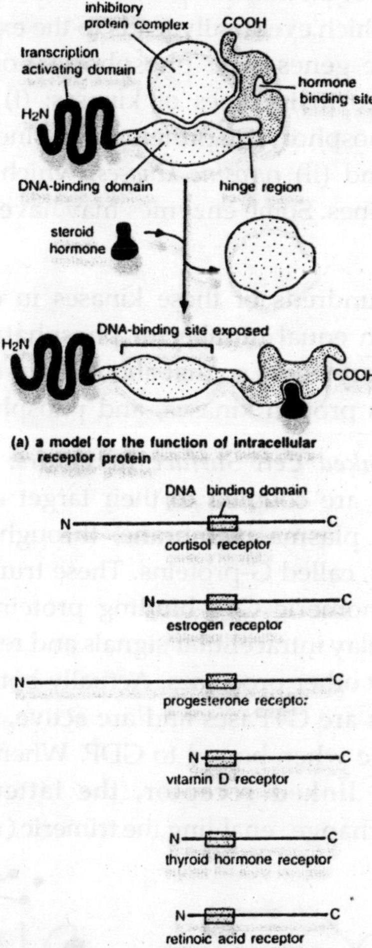

(b) six receptors, each having a DNA binding domain

The intracellular receptor superfamily; (a) a model for the mechanism of action of an intracellular protein; (b) a variety of structures known in six different intracellular receptors (DNA binding domain is shown by a box in each case) (redrawn from Alberts et al., 'Molecular Biology of the Cell' 1994).

The ion-channel-linked receptors mentioned above are dealt and therefore will not be discussed in this chapter. The other two types of receptors *i.e.* G-protein and enzyme linked receptors initiate a cascade of events. Each such event involves phosphorylation of an inactive protein, thus altering it into an active protein, which eventually leads to the expression of one or more specific genes. The phosphorylation cascades arc facilitated by two main types of kinases: (i) *serinelthreonine kinases*, which phosphorylate proteins on serines and less often on threonines and (ii) *tyrosine kinases*, which phosphorylate proteins at tyrosines. Some enzymes may have both the above activities.

There are hundreds of these kinases in each cell of an organism and an equal number of phosphatases, that bring about dephosphorylation, so that about 1% of each eukaryotic genome encodes protein kinases, and phosphatases.

G-protein-linked Cell Surface Receptors: The G-protein-linked receptors are coupled to their target enzymes or ion channels in the plasma membrane, through trimeric GTP-binding proteins, called G-proteins. These trimeric G-proteins differ from monomeric GTP-binding proteins (often called GTPases), that relay intracellular signals and regulate vesicular traffic and many other processes. Actually both of these GTP-binding proteins are GTPases and are active, when bound to GTP and inactive when bound to GDP. When a ligand binds to a G-protein linked receptor, the latter undergoes a conformational change, enabling the trimeric ($\alpha\beta\gamma$) G-protein

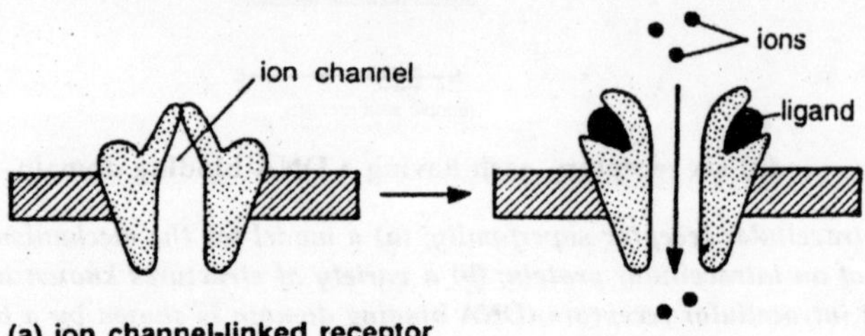

(a) ion channel-linked receptor

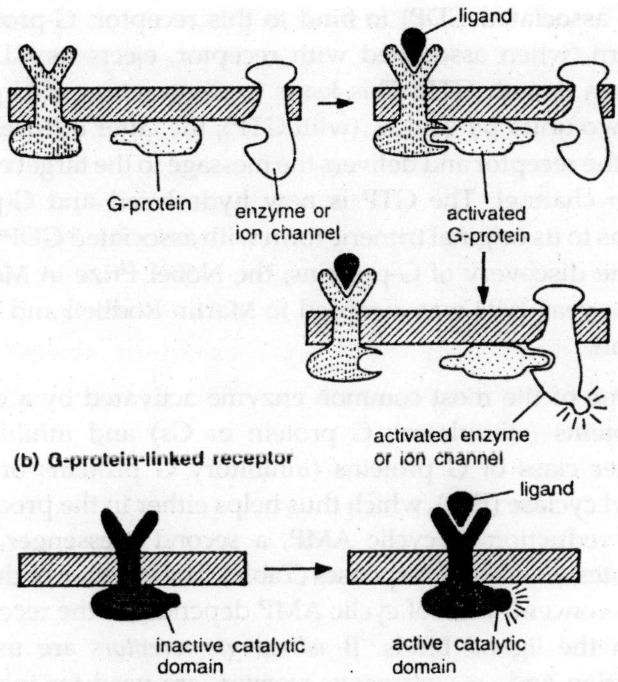

Three classes of cell surface receptors (redrawn from Alberts. 'Molecular Biology of the Cell', 1994).

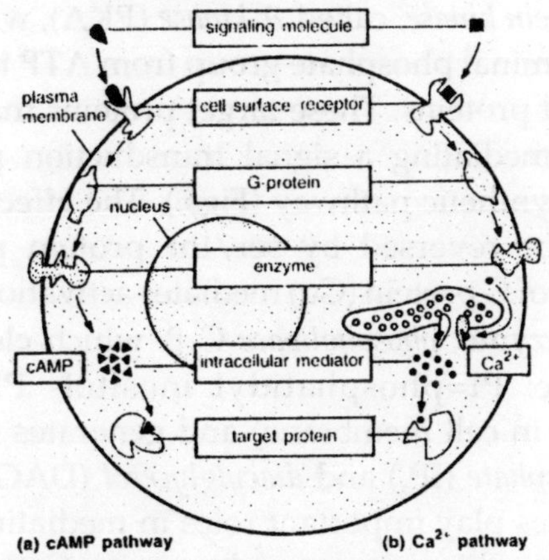

Two major pathways by which G-protein linked cell surface receptors initiate cell signalling pathways (redrawn from Alberts. 'Molecular Biology of the Cell', 1994).

(with associated GDP) to bind to this receptor. G-protein, in its turn, when associated with receptor, ejects its GDP and replaces it with GTP. This leads to dissociation of G-protein into two units βγ and α (with GTP); the latter diffuses away from the receptor and delivers the message to the target enzyme or ion channel. The GTP is now hydrolysed and G-protein returns to its original trimeric form with associated GDP (Figs.). For the discovery of G-proteins, the Nobel Prize in Medicine for the year 1994 was awarded to Martin Rodbell and Arthur Gilman.

One of the most common enzyme activated by a class of G-proteins (stimulatory G protein or Gs) and inhibited by another class of G proteins (inhibitory G proteins or Gi) is adenyl cyclase (Fig.), which thus helps either in the production or in reduction of cyclic AMP, a second messenger which mediates several cell responses (Table). The increase or decrease of the concentration of cyclic AMP depends on the receptor to which the ligand binds. β *adrenergic receptors* are used for activation and α_2-*adrenergic receptors* are used for inhibiting adenyl cyclase thus influencing the concentration of cyclic AMP. The cyclic AMP exerts its influence through *cyclic-AMP-dependent protein kinase*, called *A-kinase* (PKA), which catalyzes transfer of terminal phosphate group from ATP to specific ser/thr of selected proteins. These target proteins may themselves be enzymes mediating a signal transduction pathway or a biochemical synthetic pathway (Figs.). The effects of A-kinase (PKA) may be reversed by ser/thr protein phosphatases. Another class of G protein (Gq) mediates activation or inhibition of another enzyme, *phospholipase* C-β which cleaves PIP_2 (PI bisphosphatc; PI=phosphatidyl inositol- PI is a minor phospholipid in cell membrane) and generates two products: *inositol triphosphate* (IP_3) and *diacylglycerol* (DAG) (Figs.). Both these molecules play important roles in mediating several cell responses, through increasing or decreasing Ca^{++} concentration in the cytosol. The pathway is described as *inositol phospholipid pathway* (Table).

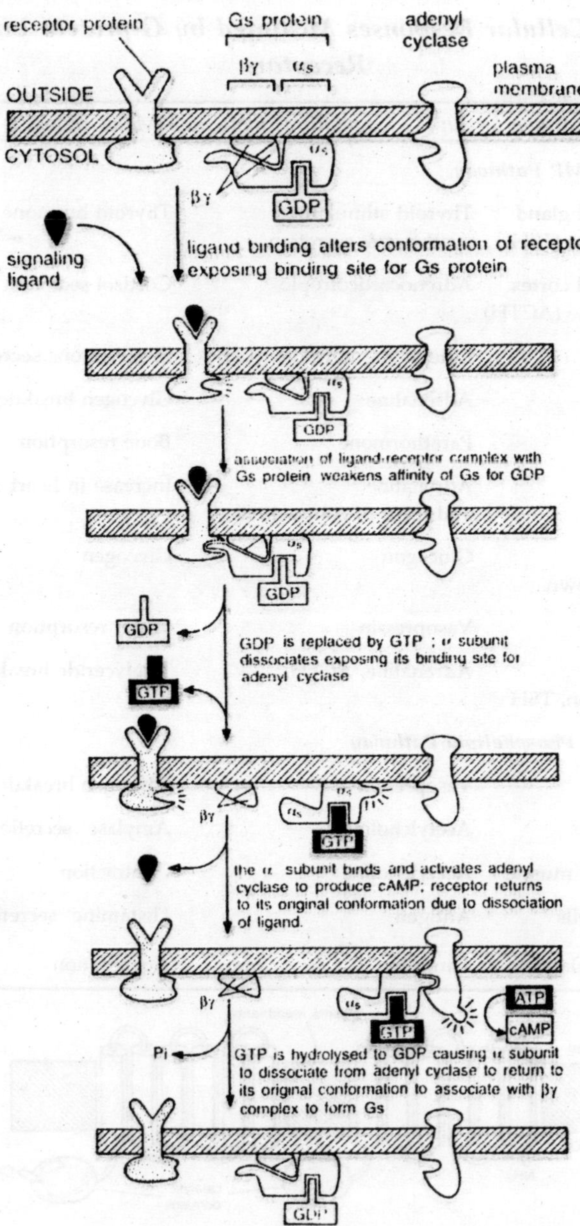

A typical trimeric (αβγ) G protein, where GTP binding releases α-subunit from βγ dimer; α-GTP subunit becomes free to act upon target protein.

Some Cellular Responses Mediated by G-protein Linked Receptors

Target tissue /organ	*Signalling molecule*	*Major response*
I. Cyclic AMP Pathway		
Thyroid gland hormone (TSH)	Thyroid stimulating synthesis & secretion	Thyroid hormone
Adernal cortex hormone (ACTH)	Adrenocorticotropic	Cortisol secretion
Ovary	Luteinizing hormone (LH)	Progesterone secretion
Muscle	Adrenaline	Glycogen breakdown
Bone	Parathormone	Bone resorption
Heart	Adrenaline and force of contraction	Increase in heart rate
Liver breakdown	Glucagon	Glycogen
Kidney	Vasopressin	Water resorption
Fat glucagon, TSH	Adrenaline, ACTH,	Triglyceride breakdown
II. Inositol Phospholipid Pathway		
Liver	Vasopressin	Glycogen breakdown
Pancreas	Acetylcholine	Amylase secretion
Smooth muscle	Acetylcholine	Contraction
Mast cells	Antigen	Histamine secretion
Blood platelets	Thrombin	Aggregation

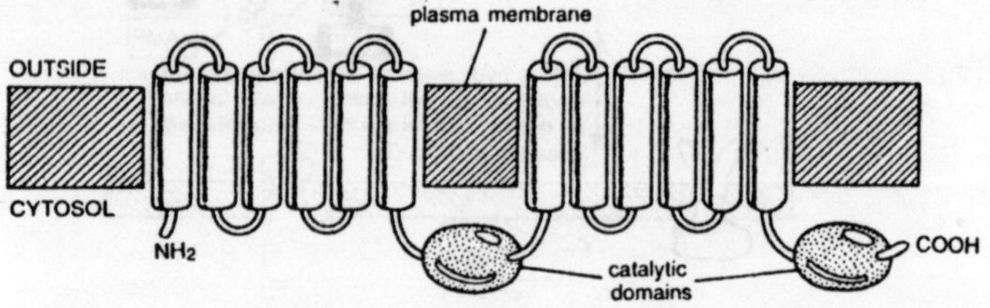

Structure of adenyl cyclase enzyme, showing two clusters of six transmembrane segments, separating two similar cytoplasmic catalytic domains (redrawn from Alberts. 'Molecular Biology of the Cell', 1994).

IP_3 released from PIP_2 due to phospholipase C-b diffuses rapidly through the cytosol, binds to *IP_3 -gated* Ca^{++}'*release channels* in ER membrane and helps in release of Ca^{++} from ER, thus increasing Ca^{++} concentration in the cytosol. Similarly diacylglycerol (the other product of hydrolysis of PIP_2) activates

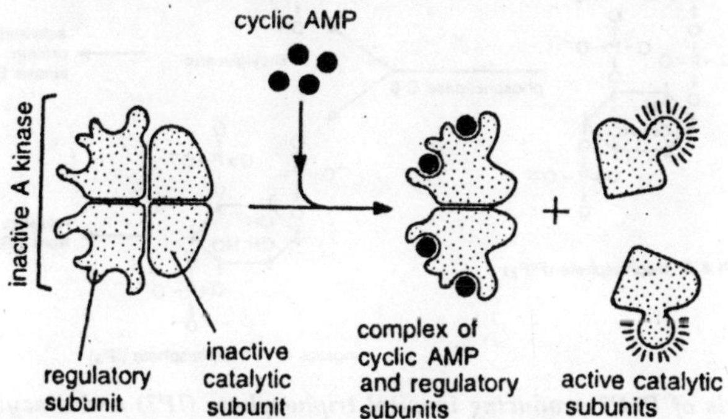

The activation of cyclic-AMP dependent protein kinase (A-kinase or PKA) due to phosphorylation of a subunit of the enzyme (redrawn from Alberts. 'Molecular Biology of the Cell', 1994).

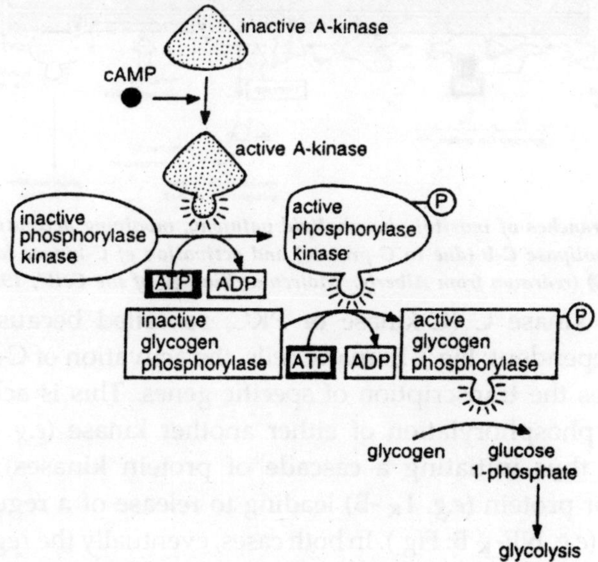

Stimulation of glycogen breakdown by cyclic AMP in skeletal muscle cells (redrawn from Alberts. 'Molecular Biology of the Cell', 1994).

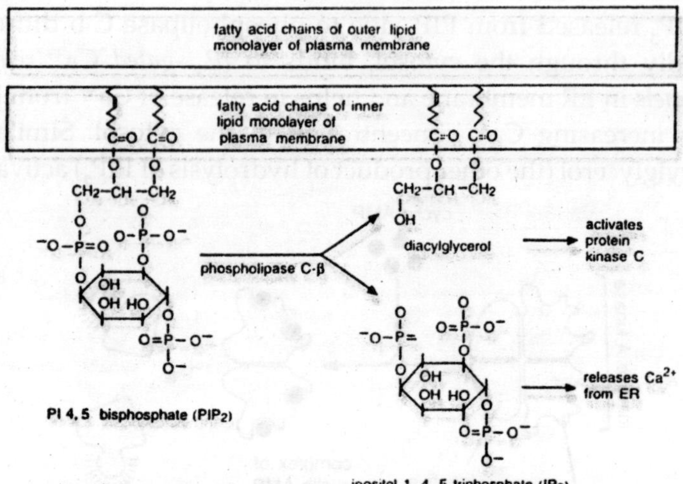

Hydrolysis of P1P2 producing inositol triphosphate (IP3) and diacylglycerol (DAG) (redrawn from Alberts. 'Molecular Biology of the Cell', 1994).

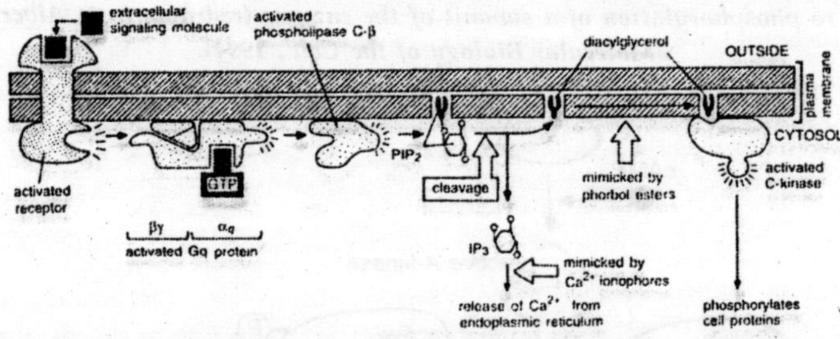

Two branches of inositol phospholipid pathway, involving activation of phospholipase C-b (due to G-protein) and activation of C-kinase (due to DAG) (redrawn from Alberts. 'Molecular Biology of the Cell", 1994).

protein kinase C (C-kinase or PKC; so called because it is Ca^{++}-dependent; Fig.). In many cells, the activation of C-kinase increases the transcription of specific genes. This is achieved due to phosphorylation of either another kinase (*e.g.* MAP-kinase, thus initiating a cascade of protein kinases) or an inhibitor protein (*e.g.* Iκ-B) leading to release of a regulatory protein (*e.g.* NF-κ B; Fig.). In both cases, eventually the regulator protein (*e.g.* a transcription factor) stimulates the transcription of specific genes.

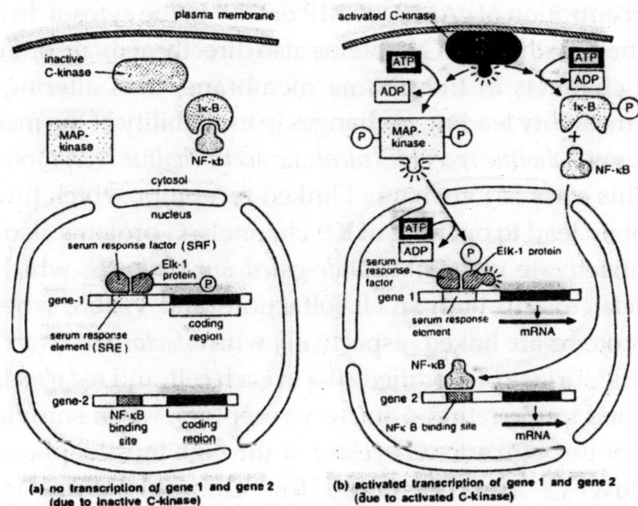

The role of C-kinase (PKC) in activation of the transcription of specific genes, through following two intracellular pathways (i) one pathway involves C-kinase phosphorylating MAP-kinase, which in its turn phosphorylates the protein Elk-1; Elk-1 along with SRF (serum response factor) activates serum response element (SRE); (ii) the other pathway involves C-kinase, phosphorylating Iκ-B, which releases NF-κB to reach the nucleus and activate transcription of specific genes (redrawn from Alberts. 'Molecular Biology of the Cell', 1994).

Cytosolic Ca^{++} acts as a second messenger, its concentration, influenced as above, regulates gene expression in a variety of ways. An important pathway is binding of Ca^{++} to calmodulin. The Ca^{++}/calmodulin complex has no enzyme activity and acts by altering the activity of other proteins (*e.g.* Ca^{++}-ATPase or CaM-kinases). The CaM kinases (Ca^{++}/calmodulin-dependent protein kinases), which mediate most of the effects of Ca^{++} in the cell, include many enzymes including the following : (i) *myosin light-chain kinase* (it activates smooth muscle contraction); (ii) *phosphorylase kinase* (it activates glycogen breakdown); (iii) CaM kinase II (Fig.; it activates secretion and resynthesis of neurotransmitter, thus helping in memory and learning).

We discussed above that the G proteins perform their function by regulating the activity of enzymes by altering the concentration of cAMP/cGMP or Ca^{++} in the cytosol. In addition to these pathways, G-proteins also directly activate or inactivate ion channels in the plasma membrane, thus altering the ion permeability leading to changes in excitability of the membrane. An *acetylcholine receptor (nicotinic acetylcholine receptors)* belongs to this class of ion-channel linked receptors, which through G-protein lead to opening of K^{++} channels. G proteins also activate or inactivate *cyclic nucleotide-gated ion channels,* which play a crucial role in both smell (olfaction) and vision, where these G proteins are linked respectively with *olfactory receptors,* located on the surface of modified cilia in each cell, and *rod photoreceptors* located in the retina. Olfactory receptors, when stimulated, act either through adenyl cyclase or through inositol phospholipid pathways and IP3-gated ion channels. Similarly, rod photoreceptors are coupled with G-protein (Gt or transducin), which activates *guanyl cyclase* or *cyclic GMP phosphodiesterase.*

For the discovery of G-proteins, the Nobel Prize in Medicine for the year 1994 was awarded to Martin Rodbell and Arthur Gilman.

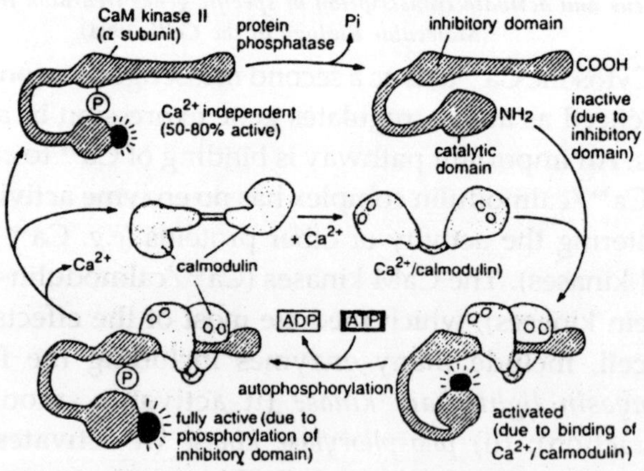

The activation of CaM kinase II; it consists of 12 subunits of four kinds (α, β, γ, δ), but only one (α), expressed in the brain, is shown in this figure; Ca2+/calmodulin helps in autophosphorylation of the enzyme, thus rendering it Ca^{2+} independent (redrawn from Alberts el al., 'Molecular Biology of the Cell', 1994).

Enzyme-linked Cell Surface Receptors: Enzyme linked receptors are also transmembrane proteins like the G-protein linked receptors. Their cytosolic domains either have an enzyme activity (as in several tyrosine and serine/threonine kinases) or are associated directly with an enzyme (as in *JAK-STAT pathway*, discussed later in some detail in this Chapter). There are five classes of enzyme linked receptors :

(i) receptor guanyl cyclases (catalyze production of cyclic GMP);

(ii) receptor tyrosine kinases (catalyze phosphorylation of specific intracellular proteins);

(iii) tyrosine-kinase associated receptors (associate with proteins having tyrosine kinase activity);

(iv) receptor tyrosine phosphalases (remove phosphate group from tyrosine) and

(v) receptor serine/threonine kinases (catalyze phosphorylation of specific serine/ theronine residues of specific intracellular proteins).

Some of the details of these five types of receptors, including suitable examples are summarized in Table. In all these cases, ligand binding usually induces the receptors to dimerize, which activates the kinase domain of the receptor or an associated non-receptor protein kinase. When the receptor is activated, two molecules of a tyrosine kinase in a dimer, cross-phosphorylate themselves on multiple tryosine residues (only dimerized receptors are active). The tyrosine phosphorylated proteins are the binding sites of intracellular signalling proteins, which, via their SH_2 domains, bind to specific phosphotyrosine residues. In this way, a multiprotein signalling complex is activated from which the signal spreads to the cell interior.

Cell surface receptors in cholesterol metabolism and drug production (1985, '88 Nobel Prizes). During last more than two decades, a study of regulation of cholesterol metabolism was

Five Different Types of Enzyme Linked Cell Surface Receptors and their Characteristics.

Ligand	*Receptor*	*Catalytic site*	*Response on ligand binding*
I. Receptor guanyl cyclasc			
1. A trial natriuretic peptides or ANPs (a peptide hormone secreted in atrium of heart, when BP rises)	ANP receptors in kidney cells & smooth muscle cells of blood vessels	Guanyl cyclase site in cytosolic domain of receptor protein (no G-protein is involved)	Cyclic GMP is produced and activates G-kinase which phosphorylates ser/thr in specific proteins
II. Receptor tyrosine kinases			
2. Epidermal growth factor or EGF (a small protein, 53 amino acids long)	EGF receptor (single pass transmembrane proteins; 1200 amino acids)	One tyrosine kinase cytosolic domain	Autophosphorylation followed by phosphorylation of a cascade of intracellular signalling proteins having SH2 and SH3 (SH stands for Src homology) domains (*e.g.* GTPase- activating protein or GAP, phospholipase C)
3. Nerve growth factor or NGF	NGF receptor	Two tyrosine kinase cytosolic domains	Phosphorylation of receptor
4. Insulin and insulin like growth factor (IGF-1)	Insulin & IGF-1 receptors (a tetramer)	Two tyrosine kinase cytosolic domains	Allosteric interaction of autophosphorylation of receptor

Contd...

Ligand	*Receptor*	*Catalytic site*	*Response on ligand binding*
5. Platelet derived growth factor or PDGF and microphage colony stimulating factors or M-CSF	PDGF receptor & M-CSF receptor	One tyrosine kinase cytosolic domain interrupted by a 'kinase insert region'	Autophosphorylation of activated receptor
6. Fibroblast growth factors or FGFs	FGF receptor		
7. Vascular endothelid growth factor (VEGF)	VEGF receptor		
III. Tyrosine kinase associated receptors			
8. Cytokines (*e.g.* interleukins like IL-2)	Cytokine receptors (*e.g.* IL-2 receptor)	Janus family of non-receptor tyrosine kinases (Jakl, Jak2, Tyk2)	Assembly of functional receptor, which activates a tyrosine kinase
IV. Protein tyrosine phorphatases			
9. Foreign antigens	CD45 protein on the surface of WBCs	Tyrosine phosphatase cytosolic domain	Activation of T and B lymphocytes
V. Receptor serine /threonine kinases			
TGF-b superfamily (*e.g.*TGF-bl to b5, activins, bone morphogenetic proteins)	TGF-P receptor (single pass transmembrane protein)	Serine-threonine kinase cytosolic domain	Activation of kinase domain

conducted by S. Brown and J.L. Goldstein (at University of Texas, U.S.A.), both of them sharing the Nobel Prize for Physiology or Medicine in 1985. Their study mainly involved the study of cell surface receptor for low density lipoprotein (LDL), which is a major transporter of blood cholesterol from blood to the interior of fibroblast and liver cells.

The receptor molecule forms a complex with LDL and enters the cell through pinocytosis. The receptor is recycled and LDL is degraded so that cholesterol can be utilized in the cell. When a patient has a defective gene for receptor, it leads to high blood cholesterol and LDL. This leads to the development of familial hypercholesterol-anemia or coronary atherosclerosis (before the age of 20 in a homozygous defective). Brown and Goldstein also synthesized the gene for enzyme HMGCoA reductase, which is responsible for cholesterol synthesis and is under the control of a feedback inhibition. These researches may eventually lead to prevention and treatment of *atherosclerosis*.

Signal Transduction Pathways

Signal transduction using 'second messengers': In animals as well as in plants, the cellular machinery for protein synthesis responds to external stimuli. These external stimuli may include a whole range of diverse factors including the following :

(i) the presence of hormones, growth factors and other signalling molecules (*e.g.* in animals);

(ii) the presence of pheromones secreted by cells of two different mating types (*i.e.* in yeast);

(iii) presence of light (different wave lengths), gravity, minerals, water, gases and specific soil structure (*e.g.* in plants) and

(iv) cell-cell contact.

In some of these cases, cell surface receptor molecules are present and transmit the signal to the interior of the cell (for water-soluble signals like proteins, peptides and neuro-transmitters). In other cases, substances like hormones interact with intracellular receptors, which transmit the signal further to the target site (for water insoluble lipophilic signals like steroids, thyroid hormones, retinoic acid, etc.). In all these cases (including prokaryotes), the external stimulus needs to be transduced and integrated with internal signals, which follow a cascade of events called- *'signal transduction pathway'*-leading to the response, which is a specific gene expression in many cases (Fig.).

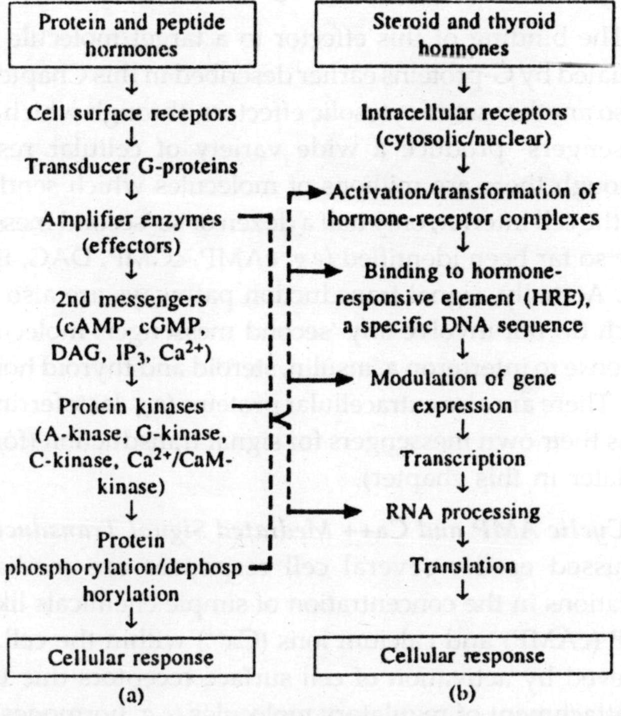

Different steps involved in 'signal transduction pathways' for regulation of gene expression due to action of (a) protein/peptide hormones; (b) steroid and thyroid hormones (also note cross-linking between two pathways).

Signal transduction pathways precede most of the mechanisms for regulation of gene expression discussed earlier. In prokaryotes, often the precursor molecule (*e.g.* lactose) or the end product (*e.g.* an amino acid) of a biosynthetic pathway takes part in regulation of gene activity. But in 'signal transduction pathways' in eukaryotes a cascade of molecules, leading to the activation of one or more specific transcription factors, is involved in regulation of gene activity. The extracellular *'signals'* in the form of ligands, are transduced across the plasma membrane to cell compartments (nucleus and other cells organelles) via a variety of molecules described as *'second messengers'*. The *'effectors' i.e.* molecules which initiate the pathways for the production of *'second messengers'*, are often described as *'first messengers'*.

The binding of this effector to a target molecule is often mediated by G-proteins earlier described in this Chapter. There is also another set of cytosolic effectors, through which 'second messengers' produce a wide variety of cellular responses. Although there are millions of molecules which send signals into the cell interior, only half a dozen or so 'second messengers' have so far been identified (*e.g.* cAMP, cGMP, DAG, IP_3, Ca^{2+}, etc.). Actually, signal transduction pathways are also known, which do not involve any 'second messenger' molecules (*e.g.* response to interferon a, insulin, steroid and thyroid hormones, etc.). There are also extracellular proteins (*e.g.* lactoferrin) which act as their own messengers for signal transduction (for details see later in this chapter).

***Cyclic AMP and* Ca++ *Mediated Signal Transduction*:** As discussed earlier several cell responses are mediated by alterations in the concentration of simple chemicals like cyclic AMP (cAMP) and calcium ions (Ca^{++}) within the cell. This is achieved by activation of cell surface receptors due to either the attachment of regulatory molecules (*e.g.* hormones) or due to other physical factors. The activation of receptor leads to activation of a G-protein, which in its turn activates the amplifier enzyme, adenylate cyclase (an effector). This enzyme converts ATP into cAMP, leading to an increased concentration (inside the cell) of cyclic AMP, Ca^{++} or whatever the internal mediator is. Since cyclic AMP conveys the message in the interior of the cell, it was described as a *'second messenger'*. In most cases, cAMP activates a protein kinase A (PKA), which phosphorylates several specific enzymes (*e.g.* lipase for mobilization of fat; phosphorylase for enzymatic breakdown of glycogen in the liver). In addition to its role in activation of several cellular enzymes, cAMP often causes an elevation of cytosolic level of Ca^{++}, which is another second messenger. Epinephrine (adrenaline hormone) also acts through cAMP to modulate the intracellular level of Ca^{++} in muscles and nerve tissues.

Lipids (DAG, IP_3) and Ca^{++} Mediated Signal Transduction: Some external signals have been found to stimulate the turnover of a membrane lipid and change the intracellular concentration of Ca^{++}. In this connection, the metabolism of phosphatidylinositides (PIs) has been shown to play an important role in the regulation of cell growth. It was also shown that increased PI metabolism is a reflection of signal transduction across the cell membrane. In 1980s, the activation of receptors could be linked to the rapid hydrolysis of PIs and mobilization of Ca^{++} in many tissues in response to a variety of external stimuli. As discussed earlier in this chapter hydrolysis of a PI (phosphatidylinositol 4,5 bisphosphate = PIP_2) utilizes an amplifier enzyme, the Pi-specific phospholipase Cβ (PLC) and produces various messenger molecules. The following three 'lipid second messengers' have been documented :

(i) *1,2-diacylglycerol (DAG);*

(ii) *inositol 1,4,5-triphosphate (Ins* 1, 4, 5, - P_3 = IP_3,) and

(iii) *archidonic acid.*

The hydrolysis of PIP_2, upon binding of extracellular ligands to cell surface receptors is also coupled to G-protein activation. Activated G-protein modulates PLC, which in turn hydrolyses PIP_2 into DAG and IP_3. Although, the signal transduction utilizing these lipid messengers has been shown to be complex, it is known that DAG serves as a co-factor and together with Ca^{++} activates *protein kinase* C (PKC).

During this process, DAG itself is metabolized to archidonic acid, a precursor of other mediators. (PKC plays a central role in the transduction of extracellular signals into a cellular response). On the other hand, IP_3 promotes the release of Ca^{++} into the cytoplasm through binding to a specific receptor on the surface of endoplasmic reticulum (ER) and/or other intracellular vesicles that store Ca^{++}.

This interaction induces a conformational change in proteins specific to Ca^{++} channel and allows Ca^{++} ions to exit from intracellular storage sites into the cytoplasm. This calcium influx following stimulation of the PI signal pathway is a widespread and physiologically most important phenomenon. Ca^{++} in the cytoplasm has several roles to play in the cellular regulation, including activation of PKC by DAG. In view of the role of IP3 as a 'second messenger' in operating Ca^{++} channels as above, these Ca^{++} channels have been described, in the past, as SMOCs (second messenger operated calcium channels). However, since the role of 'second messengers' is not fully established, the usage of the term SMOC has been questioned.

Ras-mediated Signal Transduction Pathways: Receptor tyrosine kinases and G-protein linked receptors, both initiate intracellular signalling cascades in which Ras proteins belonging to Ras superfamily of monomeric GTPases, play an important role. Like G-proteins and other GTPases, Ras proteins are active, when bound to GTP and inactive when bound is GDP. Two classes of signalling proteins, that regulate Ras activity include the following :

(i) GTPase-activating proteins (GAPs) increase the rate of hydrolysis of bound GTP by Ras, thereby inactivating Ras.

(ii) Guanine nucleotide releasing proteins (GNRPs) promote the release of GDP and uptake of GTP, thus restoring the activity of Ras (43.36).

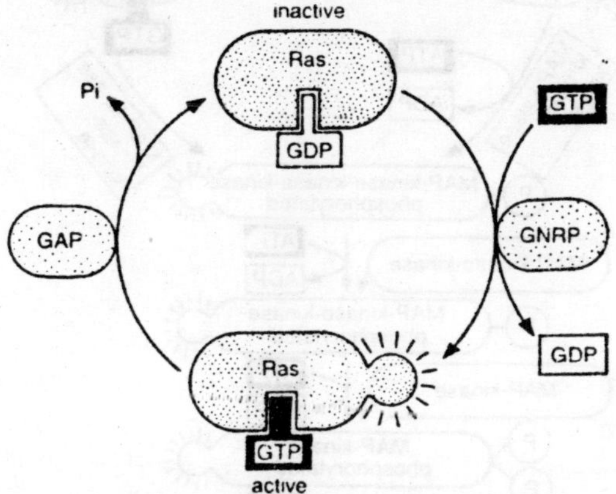

Regulation of Ras activity by GAPs (GTPase activating proteins) or by GNRPs (guanine nucleotide releasing proteins). (redrawn from Alberts., 'Molecular Biology of the Cell', 1994).

Therefore receptor tyrosine kinases activate Ras either by inhibiting a GAP or by activating a GNRP. When ras proteins are active, they initiate a cascade of ser/thr phosphorylations, that converge on MAP-kinases, which help relay the signal to the nucleus (Fig.). Many of these intracellular signalling proteins, when inappropriately activated, can cause cancer.

Ras mediated Signal Transduction for Cell Division: It has been known for some years, that signals received at the cell surface are transmitted via activated tyrosine kinase (associated with the receptor) in the cell membrane, through serine/threonine kinase in the cytoplasm, to the transcription factor in the nucleus (among other targets). One such signal transduction pathway, discovered fully in 1992-1993, involved a flow of signal via Ras (product of proto-oncogene *c-ras;* to Raf-1 kinase (product of another oncogene, *raf-1),* thence to MSP kinases and pp^{90rsk} and finally on to nuclear proto-

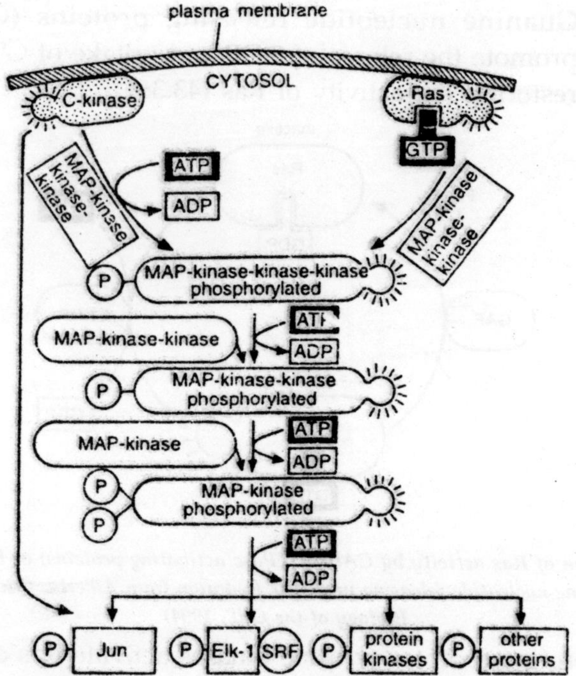

Serine-threonine phosphorylation cascade activated by Ras and C-kinase and involving a series of MAP kinases (redrawn from Alberts. 'Molecular Biology of the Cell', 1994).

oncogene products like c-Fos, c-Jun, etc. (MAP = mitogen activated protein; pp^{90rsk} = ribosomal S6 kinase).

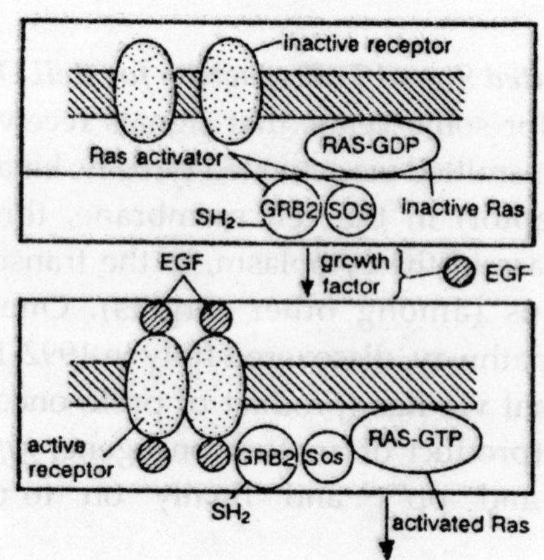

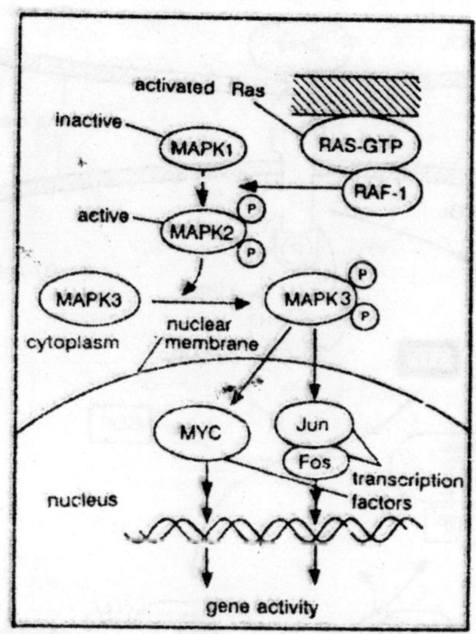

Different steps involved in Ras mediated signal transduction pathway for inducing cell division.

Following steps, shown in Figure are involved in EGF-induced Ras signal transduction pathway, leading to cell division: (i) EGF binds to receptor to form activated EGF-receptor complex, which is an active tyrosine kinase and phosphorylates itself; (ii) phosphorylation of receptor, enables its internal segment (cytosolic) to bind to GRB2 protein in the GRB2-mSos complex (GRB2 and mSos are two intermediary proteins, which remain linked together in unstimulated cells); (iii) mSos of the GRB2-mSos complex now attaches to Ras, which is anchored to the inside of the cell membrane; mSos activates Ras by changing it from inactive Ras-GDP to active Ras-GTP; (iv) Raf-1 now binds to activated Ras-GTP and gets activated; (v) Activated Raf-1 now activates MAP kinases (MAPK) forming a 'kinase cascade', (each MAPK phosphorylates, and thus activates the next member of the series); (vi) The last MAP kinase phosphorylates several cell proteins, including some transcription factors (*e.g.* Myc, Fos, Jun, etc.); (vii) Phosphorylation of transcription factors leads to expression of specific genes for cell division.

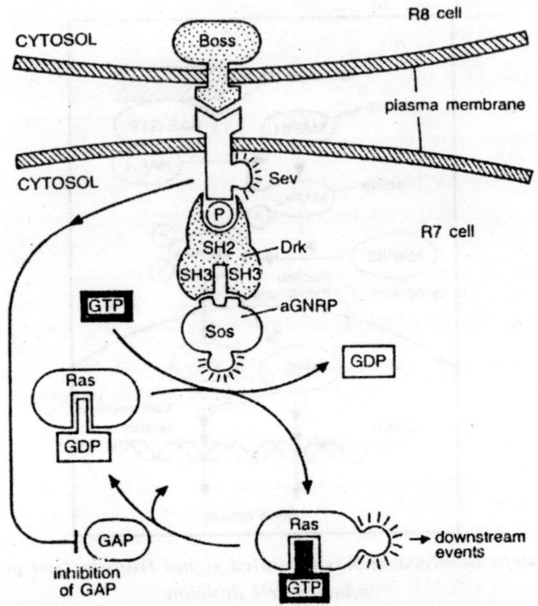

Regulation of Ras activity in early cell signalling leading to R7 development; note the presence of Boss protein (in R8 cell), which activates sev receptor protein in R7 precursor cell, leading to the activation of Ras (redrawn from Alberts. 'Molecular Biology of the Cell', 1994).

Ras mediated development of R7 photoreceptor in Drosophila: A compound eye in *Drosophila* consists of 800 ommitidia, each consisting of 8 photoreceptor cells (designated as R1-R8) and 12 accessory cells. The development of R7, required for the detection of UV light, has been studied in some detail. It involves activation of Ras, so that the activated Ras relays the signal downstream through a serine/threonine phosphorylation cascade involving a series of MAP kinases (MAP = mitogen activated protein). The activation of Ras, during development of R7 photoreceptor involves the following components (Fig.). (i) Boss protein (located in R8 cell), encoded by *bride-of-sevenless (boss)* gene works as a ligand for the Sev receptor protein (located in R7 cell); (ii) Sev receptor tyrosine kinase, once activated by Boss, activates another protein Sos (encoded by gene *son of sevenless* or *sos*), which is a guanine nucleotide releasing protein (GNRP). (iii) Another protein Drk (downstream receptor kinase), couples the Sev receptor with the Sos, using its SH2 domain for binding Sev and two SH3 domains for binding Sos. (iv) Sos is a GNRP and activates Ras by converting Ras-GDP into Ras-GTP. Yet another protein, a GTPase activating protein (GAP), facilitates hydrolysis of Ras bound GTP, thus converting active Ras-GTP into inactive Ras-GDP. If the gene for GAP mutates, R7 develops even in the absence of Sev, due to the availability of hyperactive Ras.

JAK-STAT Signal Transduction Pathway (Mediated by Cytokine Receptors): In 1994, a study of the IFN-α and IFN-γ induced transcriptional activation, led to the discovery of a new signal transduction pathway called JAK-STAT pathway (JAK stands for Janus kinases and STAT refers to cytoplasmic transcription factors called STATs; STAT stands for Signal Transducers and Activators of Transcription). Tyrosine kinases of JAK family remain closely associated with the components of transmembrane receptors and are activated after ligand-receptor attachment (ligands are polypeptides like IFN-α & IFN-γ, that are recognized by ligand binding domain of a receptor protein found in plasma membrane. The receptor consists of two domains, the ligand binding domain α and a signal transducing chain β). This leads to phosphorylation of one or more of the transcription factors called STATs. The STATs perform the following two roles: (i) they work as signal transducers by acting as substrates of the JAKs, and undergo phosphorylation leading to their own activation; (ii) after phosphorylation, homo/ heterodimerization and nuclear translocation, STATs act as transcriptional activators and activate specific genes (Fig.).

This JAK-STAT pathway was though discovered due to a study of IFN—α and IFN-γ induced activation of transcription, many other JAK and STAT proteins were later found to follow this pathway.

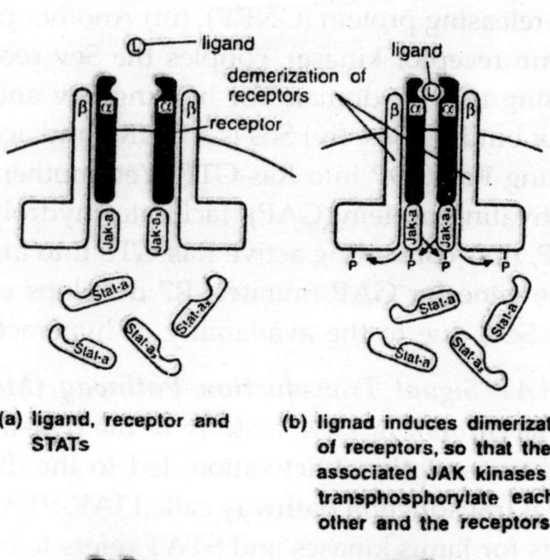

(a) ligand, receptor and STATs

(b) lignad induces dimerization of receptors, so that the associated JAK kinases transphosphorylate each other and the receptors

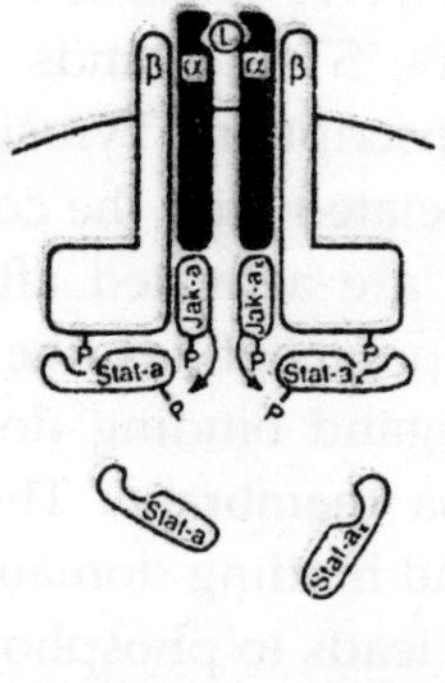

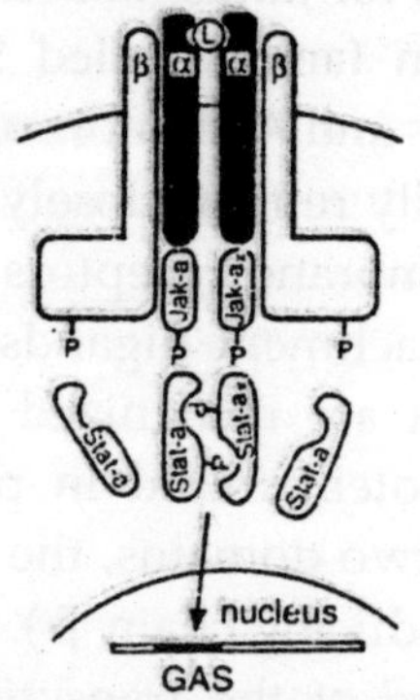

(c) STATs are recognized and phosphorylated by receptor with phosphotyrosyl residue

(d) activated STATs dimerise, translocate and activate GAS

An outline of JAK-STAT signal transduction, explaining the roles of Janus kinases (JAKs) and STATs.

Receptors and kinases of JAK-STAT Pathway: A large number of receptors, each representing a transmembrane protein are known. Following two families of these receptors are involved in JAK-STAT pathway: (i) Cytokine receptor family; (ii) *Tyrosine Kinase Receptor Family:* The ligands that interact with these two classes of receptors are listed in Table, where the JAK kinases and STATs activated by each of these ligands are also shown. There are four JAK kinases, called Tykl, Jakl, Jak2, Jak3, all of them lacking SH2 and SH3 domains. These kinases remain inactive, but remain associated with receptors, so that they are rapidly activated by a ligand stimulated phosphorylation on a tyrosine. The activated JAK kinases mediate activation of STAT proteins, leading to transcription activation of specific genes. The following steps involved in this signal transduction pathway are shown in Figure.

(i) ligand promotes dimerization of signal transducing component (β chain) of receptor;

Activation of JAK Kinases by Various Ligands

Ligands	*Receptor*	*STATs activated*
I. Cytokine receptor family		
IFN family		
IFN - /	Tyk2, Jakl	1,2,3
IFN	Jakl, Jak2	1
IL-10	Tyk2	1,3
gp 130 family		
IL-6	Tyk2, Jakl, Jak2	1,3
IL-11	Jakl	1,3
OnM	Jakl, Jak2	1,3
LIF	Jakl, Jak2	1,3
CNTF	Jakl, Jak2	1,3
G-CSF	Jakl	1,3
IL-12	Tyk-2, Jak2	3,4
-C **family**		
IL-2	Jakl, Jak3	1,3,5
IL-4	Jakl, Jak3	6
IL-7	Jakl, Jak3	5
IL-9	Jakl, Jak3	5
IL-13	Jakl	
IL-15	Jakl, Jak3	5

Contd...

Ligands	*Receptor*	*STATs activated*
gpl40 family		
IL-3	Jak2	5
IL-5	Jak2	5
GM-CSF	Jak2	5
Growth hormone family		
EPO	Jak2	5 ?
GH	Jak2	1,3,5
PRL	Jak2	1,5
II. Receptor tyrosine kinases		
EGF	Jakl, Jak2	1,3
PDGF	Jakl, Jak2	1,3
CSF-1	Jakl, Jak2	1,3

(ii) two molecules of the JAK kinase, which remain associated with specific domains of the receptor are drawn in apposition and transphosphorylate each other;

(iii) the kinases then phosphorylate a tyrosine in a distal region of receptor β-chain,

(iv) active receptor draws STAT molecules through interaction with STAT SH2 domain,

(v) STAT molecules are phosphorylated by JAK kinase, making them competent to dimerize, travel to the nucleus and bind to specific response elements in DNA.

Steroid Hormones Induced Signal Transduction: It has been demonstrated that sex steroid hormones, namely oestrogen and progesterone can regulate gene activity. Glucocorticoids, another set of steroids, found endogenously within the cells and used for treatment of leukaemia, asthma and arthritis also regulate gene expression. These glucocorticoids have a multitude of effects in various tissues. Classical example of a sex steroid inducible protein is chick ovalbumin, and two of the glucocorticoid-inducible proteins are the enzymes tyrosine aminotransferase (TAT) and tryptophan oxydase. The glucocorticoids induce synthesis of these enzymes through receptor molecules. The steroid-receptor complex interacts each with a specific DNA sequence, called steroid receptor element (SRE) or glucocorticoid receptor element (GRE) or hormone receptor element (HRE). Through recombinant DNA technology, gene constructs were prepared by fusing different lengths of DNA sequences flanking the 5' end of TAT gene with the marker gene for CAT (chloramphenicol aceryltransferasc). These cloned gene constructs were introduced into animal cells and though glucocorticoid induced expression of CAT, steroid receptor elements (SRE) were identified (Fig.)

Steroid inducible synthesis of protein, which involves the following six steps, is illustrated in Figure.

(i) Uptake of steroid hormone (S) and binding to receptor molecule in the cytoplasm (Rc) to form S-Rc complex. (Receptors for glucocorticoids are predominantly found in the cytoplasm, but those for sex steroids are found in the nucleus).

(ii) Transport of S-Rc complex to nucleus (wherever cytosolic receptor is involved), where it is now called S-Rn to specify its occurrence in the nucleus.

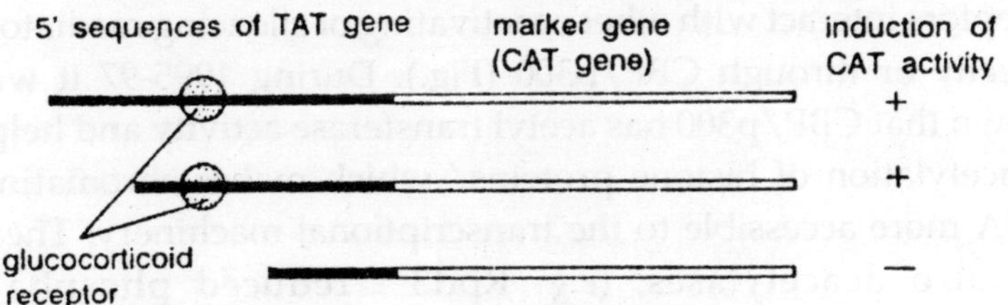

Use of deletions through recombinant DNA technology for identification of response elements.

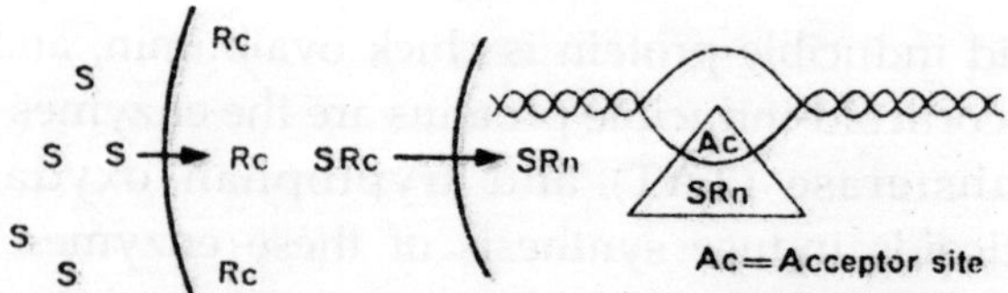

Rc
S
S S Rc SRc SRn
Ac
SRn
S
S Rc
Ac=Acceptor site
R=Receptor
S=Steroid
SRc=steroid-receptor complex in the cytoplasm
SRn=steroid-receptor complex in the nucleus

Role of steroid hormones in regulation of gene activity in eukaryotes.

(iii) Binding of S-Rn complex to specific acceptor site SRE on genome.

(iv) Activation of transcriptional apparatus in presence of specific non-histone proteins to synthesize specific mRNA.

(v) Transport of specific RNA to cytoplasm.

(vi) Synthesis of protein (like ovalbumin) and occurrence of specific steroid mediated functional response.

The nuclear hormone receptor in the above pathway needs a coactivator to initiate transcription with the help of RNA polymerase II. This coactivator is CBP or related p300 (CBP=CREB binding protein; CREB=cyclic AMP response element binding regulatory protein). CBP/p300 is a large protein, that interacts with different DNA-binding transcription factors (inducing nuclear hormone receptor) and also with the components of the basal transcription machinery. In addition to one of these coactivators (CBP/p300), nuclear hormone receptors interact with other coactivating or silencing mediators directly or through CBP/p300 (Fig.). During 1995-97 it was shown that CBP/p300 has acetyl transferase activity and helps in acetylation of histone proteins,' which makes chromatin's DNA more accessible to the transcriptional machinery. There are also deacetylases, (*e.g.* Rpd3 - reduced phosphate dependency), which remove acetyl groups from histone proteins and stabilize the repressed state of chromatin thus working as corepressors (Fig.).

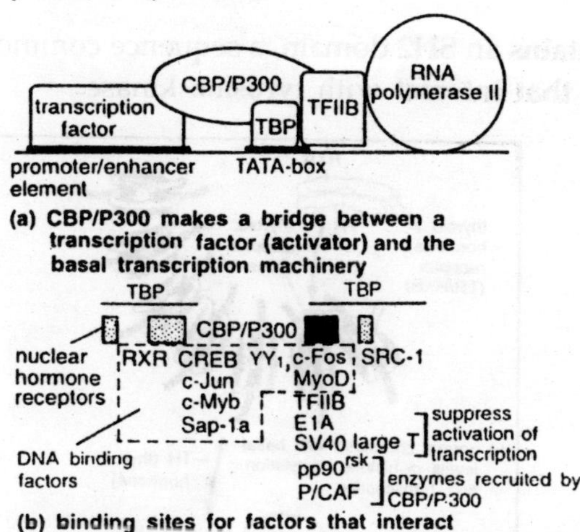

The role of CBP/p300 (a protein complex), as a co-activator, in steroid inducible synthesis of proteins.

Interferon Stimulated Gene Expression (without a Second Messenger): Interferons were once believed to have the ability to fight cancer and, therefore, received major attention of molecular biologists, even though its early clinical promise was not fulfilled. However, recent studies with interferon gave information about how cells really respond to interferon signals. In the year, 1992, several teams of researchers demonstrated that when interferon contacts its receptor on the cell surface, it immediately activates a transcription factor which moves to the nucleus, where it turns on its own particular set of genes. It was experimentally demonstrated that no *'second messengers'* are involved in this signal transduction pathway.

In mid-1980s, the genes that respond to interferon were shown to carry a common regulator sequence called ISRE (interferon stimulated response element). In late 1980s, the transcription factor that binds to ISRE was also discovered. This transcription factor consists of four protein subunits including three large subunits, and a small subunit. The fourth small subunit actually makes contact with the ISRE. It was also shown that the three large subunits of the transcription factor, each contains an SH2 domain, a sequence commonly found in proteins that interact with tyrosine kinase.

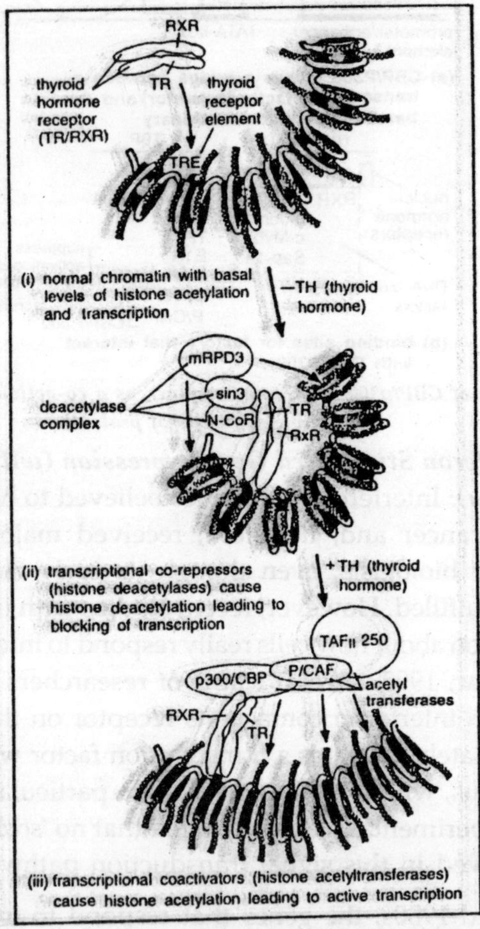

Acetylation and deacetylation of histone proteins and their role in the regulation of transcription.

In cells, which have not been stimulated by interferon-α , the three large subunits of the above transcription factor are not linked and are located only in the cytoplasm. But once interferon-α binds to its receptor, these three subunits rapidly undergo phosphorylation at the tyrosine residues due to *tyrosine kinase* associated with the receptor. The phosphorylated protein subunits associate together and move to the nucleus, where they join the smaller subunit to make the active transcription factor, thus conferring on it the ability to bind on ISRE. This leads to interferon-α stimulated expression of specific genes (Fig.).

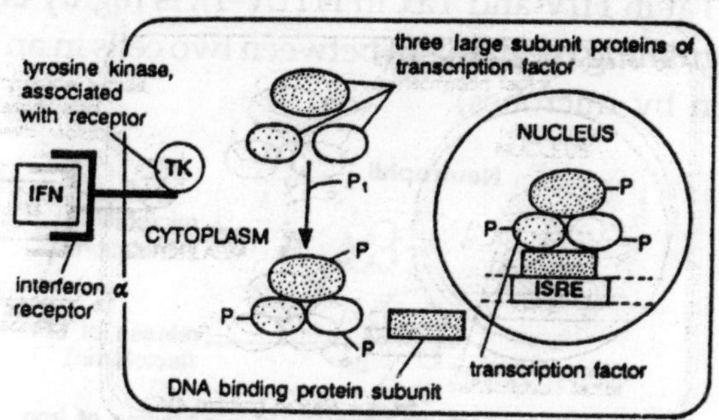

ISRE = interferon stimulated response element

A direct route for interferon induced 'signal transduction pathways' where the tyrosine kinase (TK) associated with the interferon α-receptor complex activates the transcription factor directly, thus turning on the interferon responsive genes.

Lactoferrin (an extracellular polypeptide) acts as its own messenger for signal transduction: Most of the signal transduction pathways described in this section are complex, involving as many as seven or more polypeptides, which deliver the signal from the membrane receptor to the cell nucleus. However, in Jak-STAT pathway and in the pathways utilized by steroid hormones, fewer intermediaries are used to relay the message. In 1995, an extracellular molecule, lactoferrin (a secretory protein) was found to carry the message all by itself from cell surface to the cell nucleus where, in its dimeric or tetrameric form, it binds to the sequence 5'-GGCACTTGC-3'. No cell surface receptor or second messenger is involved in this pathway. Lactoferrin, therefore, is a secretory, membrane permeable transcription factor, which is also a DNA binding protein. Lactoferrin is stored in secretory granule of neutrophils, until a pathogen stimulates its release. Although a part of the released lactoferrin is used to suppress microbial growth, most of it crosses the plasma membrane or endosomal membrane of natural and lymphokine-activated killer cells (NK and LAK cells), and there activate appropriate genes in the nucleus (Fig.). Therefore, this pathway followed by lectoferrin (perhaps also by Tat in HIV and Tax in HTLV-1), is highly effective and rapid signalling mechanism between two cells in an emergency (invasion by microbes).

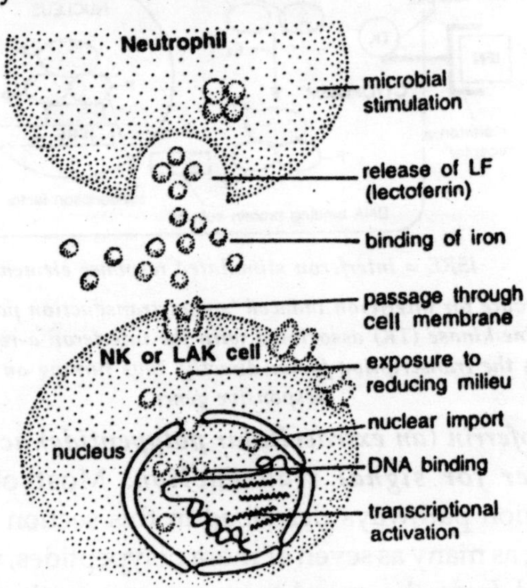

Lactoferrin induced activation of transcription, in which lactoferrin is the sole molecule carrying message from cell surface to the cell nucleus.

Nitric Oxide as a Signalling Molecule (1998 Nobel Prize): Nitric oxide (NO) is another signalling molecule, which neither needs vesicular transport (it diffuses through plasma membrane) for secretion, nor does it need any receptor at the target cell. It plays a vital role in vasodilation (dilation of blood vessels; deficiency of NO causes angina), in the control of viral/bacterial infection and in the neurotransmission (including penile erection). The work on the role of NO as a signalling molecule was recognized in 1998 by the award of Nobel Prize for Medicine to the following three scientists from USA : (i) Robert F. Furchgott, (ii) Louis J. Ignarro and (iii) Ferid Murad.

9

Circuits for Lytic Cycle

When a 'virulent' bacteriophage (like T2, T4, SPO1, etc.) infects a host bacterium, it entirely depends on the host for its multiplication. It subverts the host's function and utilizes the host machinery for producing a large number of progeny phage particles. The bacterium cell undergoes lysis and dies to liberate a large number of these phage particles, which are each then ready to start another cycle by infecting new bacterial cells. This cycle is known as lytic cycle. However, there are other phages called 'temperate' phages which have a dual existence such that they may either perpetuate through a lytic cycle as above or may take the form of a 'prophage' by integration of their DNA with the DNA of bacterial cell. The latter condition is called 'lysogeny', which provides immunity against infection by further phage particles of the same type, so that there is only one copy of phage per bacterial cell. We know that this prophage may become free, turn virulent and then take the course of lytic cycle as in virulent phages.

When the phage DNA enters the bacterial cell leaving the protein shell outside, it directs the host cell for preferential replication of phage DNA. In doing so, the phage genes are expressed in a particular order. The lytic cycle is divided into (i) 'early infection' which covers the period from the time of infection to the start of replication of phage DNA and (ii) 'late infection' which includes the period from start of phage DNA replication to the final step of 'lysis'. During this life cycle of a phage (particularly λ phage), different genes of the phage genome are expressed in a regulated manner and form a cascade of gene expression, so that the product of one gene regulates the expression of the next gene. These aspects of the regulation of gene expression in phages will be discussed in this chapter.

Regulation by a Cascade in Phages

In phages, the early, middle (or delayed early) and late genes are expressed in an order and a study of genetic map suggests, that genes are arranged on the map in the order in which they are expressed. This, therefore, needs smallest number of switches and the regulation is exercised with a maximum economy. In the very early genes, the transcription apparatus of the host is utilised and the product is then used for initiating the transcription of the next class of genes (delayed early or middle genes). The transcription of these middle genes will be necessary for expression of late genes through a regulator, such that these successive controls (where expression of each set is necessary for the expression of the next set of genes) creates a cascade, in which a group of genes are turned on or off at a particular time.

Alternative Sigma Factors in Phage SPO1

The regulation of gene expression through a cascade has been studied in phage SPO1 infecting *Bacillus subtilis.* It has been shown that for the expression of early genes of SPO1, bacterial RNA polymerase ($\alpha_2\beta\beta'\sigma^{55}$) is used, because the promoters of these early genes like the promoters of host genes can be recognized by this RNA polymerase holoenzyme. The consensus sequence in SPO1 at -35 and -10 base pairs resembles those of *E. coli* and *B. subtilis.* For the expression of middle and late genes, however, expression of at least three regulator phage genes, namely genes 28, 33 and 34 is needed. The products of these genes are respectively called gp^{28}, gp^{33} and gp^{34}. The gene 28 is an early gene, whose expression is needed for the expression of middle genes. Similarly, genes 33 and 34 are middle genes whose expression is a prerequisite for the expression of late genes.

It has been shown that gp^{28} replaced σ^{55} in the holoenzyme $\alpha_2\beta\beta'\sigma^{55}$ and therefore gp^{28} is sometimes also referred to as $\sigma^{gp^{28}}$ This RNA polymerase with gp^{28} can then transcribe the middle genes. The next transition involves replacement of gp^{28} by gp^{33} and gp^{34}, so that RNA polymerase may now transcribe late genes (Fig.).

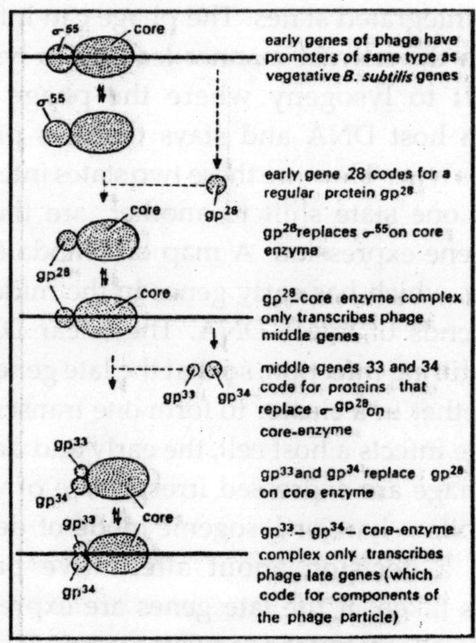

Involvement of several sigma factors, which change the specificity of host RNA polymerase for transcription of phage SPO1 genes.

Thus there are at least three sigma factors used in phages SPO1, namely σ^{55}, $\sigma^{gp^{28}}$, $\sigma^{gp^{33}-gp^{34}}$. We showed the replacement of σ^{43} by σ^{37} and σ^{29} during bacterial sporulation, suggesting that selective transcription can be achieved by replacement of sigma factor in RNA polymerase holoenzyme. We do not clearly know about the factors which cause this replacement of one sigma factor by another, but there are different promoter sequences in different genes to be recognized by different sigma factors.

Lytic Cascade in Lambda (λ) Phage **:** Lambda (λ) bacteriophage exhibits one of the most interesting, but intricate cascade circuits for the two alternative pathways (lytic cycle or lysogeny), that it can follow after infection. A brief mention of this phage and its mode of infection, where it was also explained that the temperate phages like lambda can be classified as episomes, since these are dispensable and exist in free as well as integrated states. The phage can infect bacteria and multiply in the normal manner leading to lysis, but this mode can shift to lysogeny where the phage DNA gets integrated with host DNA and stays there as prophage till again it becomes lytic. 'How are these two states interconnected, and how does one state shift to another' are the subject of regulation of gene expression. A map of lambda (λ) DNA is given in Figure, which has early genes in the middle and late genes at both ends of linear DNA. The linear DNA takes a circular shape during infection, so that the late genes from both ends come together in a cluster to form one transcription unit. When the phage infects a host cell, the early and delayed early genes of the phage are expressed irrespective of whether the phage has to follow lytic or lysogenic mode of development. At this stage, a decision about alternative pathways of development is taken. If the late genes are expressed at this stage, lytic cycle sets in, but if a regulator gene (*cI*) synthesizing a repressor is expressed, the late genes can not be expressed and lysogeny sets in (Fig.). It is interesting to note that repressor protein coded by *cI* regulates its own synthesis also by working as an activator, so that if repressor is inactivated, its further synthesis does not take place and the phage is forced to enter a lytic cycle. Mutants in the gene *cI* cannot maintain lysogeny, since then the late genes can not be stopped from being expressed, due to the absence of active repressor.

Three Sets of Genes in Phage Lambda λ

Immediate Early Cro and N Genes and their Control Regions (P_R/O_R and P_L/O_L) : As soon as the phage infects, transcription is initiated with the help of two promoters, P_R on the right side and P_L on the left side. The transcription starts on both DNA strands, one strand being transcribed on the left side and the other on the right side. There are two immediate early genes *cro* and *N*, *cro* being transcribed on the right side

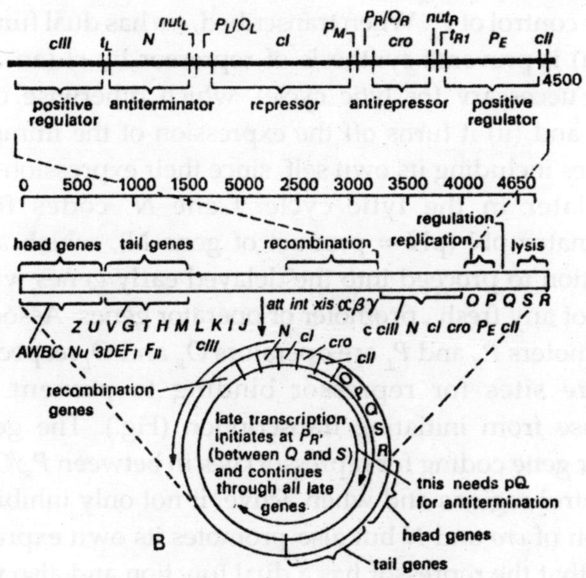

A map of lambda (λ) phage showing clustering of genes (A) and the circularization of DNA during infection (B) thus bringing together the genes on extreme left and extreme right.

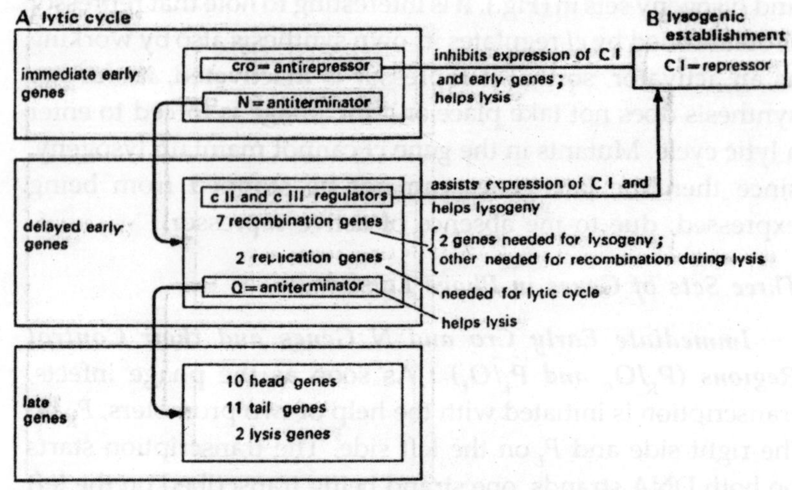

Regulation of the activity of three classes of genes in lytic cycle and in lysogenic establishment in lambda phage.

under the control of P_R and N being transcribed on the left side under the control of P_L. When transcribed, *cro* has dual function, namely (i) it prevents synthesis of repressor by *cI* (an action which is necessary for lytic cycle), which otherwise causes lysogeny and (ii) it turns off the expression of the immediate early genes including its own self, since their expression is not needed later in the lytic cycle. Gene *N*, codes for an antiterminator pN (pN = product of gene N), which allows transcription to proceed into the delayed early genes without the need of any fresh , promoter or operator genes. Associated with promoters P_R and P_L are operators O_R and O_L respectively which are sites for repressor binding to prevent RNA polymerase from initiating transcription (Fig.). The gene *cI* (regulator gene coding for repressor) lies in between P_R/O_R and P_L/O_L control regions and when active, it not only inhibits the expression of *cro* and *N* but also promotes its own expression (Fig.), so that the repressor has a dual function and also works as an activator for its own autogenous expression and regulation.

Delayed Early Genes (cII/cIII and Q) and their Control on Lysogeny and Lytic Cycle : Among the delayed early genes, regulators c*II* (on the right) and c*III* (on the left) regulate synthesis of repressor by cI, to allow phage enter lysogeny (Figs.). On the other hand, a gene *Q* (on the right) is a regulator which acts as an antiterminator, and allows the transcription of late genes meant for lysis. The two pathways (lytic and lysogenic) are so intimately related that it is difficult to predict which pathway will be followed and how is the decision for two alternative pathways is actually taken. This will be further discussed later in this chapter.

Late Genes for Lysis (right end) and, Tail and Head Genes (left end) : In the phage particle, the late genes are found at the two ends of the linear DNA molecule, genes *S* and *R* on

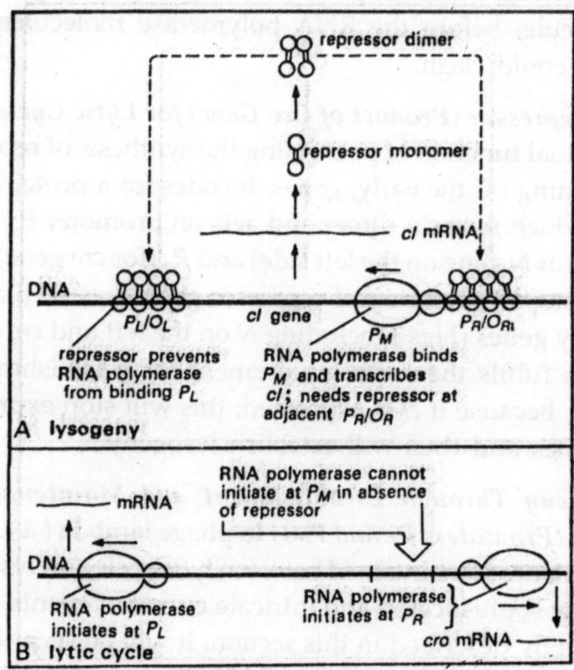

The maintenance of lysogeny by an autogenous circuit of repressor production (A), and the onset of lytic cycle due to interruption in this circuit caused by absence of repressor (B).

the right end, and the tail and head genes on the left end. Genes for recombination are also found on the left side of *cIII* and are transcribed as late genes. Soon after the infection, the two ends of DNA molecule join to form a ring, so that all late genes are arranged in a single group containing *S-R* genes from right end and head and tail genes *A* to *J* from left end (Fig.).

There is a promoter gene P_R between genes *Q* and *S*. In the absence of *pQ* (product of *Q* gene), the transcription from P_R is constitutive, but terminates at t_{R3} lying close to P_R' giving a product, 6S RNA (194 bases long). However, when *pQ* is present, it suppresses t_{R3} and 6S RNA is extended with the result that late genes are heavily expressed. Thus, *Q* gene induces transcription of late genes, which continue to be transcribed all through their length. On the left side also, late genes continue to be transcribed. Transcription from both sides stops on the ring molecule, before the RNA polymerase molecules from two sides could clash.

Anti-repressor (Product of Cro Gene) for Lytic Cycle : *Cro* gene has dual function of preventing the synthesis of repressor and of turning off the early, genes. It codes for a protein (9000 daltons) which forms a dimer and acts on promoter P_M (for *cl* gene), P_L (for *N* gene on the left side) and P_R (for *cro* gene) itself, so that it stops expression of repressor gene *cl* as well as that of the early genes (Figs.) including *N* on the left and *cro* on the right. This fulfils the early requirement for establishment of lytic cycle, because if *cl* is expressed, this will stop expression of late genes and thus will establish lysogeny.

Lysogeny Through Establishment and Maintenance of Repressor (Promoters PE and PM) : In phage lambda (λ), a very delicate balance is maintained between lytic cycle and lysogeny, with a very sophisticated and intricate circuit of events, which will be briefly described in this section. It will be seen that the repressor's synthesis is first established with the help of a promoter P_E under the control of genes *eII* and *cIII*, and is subsequently maintained by the repressor itself through a promoter P_M.

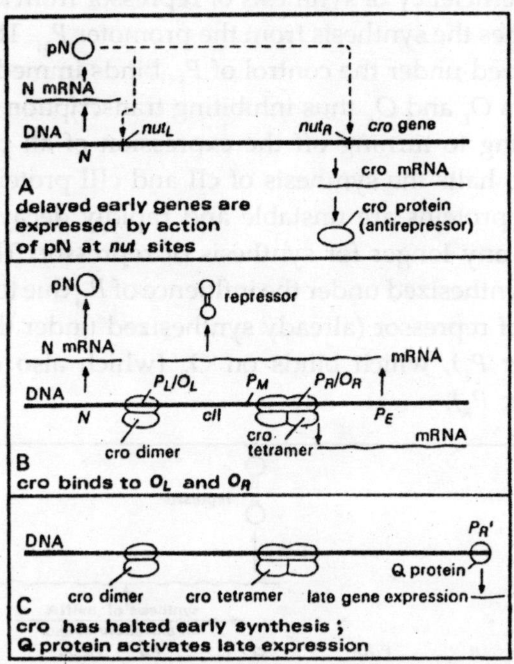

Synthesis and role of Cro protein in prevention of repressor maintenance both through P_M (directly due to binding at O_L and O_R) and P_E (indirectly due to switching off the delayed early genes), thus leading to lysis.

Promoter P_E (or P_{RE}) for Repressor Establishment and the Clear Plaque Genes (cI, cII, cIII) : There is a promoter gene P_E lying between *cro* and *cII* genes. The product of genes *cII* and *cIII* act on the promoter P_E, so that the RNA polymerase enzyme can initiate transcription towards the left thus synthesizing RNA on genes *cro* and *cI*. This transcript will consist of :

(i) the transcript from the antisense strand of *cro* gene, which can not be translated (*cro* gene is transcribed in the right direction from P_R to yield translatable mRNA) and

(ii) mRNA from *cl* gene which can be translated to give rise to repressor (Fig.).

The efficier.cy of synthesis of repressor from *cl* through P_E is 7-8 times the synthesis from the promoter P_M. This repressor synthesized under the control of P_E, binds immediately to the operators O_L and O_R thus inhibiting transcription from P_L and P_R, leading to turning off the expression of all phage genes. This also halts the synthesis of cII and cIII proteins. Since cII and cIII proteins are unstable and rapidly decay, P_E can not be used any longer for synthesis of repressor. The repressor is now synthesized under the influence of P_M due to the positive control of repressor (already synthesized under the control of promoter P_E), which binds on O_R (which also contains the promoter P_R).

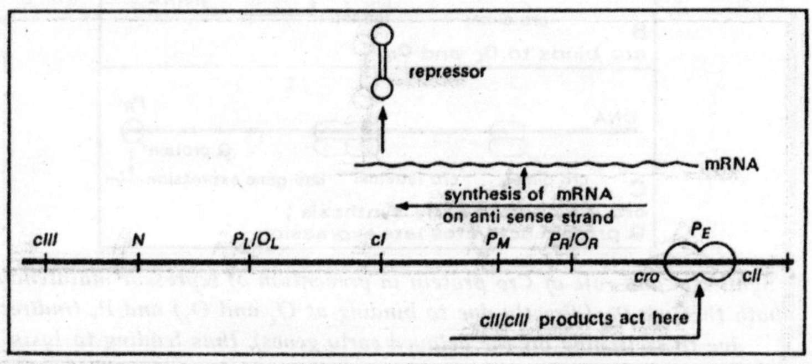

The role of cII and cIII in repressor synthesis due to their action at* PE *site leading to transcription of antisense strand.

The genes *cI, cII* and *cIII* are so named, because mutants in these genes produce clear plaques, due to the presence of only lysed cells. In the wild type phage, when the lysis occurs, due to the presence of some lysogenic bacteria, clear plaques are not obtained (plaque is a region on the culture medium on the plate, generated due to lysis of bacterial cells). Mutants in *cI* (*cI*⁻) differ from the mutants *cII*⁻ and *cIII*⁻, since the former can neither establish, nor maintain lysogeny. The mutants *cII*⁻ and *cIII*⁻ have difficulty in establishing lysogeny, but once established, lysogeny can be maintained. This suggested that the products of genes *cII* and *cIII* are needed only for the establishment of lysogeny, but not for its maintenance. The genes *cII* and *cIII* are thus needed to circumvent the difficulty of autogenous circuit of *cI*, which can not start the synthesis of repressor *de novo*. The different steps involved in establishment and autogenous maintenance of lysogeny are shown in Figure.

Promoter P_M (P_{RM}) for Autogenous Maintenance of Repressor: Once the synthesis of repressor is established

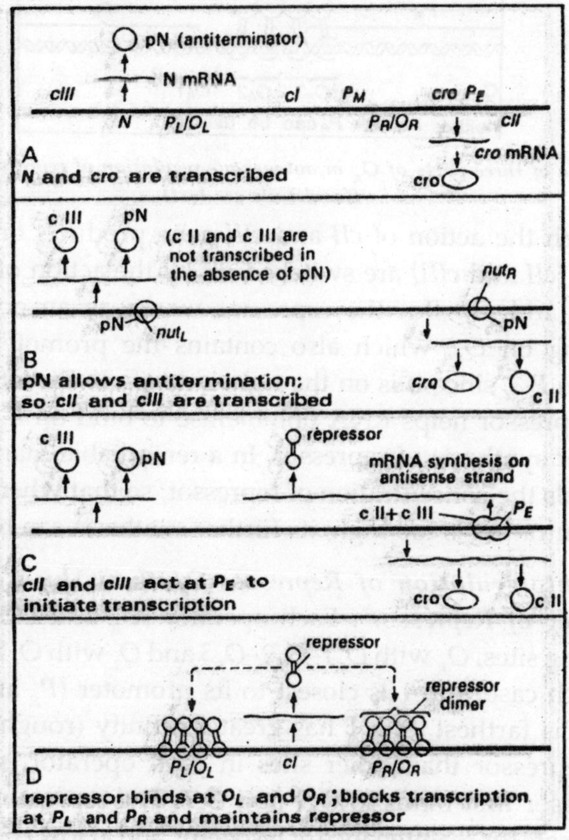

Establishment of lysogeny through a cascade involving the antiterminator protein pN and its further maintenance by the onset of an autogenous repressor maintenance circuit.

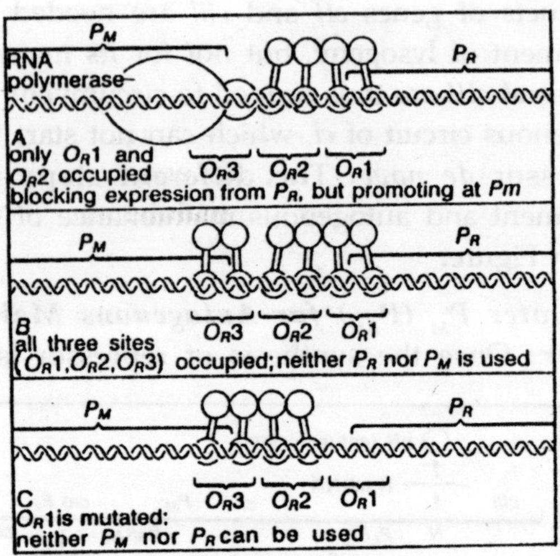

The role of three parts of O_R in autogenous regulation of repressor synthesis (for details see text).

through the action of *cII* and *cIII* gene products on $P_{E'}$ these genes *(cII* and *cIII)* are switched off by the action of repressor on O_R and O_L. But the repressor works as an activator by binding on $O_{R'}$ which also contains the promoter P_M (also known P_{RM} since it is on the right side of *cI)*. By acting on $O_{R'}$ the repressor helps RNA polymerase to bind on P_M and start further synthesis of repressor. In a remarkable manner it also controls the concentration of repressor, so that when repressor concentration is too high, its further synthesis can be stopped.

Autoregulation of Repressor Synthesis by Cooperative Binding of Repressor : Each operator (O_R and O_L) has three binding sites, O_R with $O_R1–O_R2–O_R3$ and O_L with $O_L1–O_L2–O_L3$. In each case, site 1 is closest to its promoter (P_R and P_L) and site 3 is farthest. Site 1 has greater affinity (roughly tenfold) for repressor than other sites in each operator, so that the repressor first binds to O_R1 and O_L1. The repressor functions as dimers for maintenance of lysogeny, and at low concentration it binds only to sites O_R1 and O_L1, while with increasing concentration, the repressor dimers can also bind to sites 2(O_R2

and O_L2) and 3(O_R3 and O_L3). Presence of repressor on site 1 increases its affinity for site 2 but this interaction does not proceed to site 3. When lysogeny is established, only sites 1 and 2 are occupied but site 3 is not occupied. Since P_R lies within O_R1 and P_L lies within O_L1, occupancy of O_R1–O_R2 and O_L1-O_L2 physically blocks P_R and P_L respectively and inhibit the expression of all genes thus establishing lysogeny.

The relationship of O_R and P_M is such that P_M (site for RNA polymerase binding) lies rather close to O_R2. When dimers are bound to O_L1–O_L2, the dimer at O_R2 interacts with RNA polymerase in such a way that the latter can utilize P_M for transcription (Fig.). It is interesting that a repressor dimer bound on O_R2 can inhibit transcription from P_R, but promote transcription from P_M.

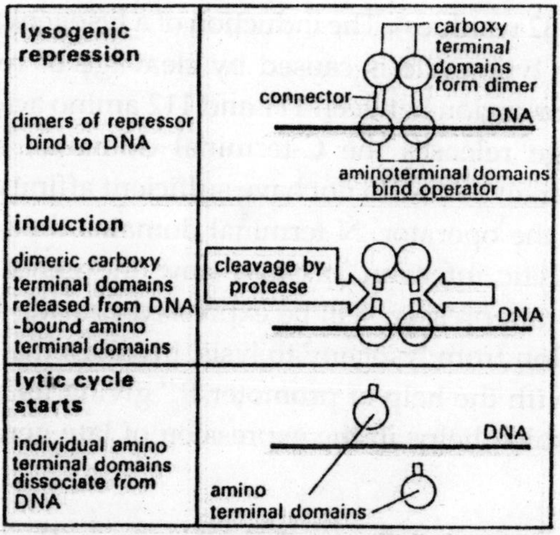

lysogenic repression dimers of repressor bind to DNA	carboxy terminal domains form dimer connector DNA aminoterminal domains bind operator
induction dimeric carboxy terminal domains released from DNA -bound amino terminal domains	cleavage by protease DNA
lytic cycle starts individual amino terminal domains dissociate from DNA	DNA amino terminal domains

The role of C-terminal domains in controlling the binding of repressor dimers to the operator site (through the N-terminal domain of repressor). (Note that due to cleavage of C-terminal domains, the N-terminal domains lose their affinity for binding).

However, when repressor concentration increases to such an extent, that it may now occupy even O_R3 in addition to O_R1–O_R2, the transcription from *cI* is prevented. This effect depends on the fact that P_M lies within O_R3 and when O_R3 is occupied by repressor, it is not available for RNA polymerase. When repressor synthesis stops due to occupancy of O_R3 by repressor, the concentration of repressor consequently falls, and O_R3 again, becomes available for RNA polymerase and synthesis of repressor starts once again due to the presence of repressor on O_R2. Thus repressor is an autogenous regulator of its own synthesis causing the inhibition by negative control at its high concentration and promoting by positive control at its low concentration.

Change from Lysogeny to Lytic Cycle by Cleavage of Repressor Dimers in the Connector Region : The repressor functions as a dimer and each unit (27,000 daltons) has (i) a N-terminal domain (1-92 residues), which binds to operator; (ii) a C-terminal end (133-132 residues), which enhances the affinity of N-terminal domain for binding and (iii) a connector region (93-132 residues). The induction of a lysogenic prophage to enter the lytic cycle is caused by cleavage of repressor in the connector region between 111 and 112 amino acid residues. The cleavage releases the C-terminal domains, so that N-terminal domains now do not have sufficient affinity to remain attached to the operator. N-terminal domains thus dissociate, leading to lytic infection, because now the genes repressed earlier due to repressor, will be expressed causing lysis (Fig.). This transition from lysogeny to lysis, involves the expression of *Q* gene with the help of promoter P_R' giving its product *pQ* regulator which helps in the expression of late genes *S* and *R* (Fig.).

DNA Binding of Cro and λ Repressor Proteins : The structure of lac repressor and CAP proteins, with respect to its DNA binding properties. A similarity of CAP with Cro and λ repressor proteins, discussed in this chapter, has also been demonstrated (Table). Cro consists of a single domain with three a helices (α,α_2,α_3) and a β sheet. It has a helix-tura-helix motif that binds the operator symmetrically (Fig.). λ repressor resembles Cro, but consists of two distinct domains and a connector. The N-terminal domain of λ repressor is shown in Figure. A helix-turn-helix model for DNA binding holds good in this case also. The amino terminal ends form two long arms, which wrap around the DNA (unlike Cro).

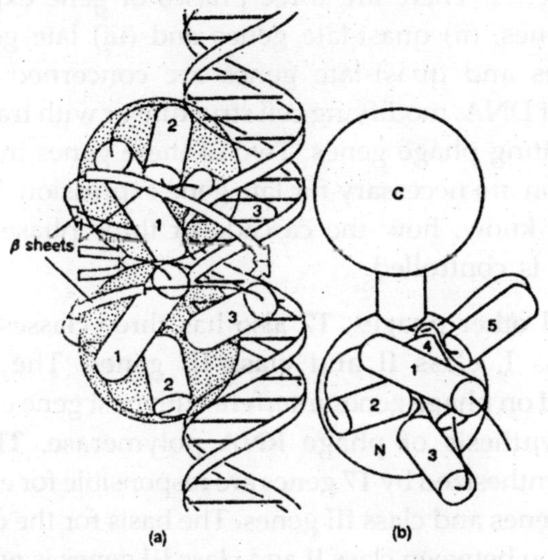

(a) Binding of Cro protein with operator (note that Cro consists of a single domain and its active form is a dimer; each monomer consists of three α helices and a β sheet); (b) N-terminal domain of X repressor; α helices 2 and 3 are involved in binding DNA.

Three Double Stranded DNA Binding Proteins and their Characteristics.

Protein	*Action*	*Monomer (no of amino acids)*	*DNA binding domains*	*Other domains*	*Actice form*	*Size of DNA site (bp)*
λ *Cro*	Repressor	66	Whole protein	none	Dimer	17
λ repressor (or CI)	Repressor or activator	236	N-terminal 92 residues	C-terminal (dimerizes)	Dimer	17
CAP	Activator or repressor	209	C-terminal 74 residues	N-terminal (binds cAMP)	Dimer	15

Functional Grouping of Genes in Phages T4 and T7 : The bacteriophage T4 has a large genome in which there are (i) essential genes and (ii) nonessential genes. These genes exhibit extensive functional grouping, but we do not fully understand the significance of the presence of so many non-essential genes. There are three phases of gene expression : (i) early genes, (ii) quasi-late genes and (iii) late genes. The early genes and quasi-late genes are concerned with the synthesis of DNA, modifying cell structure or with transcribing and translating phage genes. Two of these genes involved in transcription are necessary for late gene expression. However, we do not know, how the cascade of three phases of gene expression is controlled.

Like all other phages, T7 also has three classes of genes called class I, class II and class III genes. The enzymes synthesized on phage genes interfere with host gene expression and the synthesis of phage RNA polymerase. The phage enzymes synthesized by T7 genes are responsible for expression of class II genes and class III genes. The basis for the difference in expression between class II and class III genes is not known.

10

Operon Circuits in Bacteria

Different genes in an organism are meant for the synthesis of different proteins. All these proteins are not needed at one time. Further, the same proteins are not needed in different tissues. Specific enzymes are needed at different times and in different tissues in the life cycle of an organism. However, at all times in the life cycle, and in different tissues/organs, every cell contains same set of genes. It would be necessary, therefore, to have mechanisms which would allow only the desired genes to function at a particular time.

The activity of other genes will have to be restricted. This will require temporal (in time) as well as spatial (in space) regulation of the expression of individual genes. A variety of mechanisms are now known, which regulate gene expression at different levels including transcription, processing of mRNA and translation. Although information on regulation was available from bacteria for quite some time, only in recent years, work in this area became possible in eukaryotes. This chapter is devoted mainly to bacteria.

Induction and Repression

In *E. coli,* synthesis of β galactosidase, an enzyme meant for hydrolysis of lactose into glucose and galactose, has been studied in considerable detail.

$$\text{lactose} \xrightarrow{\beta\text{ galactosidase}} \text{glucose} + \text{galactose}$$

If β galactosides (*e.g.* lactose) are not supplied to *E. coli* cells, the presence of β galactosidase is hardly detectable, but as soon as lactose is added, production of enzyme β galactosidase increases (Fig.) as much as 10,000 times. The enzyme quantity again falls down as quickly as the substrate (*i.e.* lactose) is removed. Such enzymes, whose synthesis can be induced by adding the substrate are known as inducible enzymes and the genetic systems responsible for the synthesis of such an enzyme are known as inducible systems.

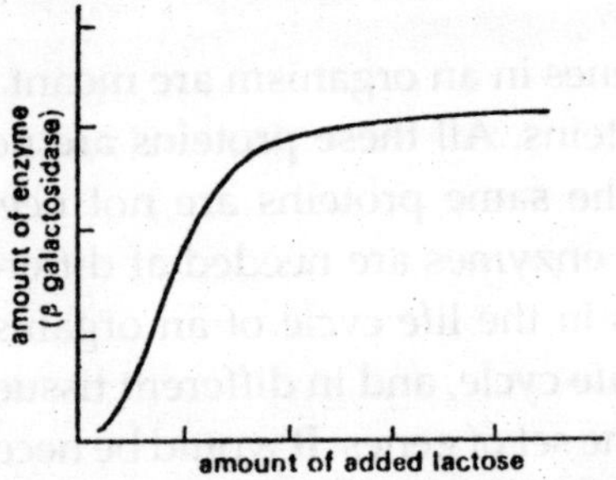

Induction of β galactosidase synthesis by lactose.

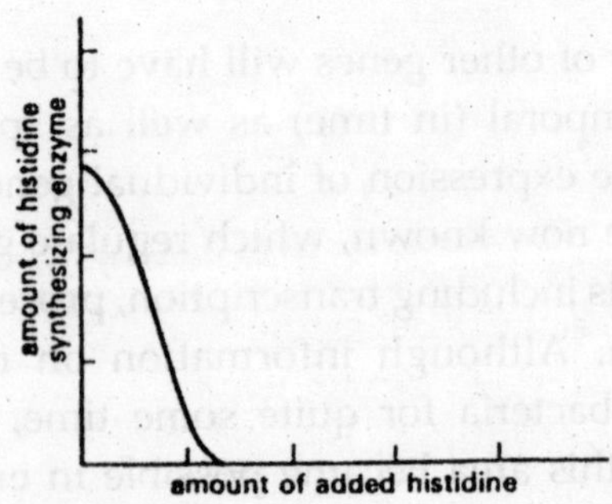

Repression of histidine synthesis enzymes by histidine.

In some other cases the situation is reverse. For instance, when no amino acids are supplied from outside, *E. coli* cells can synthesize all the enzymes needed for synthesis of different amino acids. However, if a particular amino acid like histidine is added, production of histidine synthesizing enzymes will fall down (Fig.). In such a system the addition of end product of a biosynthetic pathway will check synthesis of the enzymes needed for its biosynthesis. Such enzymes whose synthesis can be checked by adding end product are known as repressible enzymes and these genetic systems would be known as repressible systems.

Inducer and Co-repressor

The substrate whose addition induces synthesis of an enzyme, (as lactose in case of the synthesis of β-galactosidase), would be called inducer. Similarly, the end product, whose addition will check the synthesis of biosynthetic enzymes (as histidine in case of histidine biosynthetic enzymes), is known as co-repressor.

Although ordinarily only a substrate acts as an inducer, but rarely there are molecules that adequately resemble the natural inducers, but are not metabolized themselves by the enzyme. An excellent example is isopropyl thiogalactoside (IPTG), which, like several other galactosides, resembles lactose and thus has the property of induction. However, IPTG is not recognized by β-galactosidase. Such inducers which induce enzyme synthesis, without getting metabolized are called gratuitous inducers. Such inducers are very useful for conducting recombinant DNA studies, since they induce enzyme synthesis, but remain in the cell in original form. This also indicates that the substrate should have sites other than those for target enzyme, which will help the substrate in inducing enzyme synthesis.

We learnt above that in the absence of lactose no β-galactosidase is synthesized. This would mean that in the absence of 'inducer' the gene or genes responsible for the synthesis of β-galactosidase do not function. How is it brought about ? It is proposed and verified that a class of molecules called repressors are found in cells and these repressors check the activity of genes. Ordinarily, these repressors may be active or inactive. Active repressor may be rendered inactive by addition of an inducer, while an inactive repressor can be made active by addition of a co-repressor.

(1) *Inducible system :*

active repressor + inducer = inactive repressor.

(2) *Repressible system :*

inactive repressor + co-repressor = active repressor.

The Operon Model for Transcriptional Regulation **:** F. Jacob and J. Monod in 1961, on the basis of their study on the inducible system for the synthesis of β-galactosidase enzyme in *E. coli,* proposed a model in order to explain the induction or repression of enzyme synthesis. The model, popularly known as operon model, has been variously modified, ever since it was originally proposed by Jacob and Monod. In its original form 'operon model' assumed the presence of a unit of coordinated control of protein synthesis and this unit was called 'operon'. An operon consisted of (i) an operator gene which controlled the activity of (ii) a number of structural genes which take part in the synthesis of protein(s). This means that the structural genes will synthesize mRNA under the operational control of an operator gene. The operator, in its turn, is under the control of a repressor molecule synthesized by a regulator gene, which is not a part of the operon. The mechanism of regulation of protein synthesis utilizing operon model can be illustrated using the example of two different operons (lac and tryptophan operons) in bacteria.

	lac i	lac p / lac o	lac Z	lac Y	lac A	
DNA (promotes for lac i, ~40)	1040	(26)	3510	780	825	base pairs
mRNA	lac i mRNA		mRNA for lac genes (z, y, a)			
polypeptide	360 38000		1021 116,000	275 30,000	275 30,000	amino acids daltons
active protein	tetramer 152,000		tetramer 500,000	monomer 30,000	dimer 60,000	daltons
function	repressor		β-galactosidase	permease	trans-acetylase	

The lactose operon of E. coli and its regulatory gene (redrawn from Lewin's Genes IV).

The lac operon in E. coli

The *lac* operator as shown in Figure, consists of 26 base pairs and is a part of the first structural gene *lac*Z (meant for β galactosidase). There are three structural genes in *lac* operon which includes the following : (i) gene *lac*Z (3510 bp) codes for the enzyme β galactosidase, which is active as a tetramer and breaks lactose into glucose and galactose to be utilized in the cell; (ii) gene *lacY* (780 bp) codes for β galactose permease, which is a membrane bound protein and helps in transport of metabolites and (iii) gene *lacA* (825 bp) which codes for β galactose transacetylase, an enzyme that transfers an acetyl group from acetyl-CoA to β galactosides. Any mutations in *lac*Z and *lacY* genes give *lac*⁻ genotypes, which can not utilize lactose. While *lacZ*⁻ mutants abolish enzyme activity, *lacY*⁻ mutants can not take up lactose from the medium. No defect is observed in *lacA*⁻ mutants, which is a puzzle. It is possible that acetylation gives an advantage for utilizing certain non-metabolizable analogues of β-galactoside, because it results in detoxification and excretion.

In *'lac'* operon of *E. coli*, repressor is synthesized through activity of a gene *lacI*, known as regulator gene. The repressor then attaches on operator gene and checks synthesis of mRNA (Fig.). If lactose is added, the repressor is rendered inactive, so that it cannot attach on operator gene, and the synthesis of mRNA takes place (Fig.). The activity of the regulator gene is controlled by a promoter gene on its left side. It has been shown that mutations in this promoter may increase the quantity of mRNA for repressor. This promoter is different from the promoter for lac operon which is described later in this chapter.

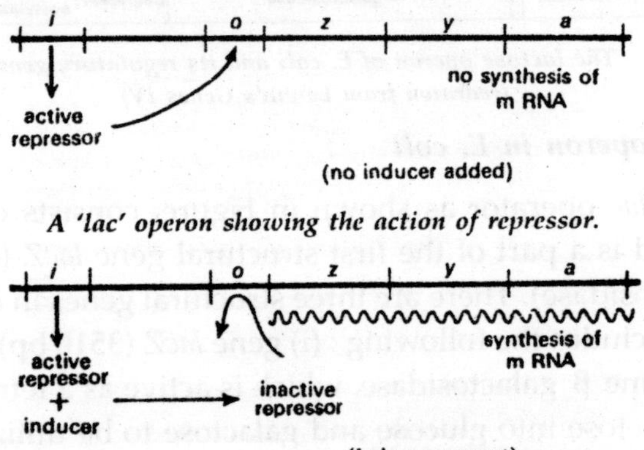

A 'lac' operon showing the action of repressor.

A 'lac' operon showing induction by inducer.

Lactose is not the Real Inducer of Lac Operon : Since in the repressed state of *lac* operon, *E. coli* cells will not synthesize β galactoside permease, one can question how lactose enters the cell. It is believed that a minimal amount (basal level) of enzyme is always present in the cell to start the metabolic pathway. It has been shown that lactose is not the real inducer of *lac* operon, but rather it is an anti-inducer, since it binds to repressor to increase its affinity for operator. On the other hand, the bound protein in inactive repressor was allolactose. It has also been shown that while β galactosidase breaks lactose into glucose and galactose, a side reaction changes galactose into allolactose and galactobiose. Therefore, in the regulation of *lac* operon, lactose is taken up and a part of it, through galactose, is converted into allolactose, which then induces the operon, and outweighs the anti-inducing effect of lactose.

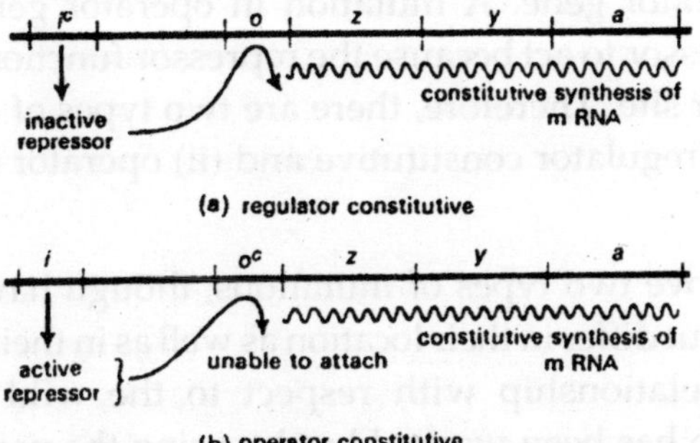

Constitutive strains resulting due to mutations in regulator gene and operator gene (redrawn from Watson : Molecular Biology of the Gene).

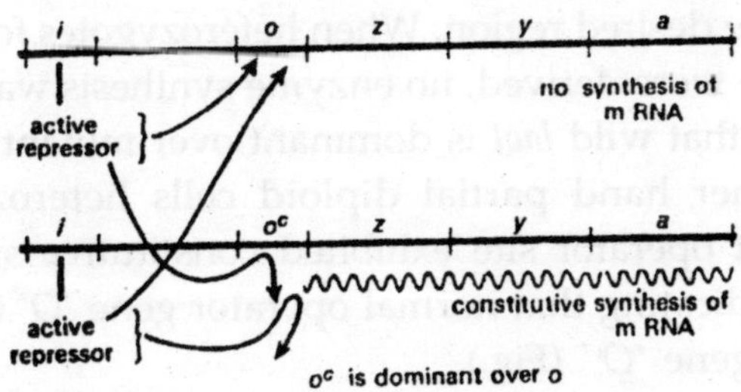

Check on enzyme synthesis in a partial diploid, heterozygous for regulator gene (I/I^c) (redrawn from Watson : **Molecular Biology of the Gene).**

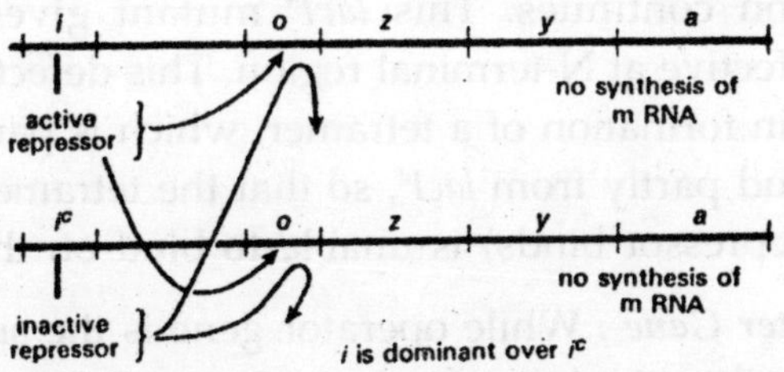

Constitutive enzyme synthesis in a partial diploid, heterozygous for operator gene (O/O^c) (redrawn from Watson : Molecular Biology of the Gene).

Constitutive Mutants : It has been found that a mutation in regulator gene *lacl* results in the absence of active repressor and consequently in uncontrolled synthesis of enzyme. Such strains which continuously synthesize enzyme independent of need are known as constitutive strains or constitutive mutants. These constitutive strains can also be obtained by mutations in the operator gene. A mutation in operator gene does not allow repressor to act because the repressor functions by acting on operator site. Therefore, there are two types of constitutive strains : (i) regulator constitutive and (ii) operator constitutive (Fig.).

The above two types of mutations, though have the same net result, but differ in their location as well as in their dominant-recessive relationship with respect to the wild type. This relationship has been worked but by using the partial diploid cells (merozygotes). Partial diploid cells are those which have one complete chromosome and a part of another chromosome covering the desired region. When heterozygotes for regulator constitutive were derived, no enzyme synthesis was observed suggesting that wild *lacI* is dominant over mutant *lacI*^{c} (Fig.). On the other hand partial diploid cells heterozygous for mutation at operator site exhibited constitutive synthesis of enzyme, indicating that normal operator gene *'O'* is recessive to mutant gene *'O*^{c}*'* (Fig.).

There is yet another class of regulator constitutive mutants, which are dominant *(lac I*^{d}*)*, so that in merozygotes *(lacI/lacI*^{d}*)*, transcription continues. This *lacI*^{d} mutant gives repressor subunit defective at N-terminal region. This defective subunit takes part in formation of a tetramer, which is partly derived from *lacI* and partly from *lacI*^{d}, so that the tetramer (the form in which repressor binds) is unable to bind on the operator.

Promoter Gene : While operator gene is the site of action of repressor, the actual site of transcription activation is different and is known as *promoter gene*, which lies to the left of operator gene. It is also believed, therefore, that perhaps RNA polymerase binds to and moves from promoter site. The repressor, when attached to operator, prevents the movement of RNA polymerase (Fig.). The promoter gene in *lac* operon was inferred from mutations at this locus, which were found to affect the maximum rate of transcription.

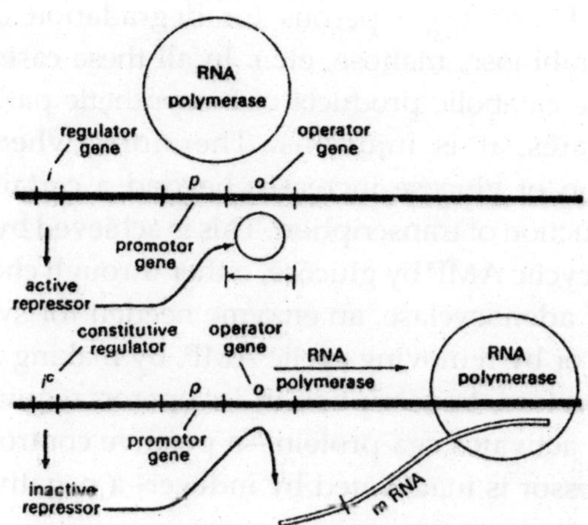

Relationship between promoter gene, repressor and RNA polymerase (redrawn from Watson : Molecular Biology of the Gene).

Thus, there are two proteins which are known to be involved in the regulation of lactose operon : (i) lac repressor and (ii) catabolise gene activator (cga) protein, also known as CAP *(cyclic AMP protein)*, or CRP *(cyclic AMP receptor protein)*. While repressor binds to operator site, cga protein binds to cga site. Moreover, repressor exercises a control by checking RNA polymerase to travel through operator, but cga protein exercises a positive control. Only when cyclic AMP molecule activates cga protein, the latter will allow RNA polymerase to bind (Fig.). A protein such as CAP, which interacts with DNA and RNA polymerase to assist in transcription initiation, is called apoinducer.

It has been shown that cyclic AMP is essential for activation of cga protein, because when cyclic AMP was removed or was absent due to mutation, transcription was reduced. In mutants lacking cyclic AMP, transcription was restored by adding cyclic AMP exogenously. It has also been shown that this cyclic-AMP—cga protein system is influenced by the level of glucose concentration. It is, therefore, sometimes described as glucose sensitive system which is common to several glucose sensitive operons in *E. coli* (*e.g.* operons for degradation of lactose, galactose, arabi ιose, maltose, etc.). In all these cases, glucose is one of the catabolic products of biosynthetic pathway and if accumulates, it is injurious. Therefore, whenever the concentration of glucose increases beyond a certain level, it leads to reduction of transcription. This is achieved by reducing the level of cyclic AMP by glucose, either through checking the synthesis of adenocyclase, an enzyme needed for synthesis of cyclic AMP or by removing cyclic AMP, by making a complex with it. Therefore, transcription in *lac* operon requires that (i) cyclic AMP activates cga protein—a positive control and that (ii) lac repressor is inactivated by inducer- a negative control.

Operator and Promoter are Cis-dominant : It should be realized that in the lac operon, operator can control only *lac* genes that are adjacent to it. If a second *lac* operon is introduced as done in partial diploids or merozygotes, none of the operators is influenced by the other operator. For instance, if in a partial diploid, one operon has normal operator *(O)* and the other has a mutant operator *(O^c)*, the operon with O^c will be expressed, but that with *O* will not be expressed.

This could be tested by the using partial diploids which not only differ in operator, but also differ in structural genes. For instance, in a partial diploid with genotype $O^cZ^+Y^-/OZ^-Y^+$, only β galactosidase will be synthesized, but not the β galactoside permease, although Y^+ is present along with O^c in the same cell, but on two different chromosomes. This phenomenon of lack of dominance extended to other chromosome (trans configuration) is called cis-dominance. This also suggests that operator does not act through a gene product like regulator gene which acts through repressor, that can diffuse and influence the other allele also. The phenomenon of cis-dominance will be expressed by all those sites, which are physically contiguous with the sequences of controls, and function by being recognized rather than by giving a gene product. It is for this reason that promoter and terminator are also cis-dominant or cis-acting, while regulator is transacting.

Uninducible Mutants (Promoter and Regulator Genes): Uninducible mutants in the *lac* operon are those which can not be induced into synthesis of enzyme even on lactose addition. These can be of two types, cis-acting and transacting.

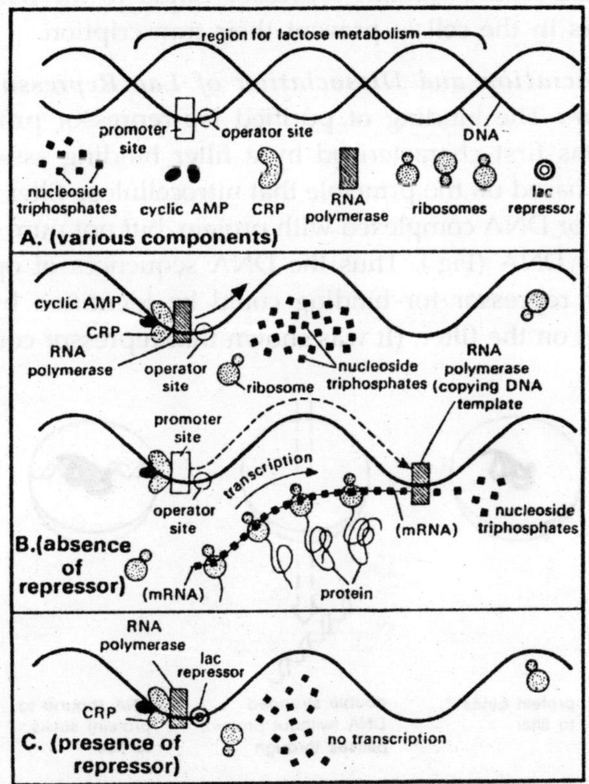

Mechanism involved in the positive control system for the regulation of gene activity in E. coli lac *operon. Note that only in the absence of the repressor. RNA polymerase enzyme can travel and transcribe* lac *operon as shown in* B. *The repressor, when present on operator site is an obstacle in the path of RNA polymerase.*

(i) Cis-acting mutations may be in the promoter site, since they act by preventing RNA polymerase from binding to the promoter, so that the operator can not function due to non-availability of RNA polymerase. (ii) Transacting uninducible mutants lie in the regulator gene, since they act by changing the repressor protein so that the altered repressor can not bind the inducer (lactose), and will always be 'locked in' to the active form for attachment to the operator. These mutations are designated as '*lac* I^s' and are dominant over wild type. This happens because mutant repressor binds to all wild type operators in the cell to prevent their transcription.

Association and Dissociation of Lac Repressor with Operator: The binding of purified lac repressor protein to DNA was first characterized by a filter binding assay. The assay is based on the principle that nitrocellulose filter retains protein, or DNA complexed with protein, but not pure double stranded DNA (Fig.). Thus the DNA sequences of operator used by repressor for binding could be identified by their retention on the filter. (It was shown that repressor could not

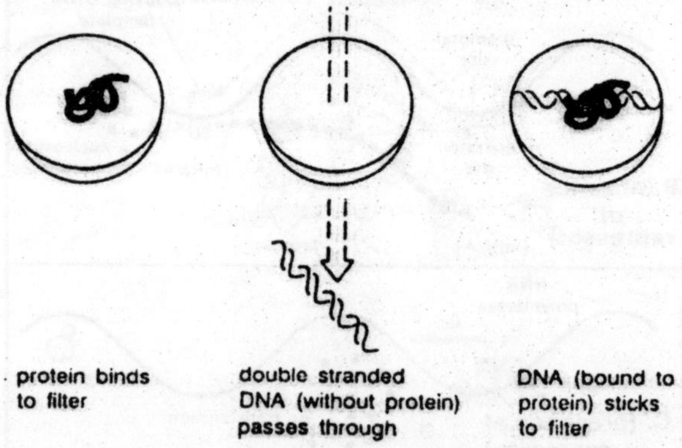

Identification of DNA sequence to which a specific protein is bound, by nitrocellulose filter binding assay.

bind on DNA obtained from an O^c mutant). The location of binding in *lac* operon extended from – 5 to + 21 within the transcription unit, thus overlapping the right end of promoter (this overlapping does not occur in all operons).' Although earlier it was believed that binding of repressor does not allow binding of RNA polymerase, now we know that both can bind simultaneously, but RNA polymerase can not initiate transcription, unless repressor is released by an inducer.

Repressor is a polypeptide of 360 amino acids and is divided into two regions : (i) 1–59 *amino acid residues* on the amino terminal end make a long headpiece, which can be cleaved at position 51, and the fragment 1–51 is called short headpiece, (ii) 60–360 amino acid residues on the C-terminal end make trypsin resistant core (trypsin cleaves the repressor at amino acid no. 59). The long and short head pieces both can bind to operator independently of the core, and therefore, it is assumed that their structure does not change in association with the core. Several models have been proposed suggesting the manner of association of repressor, but all models assume that headpieces make protrusions, which help in binding.

When lactose inducer is added, repressor will dissociate from the operator. This is explained on the basis of several models : (i) Repressor is normally in equilibrium between associated and dissociated state. When the inducer binds to dissociated repressor, it is rendered incapable of association, (ii) Inducer may bind to repressor associated with operator and changes the conformation of repressor, which allows it to get off the operator, (iii) Inducer changes the relative conformation of head and core regions of repressor, so that headpiece can not hold the DNA tightly, (iv) The repressor in

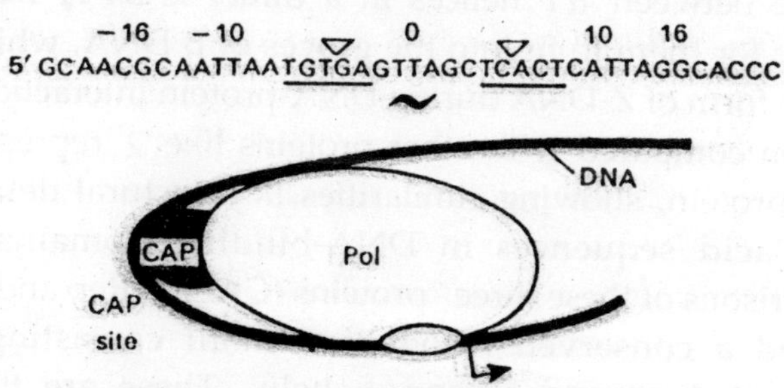

DNA bending due to binding of CAP protein.

tetrameric form may be in excess and remains associated at other random sites on DNA, but is not found free within the cell. Due to induction, when the repressor is dissociated from *lac* operator, it is stored at other random sites and used again when needed. The effect of induction, therefore, is to change the distribution of repressor , on DNA and not to release free repressor.

DNA Bending During Binding of CAP Protein : During late 1980's and 1990's, DNA-protein complexes have been analyzed to find if DNA geometry is altered, when a regulatory protein binds to it. It has been shown that DNA bends considerably, when the regulatory protein CAP binds to promoter site. Similar bending has been observed during binding of zinc finger proteins or leucine zipper proteins or TFIID. This bending of DNA is very important in activation of transcription. It has been shown that bending due to CAP alone can be about 90°, but the protein-protein interaction between CAP and RNA polymerase leads to formation of a nucleoprotein complex showing a DNA bending of upto 180° (Fig.).

The CAP protein consists of 209 amino acid residues with two domains. The carboxyl terminal domain of the protein, which is involved in DNA binding, is shown in Figure. Cyclic AMP is bound only to its larger amino terminal domain. Each CAP molecule has three a helices (α D, α E, α F) and the distance between α F helices in a dimer is 34 Å making it possible for them to fit into the groves of B DNA, which may take the form of Z-DNA during DNA-protein interaction. CAP has been compared with other proteins like λ repressor and λ cro protein, showing similarities in structural details and amino acid sequences in DNA-binding domains (Fig.). Comparisons of these three' proteins (CAP, λ Rep and λ Cro) revealed a conserved recognition motif consisting of an α-helix, a turn and second α-helix. These are therefore described as helix-turn-helix or HTH proteins.

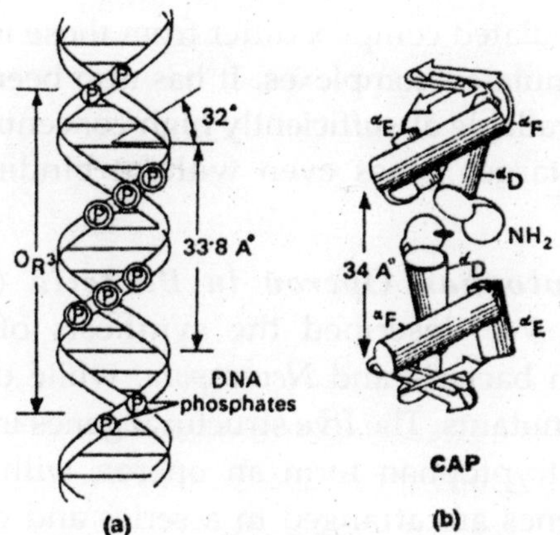

A DNA segment and its relationship (for binding site) with CAP protein involved in regulation of gene activity. Note that CAP consists of a dimer, each monomer having three polypeptides.

CAP, λ Repressor vs NTRC (Recruitment of RNAP vs Interaction with Pre-bound RNAP

As discussed earlier, CAP stimulates promoters for *lac* and *gal* operons and λ repressor stimulates promoter P_{RM} in λ phage. In contrast, NTRC is a regulatory protein (found in bacteria) that stimulates an enhancer sequence and requires ATP (*e.g. gln Ap2* gene). Further the RNA polymerase bearing σ^{70} is involved with CAP and λ repressor, while RNA polymerase bearing σ^{54} (used under nitrogen starvation conditions) is involved with NTRC. In *E. coli*, RNA polymerase bearing σ^{54} binds stably to the promoter of *glnAp2*, without the help of an activator, while that with σ^{70} binds to-*lac* promoter or λ promoter P_{RM} only with the help of an activator. It has been shown that while CAP and λ repressor recruit RNA polymerase to DNA, NTRC (and its relative NifA) works on a stably bound polymerase. As shown earlier, at most bacterial promoters, polymerase binds in at least two steps: in the first step, it binds to helical DNA to form a 'closed' complex and in the second step, the complex isomerizes to form the 'open' complex. The stability of the 'closed' complex and the requirements for the transition from 'closed' to the 'open' state in NTRC stimulated complex differ from those in CAP and λ repressor stimulated complexes. It has also been shown, that if NTRC is available at sufficiently high concentrations, it will activate the target genes even without binding any DNA sequences.

The Tryptophan Operon in Bacteria (E. coli and Salmonella): We described the synthesis of amino acid tryptophan in bacteria and *Neurospora,* while discussing the biochemical mutants. The five structural genes involved in the synthesis of tryptophan form an operon with an operator. These five genes are arranged in a series and code for three enzymes needed for tryptophan synthesis (Fig.). Transcription starts at the promoter site on the left of the structural genes. A repressor is coded by a gene *trpR,* which is not linked with the operon. The repressor has low specificity for the operator and thus remains inactive, but can become active due to association with tryptophan which functions as a co-repressor (Fig.). The repressor functions as a tetramer of four identical subunits of 12,000 daltons. The level of tryptophan, besides working as a co-repressor, also functions through feedback inhibition, where tryptophan combines with the first enzyme of tryptophan synthetic pathway and inhibits its activity through allosteric modifications (Fig.).

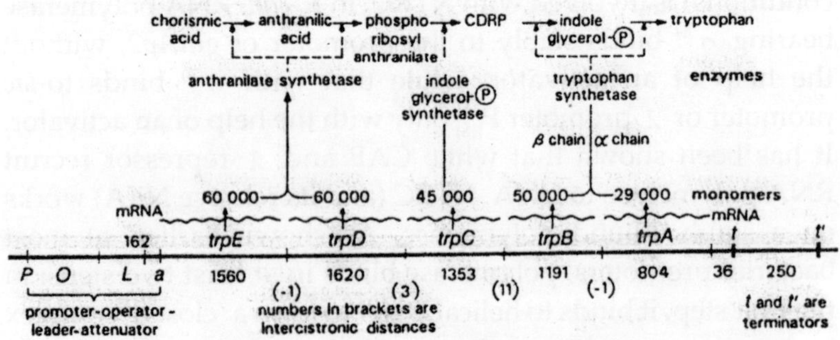

Tryptophan operon with its five genes, the promoter-operator leader-attenuator region and the corresponding biosynthetic pathway. (CDRP = carboxy phenylamino deoxyribose phosphate).

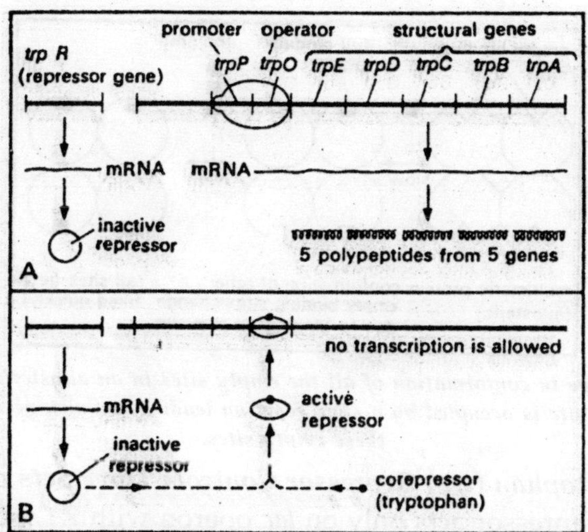

Tryptophan operon, showing transcription in the absence of corepressor (tryptophan) (as shown in A) and the prevention of transcription due to addition of corepressor, which activates the inactive repressor (as shown in B).

It is thus obvious that the synthesis of tryptophan is self regulatory, since tryptophan, when present in plenty, works firstly by feedback inhibition, and secondly through co-repression to stop further synthesis of tryptophan. The major difference between lactose operon and tryptophan operon is that tryptophan increases the affinity of repressor protein for the operator. In tryptophan operon there is yet another mechanism for regulation of tryptophan synthesis through leader sequences and attenuation.

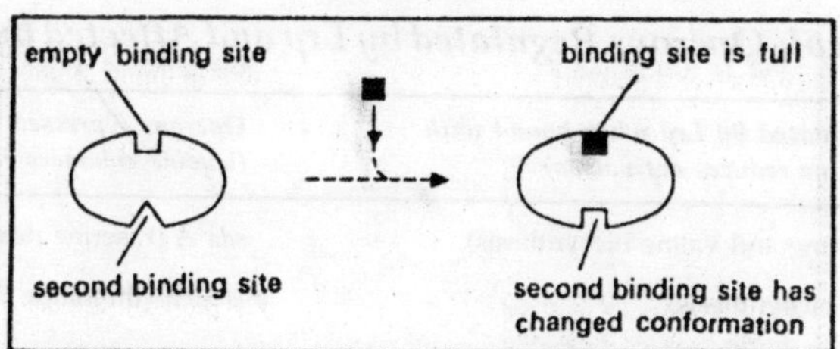

The change in conformation of the second binding site in an allosteric protein (monomer) when the first site is occupied.

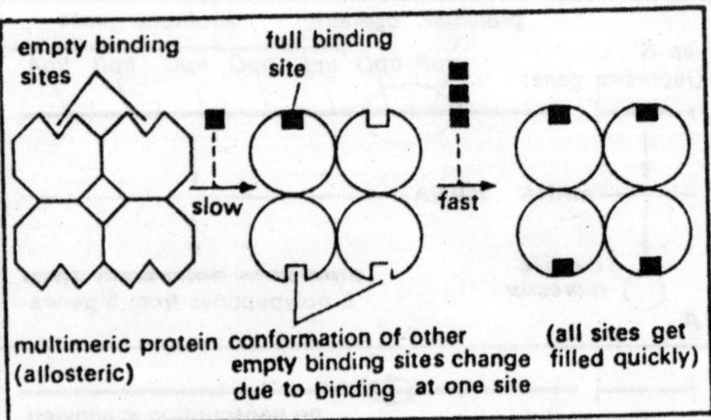

The change in conformation of all the empty sites in an allosteric protein, when one site is occupied by a slow reaction leading to quick occupation of these empty sites.

Tryptophan (trp) Repressor Controls Three Sets of Genes: The *lac* repressor acts only on *lac* operon with *ZYA* cluster of genes. However, a repressor can control several groups of genes dispersed in the genome. For instance, *trp* repressor controls following three sets of unlinked genes : (i) *trpEDBCA* meant for tryptophan synthesis enzymes; (ii) *aroH* gene coding for an enzyme needed for common pathway of aromatic amino acid biosynthesis, and (iii) *trpR* regulator gene (repressed by its own gene product = autogenous control). A related operator sequence (27bp long) is present at each of these three genes/ gene clusters, although the location of this sequence relative to start point may differ (as shown earlier). The operator may actually lie within the promoter (*e.g. trpP* locus), or downstream (*e.g. lac)* or upstream (*e.g. gal).*

Some E. coli Operons Regulated by Lrp and Affected by Leucine.

Operons activated by Lrp when bound with leucine (leucine reduces activation)	*Operons repressed by free Lrp (leucine enhances repression)*
ilvIH (isoleucine and valine biosynthesis)	*sda A* (L-serine deaminase)
serA (serine biosynthesis)	*kbl-tdh* (threonine degradation)
leuABCD (leucine bioynthesis)	*lysU* (lysyl-tRNA synthetase)
getBDF (glutamate synthase)	*livJ* (leucine transport)

Leucine Responsive Protein (Lrp) as a Global Regulator: In *E. coli,* in recent years (1990s), a protein called leucine regulatory protein (Lrp) was discovered, which responds to the presence of leucine and regulates the expression of many operons, which are collectively described as leucine/Lrp regulon. While it controls expression of many genes that are sensitive to the presence/absence of leucine, it also controls expression of other genes not affected by leucine. Therefore, sometimes Lrp is described as a global regulator of gene expression in *E. coli* and several other bacteria.

Lrp is a DNA-binding protein, which does not bind single stranded DNA. It consists of two subunits (each 1800 daltons) which make a dimer. The native Lrp protein has 163 residues and a sequence centred at 40 may represent helix-turn-helix (HTH) motif. In a cell, there are as many as 3000 molecules of this protein. This number is much higher than the number of any other regulatory protein like *lac* or λ repressor. This abundance indicates that Lrp may also function as a chromosome (DNA) organizer.

Lrp is a transcriptional regulator; when bound with leucine it functions as an activator, and when free (devoid of leucine) it functions as a repressor. Further, when it functions as an activator, its effect is reduced by leucine and when it functions as a repressor, its effect is enhanced by the presence of leucine. It is also known that some of the genes that are activated by Lrp are involved in biosynthetic pathways, and some other genes that are repressed, are involved in degradation (Table). Lrp has been shown to affect at least 11 operons directly (*e.g. gltBDF)* and several others indirectly (*e.g. glnA).* It also regulates its own synthesis and 'therefore, is also an example of an autogenously regulated protein.

Negative and Positive Controls of Transcription : The system of regulation in lactose and tryptophan operons outlined above is essentially a negative control in the sense that the operon is normally *'on'* but is kept *'off'* by the regulator gene. In other words, the gene is not allowed to express unless required. However, it is shown that through cga protein (or CAP) and cyclic AMP, a positive control is also exercised in *lac* operon (Fig.).

In a positive control the regulator gene will stimulate the production of enzyme as in case of *arabinose* operon of *E. coli.* The gene without the activator will be inactive in this case (Fig.). Most systems on regulation, studied initially, involved a negative control. The positive control systems were studied only later.

In arabinose operon (also known as OBAD operon) shown in Figure, *O, B* and *A* are structural genes, i is the initiating site, *O* is the operator site and *C* is a gene comparable to regulator gene of *'lac'* operon. Through the use of two mutations, C^- (which is characterized by reduction in transcription) and C^c (which is characterized by the constitutive synthesis of enzymes), it could be shown that the gene *C* gives rise to a protein P_1 which works as a repressor in the absence of arabinose. In the presence of arabinose, P_1 is converted into P_2, a protein working as a stimulator in transcription. The role of P_1 as repressor was confirmed when it was observed that *ara*C^+ is dominant over *ara*C^c, because if P_1 is not a repressor, then *ara*C^c should have been dominant over *ara*C^+. In this manner, arabinose operon, which was once thought to have only positive control, is now known to have both a positive and a negative control system.

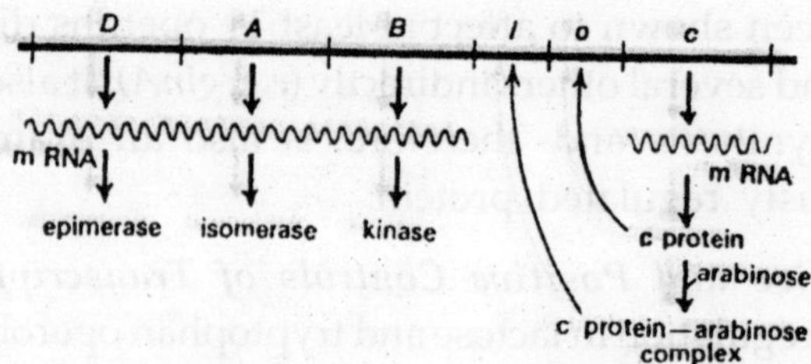

A map of arabinose operon showing positive control of protein synthesis in E. coli.

Substitution of Sigma Factor and Control of Transcription : We earlier discussed the role of sigma (σ) factor in recognition of promoter site. Therefore, whenever more than one sigma factors are available for the same core enzyme, they can be used for recognition of different promoters. Substitution of sigma factor has been shown to result in switching 'on' certain genes in *E. coli.* In some other bacteria, it leads to change in the style of life from vegetative phase to sporulation phase. Substitution of sigma factors is also used for regulation of gene expression in bacteriophages during transition from lysogenic phase to lytic phase.

Multiple Sigma Factors in E. Coli : In *E. coli,* no change from vegetative phase to sporulation phase occurs, and therefore for a long time it was believed that it has a single sigma factor, σ^{70}. However, in recent years, new sigma factors in *E. coli* have been discovered, which operate under conditions of nutrient starvation ($\sigma^{41.5}$ or σ^{S}), heat shock (σ^{32} or σ^{H}), nitrogen starvation (σ^{54} or σ^{N}) or for chemotaxis and flagellar structure (σ^{28} or σ^{F}). The details of these sigma factors are given in Table.

Multiple Sigma Factors in B. Subtilis : In bacteria like *Bacillus subtilis,* at the end of the vegetative phase, nutrients in the medium get depleted and a series of morphological changes are initiated. This involves segregation of a genome at one end of the cell, where eventually it gets surrounded by a tough spore coat. This process takes about eight hours and involves drastic changes in the biosynthetic activities of the bacterium. Synthesis of RNA continues throughout sporulation and its inhibition brings the process to an end. During this process some of the genes functioning during vegetative phase are turned off, although most genes still function. But a large number of genes specific for sporulation are expressed only during this period. At the end of sporulation 40% of mRNA is specific to sporulation.

Genes specific for sporulation can be identified by *spo⁻* mutations, which will block sporulation. These mutants may be *spo0,* where sporulation does not start at all; others could be *spoil,* etc. where later stages are affected. These genes either encode enzymes or structural proteins or are involved in transition from vegetative phase to sporulation in some other manner (see below).

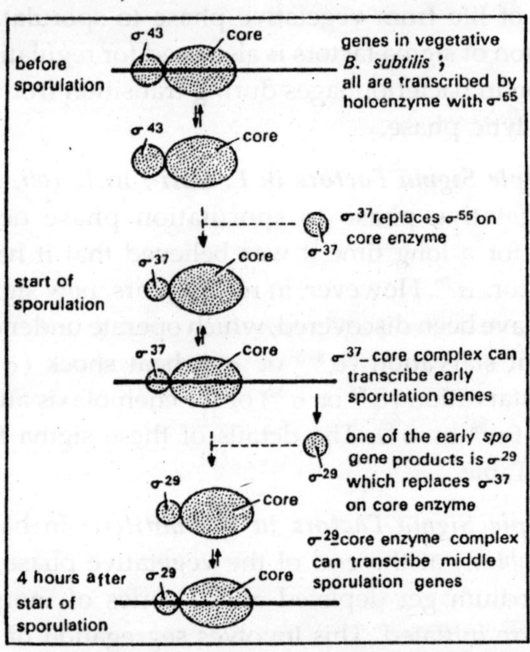

Changes in the initiation specificities of RNA polymerase by replacement of sigma factors during sporulation in Bacillus subtilis.

Some of the changes which occur during change of life style from vegetative to sporulating phase are now known and have been studied to indicate that alterations in RNA polymerase are involved in the initiation of sporulation. In RNA polymerase of *B. subtilis,* at least three sigma factors namely σ^{43} (43,000 daltons), σ^{37} (37,000 daltons) and σ^{9} (29,000 daltons) have been identified, σ^{43} is involved in transcription during vegetative phase, while σ^{37}, and σ^{39} are involved during sporulation (Fig.).

***E. coli* Sigma Factors, which Recognize Promoters with Different Consensus Sequences**

Sigma factor	*Gene*	*Mass (daltons)*	*Use*	*Consensus sequences* -35 sequence	*Consensus sequences* -10 sequence	*Separation (bp)*
σ^D or σ^{70}	*rpo D*	70,000	general	TTGACA	TATA AT	16—18
σ^S or $\sigma^{41.5}$	*rpo S*	41,500	starvation	GTTAAGC	TATACT	—
σ^H or σ^{32}	*rpo H*	32,000	heat shock	CNCTTGAA	CCCCATNT	13—15
σ^N or σ^{54}	*rpo N*	54,000	nitrogen	CTGGNA	TTGCA	6
σ^F or σ^{28}	*flal*	28,000	flagellar	TAAA	GCCGATAA	15

At the beginning of sporulation σ^{43} is replaced by σ^{37} (perhaps due to the activity of any of the five *spoO* genes), which either removes σ^{43} or modifies the core enzyme. The displaced σ^{43} can be recovered from sporulating cells. Therefore, $\alpha_2\beta\beta'\sigma^{37}$ initiates the transcription of gene needed for sporulation.

About four hours after sporulation, another protein σ^{29} appears and RNA polymerase $\alpha_2\beta\beta'\sigma^{29}$ functions for transcription (Fig.). Several other sigma factors have been isolated and characterized in *B. subtilis*. Thus, it is obvious that selective expression of genes can be caused by changes in sigma factor of RNA polymerase enzyme.

DNA Sequences Controlling Transcription

***Footprinting for Identification of DNA Sites used for Protein Binding* :** The binding site on DNA for lac repressor was characterized by filter binding assay (Fig.). However, the technique of footprinting has been extensively used for identification of DNA sites used for binding specific proteins leading to regulation of gene expression.

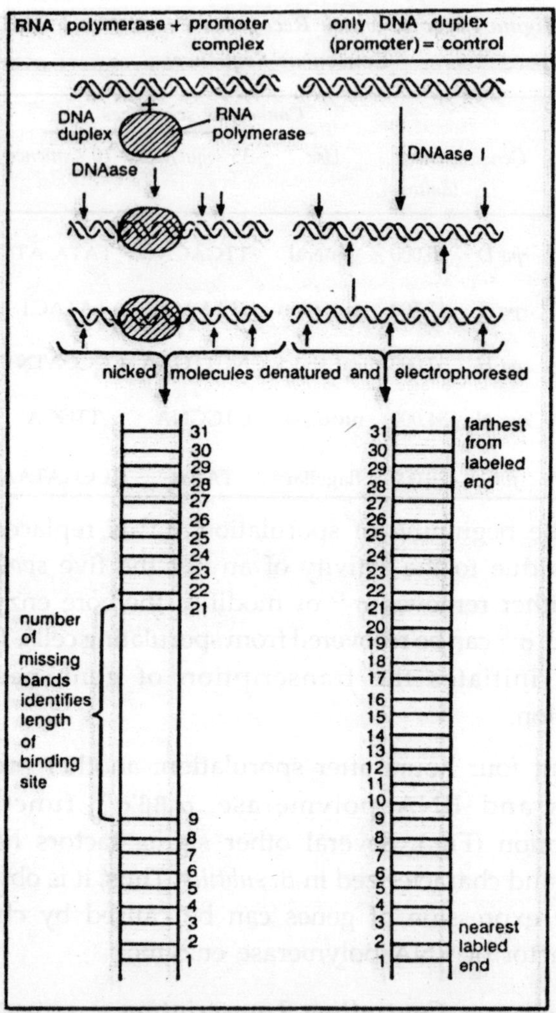

Identification of DNA binding sites for protein by the technique of DNA footprinting.

Following steps are followed using two samples of DNA to be examined (one of them used as control and the other complexed with protein, whose DNA binding sites need to be determined) : (i) DNA is end labelled; (ii) labelled DNA samples are subjected to partial cleavage (so that all cleavage sites are not attacked) with DNAaseI; (iii) population of nicked DNA molecules in both samples recovered by removing protein from the complex; (iv) DNA segments are denatured; (v) DNA segments in both samples are run in parallel for electrophoretic separation. A comparison of electrophoretic patterns in two samples will show a number of bands missing in the gel, where DNA-protein complex was used as a sample. The missing bands represent the position of DNA sites protected by protein and will allow the determination of DNA region used by protein for binding (Fig.). If above samples are treated for sequencing, the sequence of binding site can be directly 'read off from the gel.

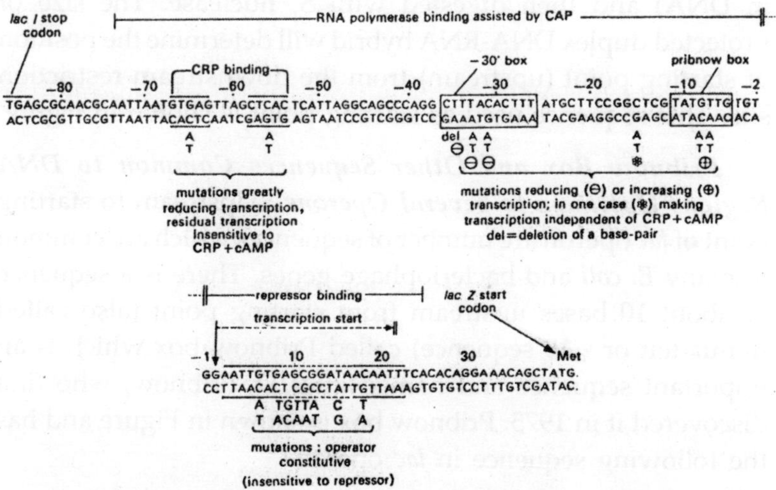

Nucleotide sequence of DNA in* E. coli lac *transcription control regions, showing functions of different regions and the effects of mutations in these regions.

DNA Sequences for CAP, RNA Polymerase and Lac-repressor : Entire DNA segment has been sequenced from transcription termination (downstream) end of *'lac I'* gene to beyond the transcription start of *Z, Y, A* genes of *lac* operon. This sequence contains the promoter and operator segments. Different parts of this sequence have been studied for their functions. For instance, sequences of three binding sites for

CAP, RNA polymerase enzyme and *'lac'* repressor could be identified through the technique of DNA footprinting. These sequences are shown in Fig.

Identification of Starting Point : In the DNA segment sequenced as above, starting point of transcription in *'lac'* operon could be identified with the help of S_1 nuclease. This enzyme has the ability to degrade single stranded DNA or RNA, but not DNA-RNA hybrids. Therefore, a DNA restriction fragment (obtained due to treatment with a specific restriction endonuclease) containing the starting point and some transcription region of DNA, was hybridized with excess of *lac* mRNA (5' end of mRNA will be complementary to start-point in DNA) and then digested with S_1 nuclease. The size of protected duplex DNA-RNA hybrid will determine the position of starting point (upstream) from the downstream restriction site. This sequence used as starting point is shown in Figure.

Pribnow Box and Other Sequences Common to DNA Regions Upstream to Several Operons : Upstream to starting point of *lac* operon are number of sequences which are common to many *E. coli* and bacteriophage genes. There is a sequence of about 10 bases upstream from starting point (also called minus ten or - 10 sequence) called Pribnow box which is an important sequence and named after D. Pribnow, who first discovered it in 1975. Pribnow box is shown in Figure and has the following sequence in *lac* operon.

–10 –5

TAT GTTG

ATACAAC

This sequence differs slightly from the consensus sequence for Pribnow box (TATAAT), where it was shown that among prokaryotes the range of conservation of bases varies from 45% to 96%. There are other common sequences located at –30 and –60 bases (upstream), the latter being CAP site, which is marked by an interrupted three base pair inverted repeat, flanked by non-inverted T-A base pair on either side as follows.

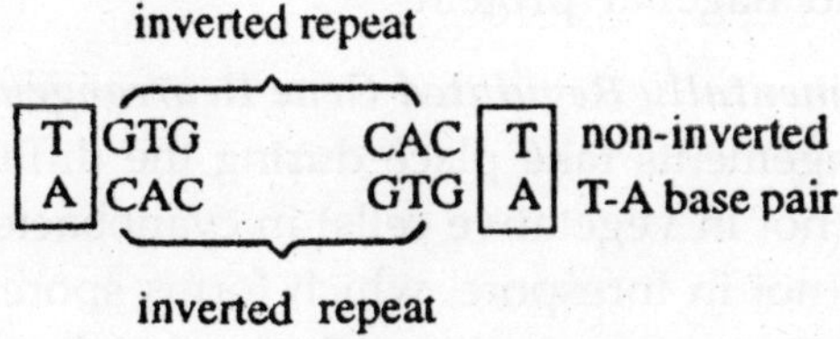

The TGTGC pentamer may lie in either orientation. This CAP binding site is common to *lac, gal* and *ara* operons and may lie at different locations relative to start point in various operons. In *gal* operon, this site may be within the promoter (between -53 to -23 bp), in *lac* operon this site lies between -72 to -52 bp and in *ara* operon it lies between -107 to -78 bp. Several of these above DNA sequences were found to be important, because mutations in these sequences influence promoter function (Fig.).

Regulation by DNA Rearrangements : Expression of certain genes in prokaryotes are also regulated through DNA rearrangements (for DNA rearrangements in eukaryotes,). These DNA rearrangements can be broadly classified into two groups.

Stochastic (Random) Gene Rearrangements : These rearrangements take place without any need, so that a small fraction of the cell population carrying rearranged genes behaves like boy scouts, always ready (if the eventuality for which they are prepared does not occur, no problem). One of the best studied examples of these gene rearrangements involves inversion of the DNA elements carrying the promoter for one of the flagellar protein operons in *Salmonella typhimurium.* In one orientation, the element promotes transcription of a gene encoding one 'flagellar protein' and 'repressor' of a distant gene encoding an alternative flagellar protein. Inversion of the elements blocks transcription of both, the 'first flagellar protein gene' and the 'repressor gene' of the second alternative flagellar protein, thus permitting expression of the second flagellar protein.

Developmentally Regulated Gene Rearrangements : These gene rearrangements take place during the differentiation of heterocysts (not in vegetative cells) in cyanobacteria or in the mother cell (not in forespore, which forms spores) of *Bacillus subtilis* during sporulation. Both the heterocyst (in cyanobacteria) and the mother cell (in *B. subtilis)* are terminal cells and do not give rise to cells of next generation, so that the gene rearrangements are not transmitted. These developmentally regulated gene rearrangements involve excision of DNA segments that interrupt coding regions of the genes. Moreover, these gene rearrangements are not random and require a series of signal transductions, gene activations and morphological differentiation.

Many species of filamentous nitrogen fixing cyanobacteria like *Nostoc* and *Anabaena* have specialized cells called heterocysts, where nitrogen fixation occurs during conditions of nitrogen limitation. Heterocysts occur at regular intervals making upto 10% of the cells in free living *Anabaena* and *Nostoc*, this frequency being much higher in symbiotic associations. The nitrogen fixation is facilitated by the presence of a 'nitrogenase complex' with three polypeptide components, that are encoded by three linked genes, *nifHDK*. In *Anabaena, nifK* gene is 11kb downstream from *NifHD* genes. This 11kb element is excised and the *nifHDK* operon is reconstructed. A second excision removes a 55kb element, which interrupts the coding region of a gene called *fdxN*. Both excisions are catalyzed by site-specific recombinases encoded within the interrupting elements themselves. In the absence of these excisions, it is impossible to transcribe the genes downstream of the interrupting elements.

Another example of DNA rearrangement is found in *Bacillus subtilis*, where some of the mother cell-specific transcriptions (including those for spore coat proteins) require a form of RNA polymerase carrying σ^K (sigK, a 27kD protein). The sigK is encoded by *sigk* gene resulting from the DNA rearrangements between two open reading frames (ORFs), *spoIVB* and *spoIIIC* (formerly believed to be separate genes). The DNA rearrangement, in this case, involves excision of a 42kb element. It was shown that the strains without this 42kb element are unable to sporulate (they sporulate if a copy of *sigk* gene is supplied) and the strains with this 42kb element sporulate normally. (For more details on DNA rearrangements in prokaryotes consult *Ann. Rev. Genet* 26 : 113-130, 1992).

Post-Transcriptional Regulation

Leader Sequences and Attenuators : While there are mechanisms controlling transcription of DNA as discussed above, there are also mechanisms, which operate- after the transcription is initiated or sometimes even when the transcription is over. For instance, in tryptophan operon of *E. coli*, the mRNA of gene *trpE* (gene for enzyme anthranilate synthetase) has a sequence of 162 bases between its initiation codon AUG and the promoter-operator region of this gene. This sequence of 162 bases is called leader' sequence and has several features of interest, which determine post-transcriptional control. (This post-transcriptional control is in addition to the normal transcriptional control provided in operon model). The leader sequence at its 3' end, but 8-12 bases before the initiating codon, has a sequence 5'AGGAGGA3', which is common to several *E. coli* genes and was discovered by J. Shine and L. Delgarno. This sequence 5'AGGAGGA3' (known as Shine Delgarno sequence) base pairs with the sequence 5'-ACC UCC UUA-OH-3' found at 3 end of 16S rRNA and thus helps in mRNA binding to ribosome.

Besides the sequence 5'AGGAGGA3' for ribosome binding, there are other sequences of interest in the leader sequence. In leader sequence of *trpE* discussed above, a mechanism is available, which allows further transcription when synthesis is needed, and stops the transcription whenever the synthesis is not needed. For instance, when tryptophan is in short supply, so that its further supply is needed, the transcription will take place and when tryptophan is in adequate supply, so that its further supply is not needed, the transcription will stop. This is achieved by a remarkable mechanism, which becomes clear by a unique feature common to leader sequences of several operons each meant for synthesis of a specific amino acid, *e.g.*, his, leu and val in *E. coli.* This unique feature includes the presence of two successive tryptophan codons at a specific site in *trpE* leader sequence. Similarly, in leader sequence of phenylalanine operon, there are three consecutive phenylalanine codons; in leader sequence of leucine operon, there are four consecutive leucine codons and in leader sequence of histidine operon, there are eight consecutive histidine codons. These regions of leader sequences each having consecutive codons for the same amino acid, for the synthesis of which the operon is meant, gives rise to mRNA for a small polypeptide about 15 amino acids long. Significance of this polypeptide synthesized by leader sequence has been explained as follows.

```
5'        10        20        30        40        50
pppAAGUUCACGUAAAAAGGGUAUCGACAAUGAAAGCAAUUUUCGUACUGAAAGG
   60        70        80        90        100
UUGGUGGCGCACUUCCUGAAACGGGCAGUGUAUUCACCAUGCGUAAAGCAAUC-
 110       120       130       140       150       160
AGAUACCCAGCCCGCCUAAUGAGCGGGCUUUUUUUUGAACAAAAUUAGAGAAUAA
   170       180
CAAUGCAAACACAAAAACCGACU.......3'
```

Met Gln Thr Gln Lys Pro Thr .....(trp E polypeptide sequence)

The leader sequence of mRNA transcribed from E. coli trp (tryptophan) *operon. Specific regions for signals are underlined. Probable ribosomes binding Shine-Delgarno sequences at 15-19 and 152-155 nucleotides; at 27-29 and 165-167, there is initiating codon AUG; two adjoining* trp *codons (UGG UGG) at 54-59; sequence 114-134, underlined by broken line, is termination structure* (attenuator) *shown in Fig.*

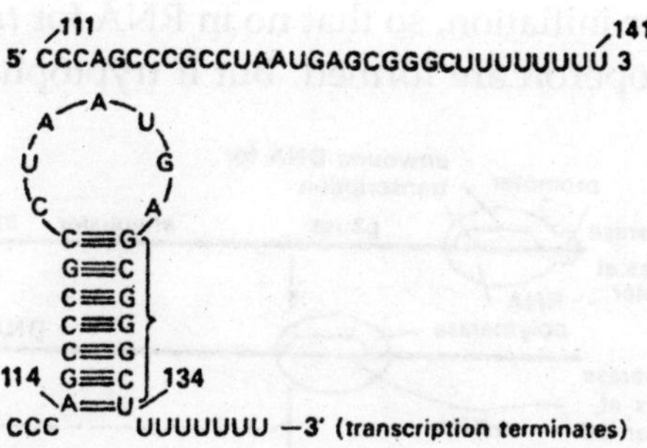

Termination structure (stem and loop), which is shown by broken underline in Fig.

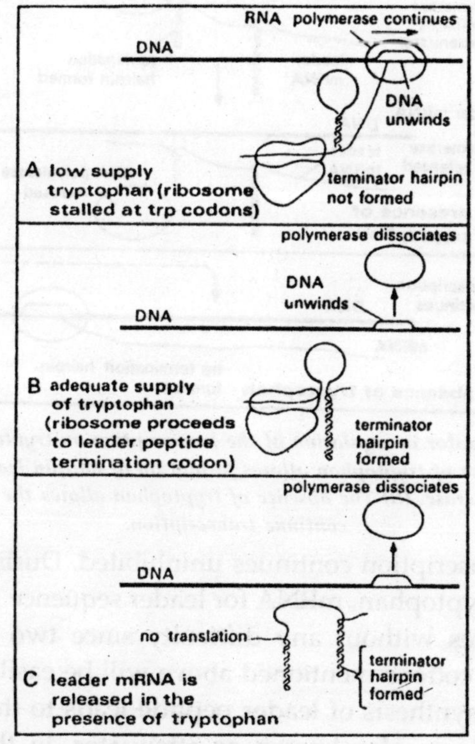

Different alternatives for RNA polymerase due to position of ribosome, which depends on hairpin formation.

In tryptophan operon, it has been shown that when tryptophan supply is adequate, transcription stops at about

147 bases after initiation, so that no in RNA for *trpE* and other genes of the operon are formed. But if tryptophan is in short

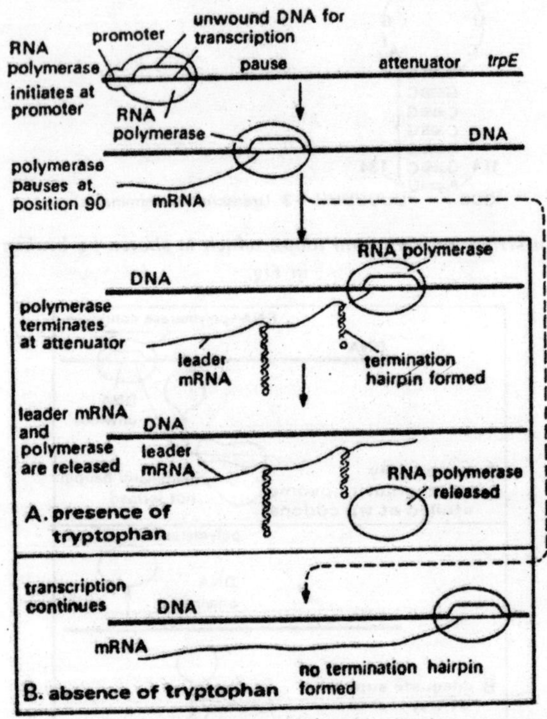

Role of attenuator in regulation of the transcription on tryptophan operon. (A) The presence of tryptophan allows formation of hairpin leading to release of RNA polymerase. (B) The absence of tryptophan allows the polymerase to continue transcription.

supply, transcription continues uninhibited. During adequate supply of tryptophan, mRNA for leader sequence is translated on ribosomes without any difficulty since two consecutive tryptophan codons mentioned above will be easily read. This favourable synthesis of leader peptide leads to the formation of hairpin loop, also known as attenuator, in the sequence between bases 114 and 134 (Figs.). This stops further synthesis of mRNA. On the other hand, if tryptophan is in short supply, the two *trp* codons (UGG, UGG) can not be translated due to lack of availability of amino acid 'trp' and thus ribosome becomes stalled at this point (Fig.) and makes the formation of terminator hairpin loop (in sequence 114-134) impossible, thus indirectly permitting transcription of *trpE* and other genes (Fig.).

The leader sequences, thus seem to have a function of attenuating (reducing) the transcription in a controlled way. It should, however be realized that such a transcriptional regulation is not possible in eukaryotes since it depends on coupling of translation with transcription and in eukaryotes, translation and transcription can not be coupled due to separation by nuclear envelope.

Autogenous Regulation of Translation : A number of examples are now known, where a protein or RNA regulates its own production. Several proteins work as repressors for repression of their own production (Table). This is achieved by binding of repressor protein to ribosome binding site (or Shine-Delgarno sequence), or initiating codon (AUG) of mRNA.

Proteins that Work as Repressor for Autogenous Regulation of Gene Expression.

Repressor	*Target gene*	*Sites of action*
R17 coat protein	R17 *replicase*	Ribosome binding site
T4 Reg A	T4 early genes	Initiation codon
T4 DNA	T4 *DNA*	Shine-Delgarno
polymerase	*polymerase*	sequence
T4 p32	T4 *gene 32*	Single-stranded 5' leader

An important example of autogenous regulation is p32 protein, which is coded by T4 *gene32* and needed for recombination, repair and replication of DNA. Whenever this protein p32 is in excess, it binds to its own mRNA and prevents initiation of translation. Several genes for ribosomal proteins, protein synthesis factors and RNA polymerase subunits (which are organized in a few operons), are also autogenously regulated at the translation level. In these cases, a regulatory protein binds to a polycistronic mRNA and inhibits expression of a contiguous sets of genes within an operon, always including its own gene. This is true about ribosomal proteins (r proteins). These r proteins are generally utilized for binding with rRNA for ribosomes assembly, but if there is excess of r proteins, one of the r proteins may bind to polycistronic mRNA and inhibit synthesis of r proteins.

In the above cases, the mDNA remains intact but can not be translated. However, there are other systems, where mRNA may be degraded as in case of synthesis of tubulin in eukaryotic cells. In this case tubulin molecules are normally assembled as microtubules, but when present in excess these may bind at a short specific sequence of its mRNA causing its degradation.

Regulation by Alternative Splicing : We discussed the phenomenon of alternative splicing, which means that a transcript or a precursor mRNA may follow one of the several paths of alternative splicing, so that the same transcript may lead to the synthesis of different protein 'isoforms' in time and space. This mechanism has been shown to be widespread in organisms ranging from viruses to mammals and is used for regulation of gene expression at the post-transcriptional level.

Regulation by Anti-sense RNA : During late 1980's and early 1990's, it has been shown that the translational control of protein synthesis can be exercised by using RNA, which is complementary to mRNA, so that this introduced RNA may form RNA-mRNA hybrids and prevents mRNA from being translated. Since this RNA will be complementary to mRNA and will interfere with its translation, it is called anti-sense RNA or mic RNA (mic = mRNA interfering complementary RNA). Such anti-sense RNA can be used to stop translation of mRNA in cells at will, and therefore is reported as a method analogous to production of mutations at will. But more important is the use of this phenomenon by cells to exercise control of gene expression at the level of translation.

In transposon Tn10, translation leading to the production of transposase (a protein required for transposition) is inhibited through the production of *anti-sense* RNA. The production of *anti-sense* RNA is controlled by a promoter which directs transcription in opposite direction to the transposase gene, producing *anti-sense* RNA, which overlaps 36 bases of mRNA at its 5' end, and is complementary to it. There is definite evidence that the synthesis of transposase enzyme in Tn10 is regulated through the synthesis of anti-sense RNA.

In *E. coli* also, the role of *anti-sense* RNA in regulating gene expression has been demonstrated in case of two genes for outer membrane protein. These two genes are *ompC* and *ompF* (Omp = outer membrane protein). The total amount of these two proteins remains constant, although their relative quantities may differ according to requirement.

It has been shown that a DNA fragment upstream of *ompC* when introduced into *ompF*$^+$ cells, inhibited the production of *OmpF* protein. This DNA fragment coded for a small 174 base RNA and is termed mic RNA (mRNA interfering complementary RNA), which was transcribed in a direction opposite to *ompC* gene. Since no ribosome binding site is present on *mic* RNA, it can not be used for protein synthesis.

But it has extensive homology with 5' end of *ompF* mRNA, and therefore, can form a hybrid with it. Since the sequences present in mic RNA include sequences complementary to ribosome binding site of *ompF* mRNA, hybrid formation will inhibit the translation initiation. Such a mechanism is believed to result in coordinate regulation of the production of OmpC and OmpF proteins.

The above two genes *ompC* and *ompF are* known to be involved in controlling the response to changes in osmolarity. Another locus *envZ* codes for a receptor protein (EnvZ) that serves as an osmosensor. EnvZ activates another protein (OmpR), the product of *ompR*. The protein OmpR regulates the transcription of *ompF* and *micF* genes. Increase in osmolarity leads to synthesis of *micF* RNA, which turns off translation of *ompF* mRNA through complementary base pairing" (antisense RNA mechanism).

Another example of the regulation of translation through anti-sense RNA is available in the nematode, *Coenorhabditis elegans*. In this case, expression of the gene *lin4* (regulating larval development) is controlled by another gene *linl4* coding for two transcripts (22 and 61 nucleotides long) having complementarity with a sequence transcribed from *lin4*. It is possible that these transcripts derived from *linl4* work as anti-sense RNA molecules.

Even in eukaryotes, it is suggested that anti-sense RNA (or even anti-sense DNA) may be used to inhibit gene expression. The presence of complementary RNA (or DNA) sequence was actually shown to reduce gene expression. The extent of utilization of this mechanism for control of gene expression in cell is not known, but it is being definitely utilized now for artificial manipulation of gene expression.

Inhibition of cloned genes by antisense RNA (or DNA) technology has been achieved in several cases, using recombinant-DNA technology through the construction of antisense expression vectors. These genes whose expression was thus manipulated, included the following :

(i) thymidine kinase (TK) gene from herpes simplex virus (HSV),

(ii) actin gene (actin protein is a major component of cytoskeleton),

(iii) genes for pigment synthesizing enzyme in petunias; (inhibition of this enzyme led to unusual pigmentation in flowers),

(iv) genes for enzymes responsible for fruit ripening in tomatoes (inhibition of these genes will help in slow ripening) and

(v) plant virus genes causing diseases. In all these cases, an antisense expression vector was constructed, so that the orientation of the gene of interest is reversed with reference to promoter sequence. These will lead to the synthesis of antisense RNA inhibiting the translation of mRNA. Using this technology inhibition of the expression of several genes as above was achieved successfully.

Not only antisense RNA, but artificially synthesized oligonucleotides (both DNA and RNA) have also been used to inhibit the translation of mRNA. This technology will be increasingly used in future for study of gene function and for treatment of diseases.

Repression and Activation of Translation : The number of regulatory circuits identified at the translational level increased steadily during 1980s and 1990s. Besides, the mechanisms of regulation discussed above, control at the translational level is also achieved by formation of secondary and tertiary structures, which either inhibit or facilitate translation.

These secondary or tertiary structures are dependent on intramolecular base pairing in mRNA, and can be variously modified due to effect of molecules acting on specific target sites. The regulatory circuits at the translational level may either involve repression or activation.

Repression of Translation : At least three different situations of repression of translation have been proposed (Fig.) :

(i) A repressor-effector molecule (*e.g.* phage R17 coat protein) may recognize and bind to a specific sequence or to a specific secondary structure (involving SD region and AUG codon), thus blocking initiation of translation, through blocking of the ribosomal binding region,

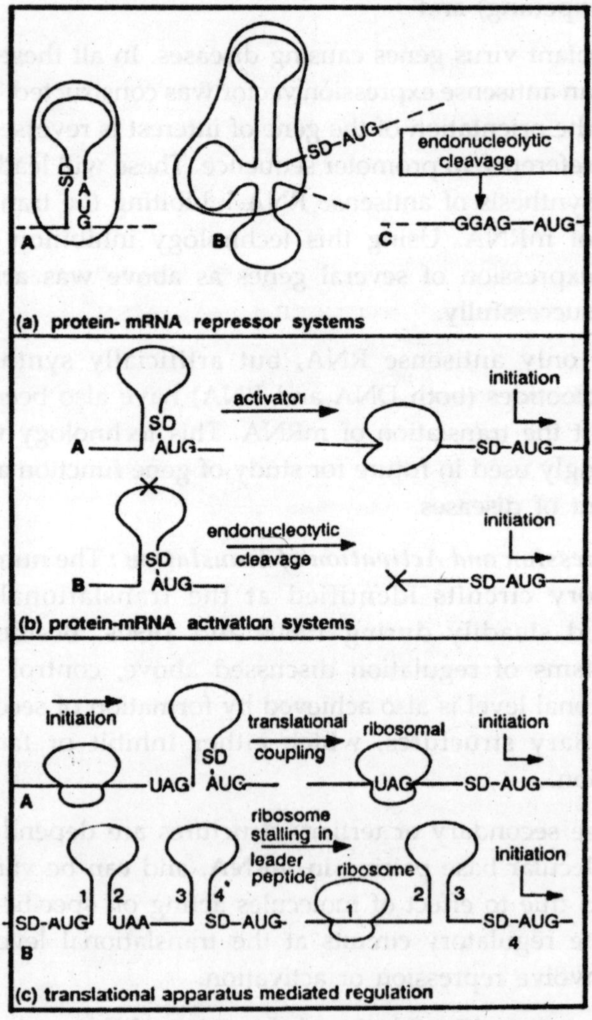

Some models of translational regulation in prokaryotes (repression and activation.

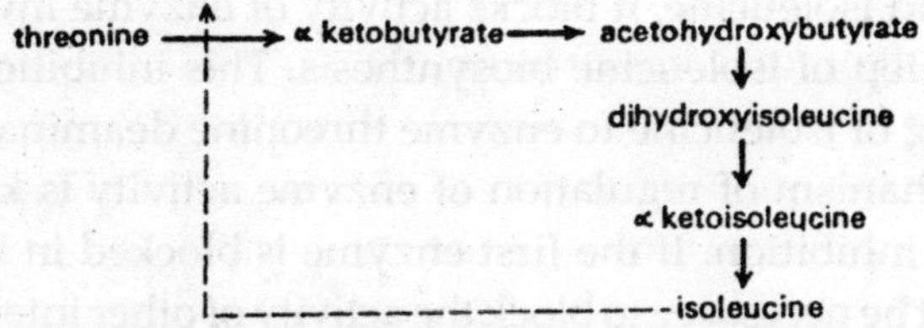

***Feedback inhibition in the synthesis of isoleucine in* E. coli.**

(ii) A repressor effector molecule may bind to an operator (not involving SD region and AUG codon), thus stabilizing an inhibitory mRNA secondary structure (*e.g.* threonyl-tRNA synthetase; phage T4 RegA protein, which recognizes TIR sequence domains of a number of T4 genes),

(iii) An effector molecule (an endonuclease) can inhibit initiation of translation by endonucleolytic cleavage of SD region (*e.g.* phage T4 *motA* mRNA SD region).

Activation of Translation : Cases of translational activation have also been reported, even though they are less well understood. The most likely mode of action of these positive effectors or activators is through destabilization of inhibitory secondary structures in mRNA either through simple binding or by endonucleolytic cleavage (Fig., *e.g.* phage λ and phage T4 RNase III). Translation of certain genes is also influenced by the translation of neighbouring genes, a phenomenon described as translational coupling. Under this mechanism, the ribosome terminating the translation on a preceding gene of a polycistronic mRNA, may sometimes release the inhibitory secondary structure found in the mRNA of the following gene (*e.g.* 6-phosphogluconate dehydrogenase gene and erythromycin methylase gene; see Fig.).

Feedback Inhibition : In some cases, the end product of a particular biosynthetic pathway gets accumulated if not utilized. Sometimes such an accumulation may stop further synthesis of this substance. For instance, if *E. coli* cells are supplied with amino acid isoleucine, it blocks activity of enzyme involved in the first step of isoleucine biosynthesis. This inhibition is due to binding of isoleucine to enzyme threonine deaminase (Fig.). This mechanism of regulation of enzyme activity is known as feedback inhibition. If the first enzyme is blocked in this case, it may not be necessary to block the activity of other intermediate enzymes, since there will be no accumulation of intermediate products.

The end product in the above case perhaps acts through allosteric transformation of the enzyme, as a result of which the site of activity of the enzyme gets transformed in such a way that it can not perform its catalytic function. This would mean that in feedback inhibition the enzyme does not function, while in repression no mRNA synthesis and hence no enzyme synthesis takes place. Therefore, this is not a case of control of gene activity, but gene expression is controlled only at a post-translational level.

Signal Transduction and 'Two Component Regulatory Systems' : In prokaryotes, a wide variety of stimuli are sensed and transduced leading to altered gene expression. Important examples in *E. coli,* include chemotaxis, nitrogen assimilation and outer membrane protein expression. Two sets of proteins, bearing sequence similarities, are known to be involved in these regulatory *systems,* which are, therefore, described as 'two component regulatory systems'. The first component is a protein kinase, described as 'kinase', that autophosphorylates (using ATP) a histidine residue, and transfers its phosphoryl group to an aspartate residue of the second component described as 'regulator'. Finally, a phosphatase activity may reset the system by restoring the regulator to its unphosphorylated state. Regulator phosphate levels are controlled in response to environmental condition. The above signal transduction, using 'two component systems' can be illustrated using the following three examples.

Bacterial Chemotaxis : In homogenous environment, *E. coli* swims smoothly, but it moves towards or away from wide variety of compounds classified as 'attractants' or 'repellents' respectively. This mechanism involves a post-translational regulation of gene expression, involving reversible phosphorylation of two proteins, CheA (kinase) and CheY (regulator).

Nitrogen Assimilation :Many species of bacteria regulate both the amount and the activity of 'glutamine synthetase' enzyme in response to nitrogen availability. The gene for this enzyme is *glnA* and its major promoter is *glnAp2*. Transcription of the gene is regulated by a two component system (NtrB = kinase, NtrC = regulator). NtrB autophosphorylates and transfers its phosphate group to NtrC, which in its phosphorylated form activates *glnAp2*. The phosphorylation of NtrC is regulated by nitrogen availability through another protein called 'PII'.

Outer Membrane Protein Expression : Earlier in this chapter, we discussed regulation of expression of genes *omp*C and *ompF* (meant for outer membrane proteins) at the translational level using antisense RNA. Another mechanism at the transcriptional level involves a two component system (EnvZ = kinase; OmpR = regulator). EnvZ autophosphorylates and transfers its phosphate group to OmpR, which in its phosphorylated form, stimulates expression of *ompF* under 'low osmolarity' conditions and that of *omp*C under 'high osmolarity' condition. Phosphorylation of OmpR is also regulated due to hydrolysis in the joint presence of EnvZ and ATP.

Bibliography

Ahmad, K. Henikoff, S. : *Epigenetic Consequences of Nucleosome Dynamics*, MIT Press, Cambridge, 2002.

Altaian, S. : *Transfer RNA*, MIT Press, Cambridge, 1978.

Anderson, S. : *Sequence and Organization of the Human Mitochondrial Genome Nature*, Oxford University Press, London, 1981.

Andre, C. A. Levy and V. Walbot : *Structure of Plant Mitochondrial Genomes*, John Wiley, London, 1992.

Atkins, J. F. : *Recoding : Dynamic Reprogramming of Translation*, Oxford, New York, 1996.

Bennett, R. H. : *Transcription, Processing and Editing in Plant Mitochondria*, Oxford, New York, 1992.

Bloom, L. B. : *Biochemical Basis of DNA Replication Fidelity*, MacMillan, London, 1993.

Branden, C. and J. Tooze : *Introduction to Protein Structure*, Garland, New York, 1991.

Darnell, J. E., H. F. Lodish and D. Baltimore : *Molecular Cell Biology*, Scientific American Books, New York, 1995.

Das, A. : *Control of Transcription Termination*, Routledge, New York, 1993.

Davidson. J. N. : *Biochemistry of Nucleic Acids*, Cox and Wyman, London, 1965.

Demidov, V. V. : *PNA and LNA throw Light on DNA, Trends,* Oxford, New York, 2003.

DuPraw, E. J. : *Cell and Molecular Biology,* Academic Press, New York, 1969.

Ebby, S. R. : *Non Coding RNA Genes and Modern RNA,* Oxford, New York, 2001

Echols, H. : *Fidelity Mechanisms in DNA Replication,* Oxford, New York, 1991.

Ellis, R. J. : *Chloroplast Protein Synthesis,* Oxford, New York, 1975.

———— : *Molecular Chaperones,* Oxford, New York, 1991.

———— : *The Biogenesis of Chloroplasts Protein,* Oxford, New York, 1974.

Fischer, L.M. : *DNA Unwinding in Transcription and Recombination,* Oxford, New York, 1982.

Friedberg, E. C. : *Relationships Between DNA Repair and Transcription,* Oxford, New York, 1996.

Gesteland, R. F. and J. F. Atkins : *The RNA World,* Oxford, New York, 1993.

Gilbert, W. and B. Muller : *Isolation of the Lac Repressor,* Oxford, New York, 1966.

Gilham, N. W. : *Genetic Analysis of the Chloroplast and Mitochondrial Genomes,* Oxford, New York, 1974.

Gilmour, R. S. and J. Paul : *Tissue Specific Transcription of the Globin Gene in Isolated Chromatin,* Oxford, New York, 1973.

Glick, B. and G. Schatz : *Import of Proteins into Mitochondria,* Oxford, New York, 1991.

Gray, M. W. : *The Evolutionary Origin of Organelles,* Oxford, New York, 1989.

Grunstein, M. : *Histones as Regulators of Genes,* Oxford, New York, 1992.

Gupta, P. K. : *Cell and Molecular Biology,* Rastogi Publications, Meerut, 2000.

Hadjiolov, A. A. : *The Nucleolus and Ribosome Biogenesis,* Oxford, New York, 1985.

Hoagland, M. G. : *The Present Status of the Adaptor Hypothesis,* MacMillan, London, 1959.

Hoeijmakers, J. H. : *Nucleotide Excision Repair from Yeast to Mammals,* MacMillan, London, 2001.

Holley, R. W. : *Structure of a Ribonucleic Acid,* McGraw Hill, London, 2002.

Ingram, V. : *The Biosynthesis of Macromolecules,* Benjamin, New York, 2001.

Jacob, F. : *Genetic Regulatory Mechanisms in the Synthesis of Proteins,* Oxford, New York, 1961.

Jukes, T. H. : *Mitochondrial Codes and Evolution,* Oxford, New York, 2000.

Kaguni, L. S. : *Eukaryotic DNA Polymerase-primase : Structure, Mechanism and Function,* MacMillan, London, 1999.

Kaziro, Y: *Structure and Function of Signal Transducing GTP Binding Proteins,* Blackwell, London, 2001.

Kendrew, J. C. : *The Three-dimensional Structure of a Protein Molecule,* Sinauer, New York, 1991

Kornberg, A. and Baker, T.A. : *DNA Replication,* Freeman, New York, 2002.

Kramer, A. :*The Structure and Function of Proteins Involved in Mammalian Pre-mRNA Splicing,* Grandada, New York, 2001.

Kurland, C. G. : *Structure and Function of the Bacterial Ribosomes,* Oxford, New York, 2000.

Lake, J. A. : *The Ribosome,* Oxford, New York, 1981.

Lehninger, A. L. : *The Mitochondrion,* Academic Press, New York, 2001

Lewin, Benjamin : *Genes,* Oxford University Press, New York, 2000.

Maxwell, E. S. and Fournier, M.J. : *The Small Nucleolar RNAs,* MacMillan, London, 1995.

McKnight, S. L. : *Molecular Zippers in Gene Regulation,* Princeton, New York, 2002.

Merrick, W. C. : *Overview : Mechanism of Translation Initiation in Eukaryotes,* Longman, London, 2001.

Moore, P. B. : *Molecular Mimicry in Protein Synthesis,* Cambridge, London, 2002.

Noller, H. F. : *Ribosomol RNA and Translation,* Oxford, New York, 1999.

Olson, R. E. : *Enzymatic Synthesis of DNA,* John Wiley, New York, 2001.

Owen, L. :*DNA Synthesis,* Freeeman, San Francisco, 2001.

Pabo, C. O. and R.T. Sauer : *Transcription Factors : Structural Families and Principles of DNA Recognition,* Printice Hall, London, 1993.

Pardee, A. B. : *Molecular Basis of Biological Regulation: Origins of Feedback Inhibition and Allostery,* Arnold, London, 2003.

Pardue, M. L. and J. G. Gall : *Molecular Hybridization of Radioactive DNA to the DNA of Cytological Preparations,* Dekker, New York, 2001.

Paszkowski, J. : *DNA and Histone Methylation in Plants,* MacMillan, London, 2002.

Razin, A. and A. D. Riggs : *DNA Methylation and Gene Function,* Longman, New York, 2002.

Rich, A. and S. H. Kim : *The Three Dimensional Structure of Transfer RNA,* Oxford, New York, 1999.

Robertson, M. : *RNA Processing : the Post-RNA World,* Oxford, New York, 1999.

Rosenberg, M. and D. Court : *Regulatory Sequences Involved in the Promotion and Termination of RNA Transcription,* Cambridge, New York, 2003.

Sager, R. : *Cytoplasmic Genes and Organelles,* Academic Press, New York, 2003.

Schmid, M. B. : *Structure and Function of the Bacterial Chromosome,* Granada, London, 1990.

Spector, D. L. : *The Dynamics of Chromosome Organization and Gene Regulation,* Oxford, New York, 2005.

Stryer, L. : *Biochemistry,* W.H. Freeman & Co., New York, 2002.

Temin, H. M. : *RNA Directed DNA Synthesis,* MacGraw-Hill, New York, 2002.

Turner, B. M. : *Histone Acetylation and Control of Gene Expression,* MacMillan, London, 2002.

Ulrich, B. : *Mechanism and Regulation of Eukaryotic Protein Synthesis,* Oxford, New York, 2002.

Uptain, S. M. : *Basic Mechanisms of Transcript Elongation and its Regulation,* MacMillan, London, 1997.

Vernadsky, V. I. : *Organellar Heredity,* Revan Press, New York, 1978.

Voinnet, O. : *RNA Silencing : Small RNAs as Ubiquitous Regulators of Gene Expression,* MacMillan, London, 2002.

Wallace, D. C. : *Diseases of the Mitochondrial DNA,* Blackwell, New York, 2002.

Wang, T. S. F. : *Eukaryotic DNA Polymerases,* MacMillan, London, 2002.

Watson, J. D. : *Molecular Biology of the Gene,* Oxford, New York, 2002.

———— : *Molecular Structure of Nucleic Acids : a Structure for Deoxyribose Nucleic Acid,* Oxford, New York, 2001.

Wilkins, M. H. F. : *Physical Studies of the Molecular Structure of Deoxyribose Nucleic Acid and Nucleoprotein,* MacMillan, London, 2002.

Williams, R. C. : *Reconstitution of Active Tobacco Mosaic Virus from its Inactive Protein and Nucleic Acid Components,* Oxford, New York, 1955.

Williamson, B. : *Split Gene Transcription,* Oxford, New York, 2001.

Woese, C. R. : *The Genetic Code : The Molecular Basis for Genetic Expression,* Harper & Row, New York, 2003.

Wuthrich, K. : *Protein Structure Determination in Solution by Nuclear Magnetic Resonance Spectroscopy,* Oxford, New York, 2001.

Yamamoto, T. : *A Bipartite DNA Binding Domain Composed of Direct Repeats in the TATA Box Binding Factor TFHD,* Oxford, New York, 2001.

Yamashita, S. : *Transcription Factor TFHB Sites Important for Interaction with Promoter-bound TFIID,* Springer, London, 2000.

Yams, M. : *Translational Efficiency of Transfer RNAs : Uses of an Extended Anticodon,* MacMillan, London, 2003.

Yanofsky, C. : *The Tryptophan Synthetase System,* Oxford, New York, 2001.

Zawel L., and D. Reinbcrg : *Common Themes in Assembly and Function of Eukaryotic Transcription Complexes,* MacMillan, London, 2002.

Zuckerman, B. : *Initiation of mRNA Translation in Prokaryotes,* Oxford, New York, 1990.

Zurick, D. : *Origin and Evolution of Mitochondrial DNA,* Oxford, New York, 1989.